M000195881

CHAMBER DIVERS

ALSO BY RACHEL LANCE

*In the Waves: My Quest to Solve the Mystery of
a Civil War Submarine*

CHAMBER DIVERS

The Untold Story of
the D-Day Scientists Who
Changed Special Operations Forever

RACHEL LANCE

DUTTON

DUTTON
An imprint of Penguin Random House LLC
penguinrandomhouse.com

Copyright © 2024 by Rachel M. Lance
Penguin Random House supports copyright. Copyright fuels creativity, encourages diverse voices, promotes free speech, and creates a vibrant culture. Thank you for buying an authorized edition of this book and for complying with copyright laws by not reproducing, scanning, or distributing any part of it in any form without permission. You are supporting writers and allowing Penguin Random House to continue to publish books for every reader.

DUTTON and the D colophon are registered trademarks of Penguin Random House LLC.

Permissions appear on page 419–420 and constitute an extension of the copyright page.

LIBRARY OF CONGRESS CATALOGING-IN-PUBLICATION DATA

Names: Lance, Rachel, author.
Title: Chamber divers : the untold story of the D-day scientists who changed special operations forever / Rachel Lance.
Description: New York : Penguin Random House, 2024. |
Includes bibliographical references and index.
Identifiers: LCCN 2023021745 (print) | LCCN 2023021746 (ebook) |
ISBN 9780593184936 (hardback) | ISBN 9780593184950 (ebook)
Subjects: LCSH: Deep diving—Great Britain—History—20th century. |
World War, 1939-1945—Commando operations—Great Britain. | Great Britain.
Royal Navy—Commando troops. | World War, 1939-1945—Naval operations—Submarine. |
World War, 1939-1945—Amphibious operations. |
Midget submarines—Great Britain—History—20th century. |
World War, 1939-1945—Naval operations, British. |
Haldane, J. B. S. (John Burdon Sanderson), 1892-1964. |
Military research—Great Britain—History—20th century. |
World War, 1939-1945—Campaigns—France—Normandy.
Classification: LCC D784.G7 L36 2024 (print) | LCC D784.G7 (ebook) |
DDC 940.54/51—dc23/eng/20231019
LC record available at https://lccn.loc.gov/2023021745
LC ebook record available at https://lccn.loc.gov/2023021746

Printed in the United States of America
1st Printing

Book design by Nancy Resnick
Title page photograph by katatonia82/shutterstock.com

While the author has made every effort to provide accurate telephone numbers, internet addresses, and other contact information at the time of publication, neither the publisher nor the author assumes any responsibility for errors or for changes that occur after publication. Further, the publisher does not have any control over and does not assume any responsibility for author or third-party websites or their content.

To my grandpa
Earl Strawbridge Lance,

1924–1999
NCDU team 10,

who specialized in rocket launchers and
underwater explosives
and pushed me for hours on the rope swing
he built in his yard

Contents

PROLOGUE

"A Piece of Cake"

DIEPPE, FRANCE
AUGUST 19, 1942
657 DAYS UNTIL D-DAY

The operation started with a simple cup of stew. By 0400, everyone had finished their hot, hearty breakfasts and begun to prep grenades. Thousands of warriors dressed in muted greens swayed with the rolling gray decks of the ships of the convoy beneath the meager light of the stars. Closer to the coast, the large ships split apart. The troops would attack at six separate locations.

Once nearer their location, British officer Captain Patrick Anthony Porteous helped his soldiers into position in the small, flat-bottomed boats that would take them the rest of the way. The propellers churned into motion. They headed for the chalky cliffs forming the dark skyline, a slight breeze from the stern nudging them landward.

The destination for Porteous's group was a sloping, pebbly shore code-named "Orange Beach II," located inside a curving hook of land in the northeast of Normandy, France. Their target cliffs tapered down to the west of the small resort town of Dieppe, and the other landing beaches—code-named "Yellow," "Red," "White," "Green," and "Blue"—lay to their left, in the east.

Porteous and his soldiers had watched tracer rounds briefly track across the skies during an unplanned encounter with armed trawlers during the Channel crossing, but the continued welcoming blink of a lighthouse onshore indicated that the scuffle had not triggered any alarms among the Germans. It was lucky because, for this group, that lighthouse was the primary means for navigation. They had limited maps, and no other visible beacons.

The Orange Beach party was divided into Groups I and II, and their joint objective sounded simple: silence the big guns in a reinforced artillery battery high on the cliff. If the battery was left intact, the German troops inside could lob mortar after gigantic, explosive mortar at the other five Allied beach locations, including the main landing sites. Group I would shimmy up a rain-carved gully in the cliff face and distract the enemy from the front. Group II would land to the west and sneak around behind. Porteous and his thirty soldiers, who formed half of F Troop of the British Number 4 Commando, had been designated part of Group II.

Captain Porteous had no idea what the true mission of the entire raid was. But like most soldiers, he knew that his role was not to reason why. His role was to destroy his assigned battery.

Group I landed without enemy opposition, guided in by the lighthouse. They used handheld tubes of explosives called Bangalore torpedoes to create a passageway up the cliff, blasting apart the thick nests of barbed wire that stuffed the narrow slot. A recent downpour had washed away the layer of dirt hiding most of the land mines, and under the moonless sky the soldiers picked their way over the safer patches of soil.

Group II was not as fortunate. As soon as Porteous and his commandos touched ground, the enemy opened fire. Machine guns tore through the night. Lightning-like bolts from tracer rounds showed that most bullets went above the commandos' heads, but still they dashed across the pebbly shoreline.

To be trapped on an exposed, featureless beach was death. The longer the soldiers lingered without cover, the longer the Germans had to aim. The warriors of Number 4 Commando unfurled coils of chicken wire over the barbed wire blocking their exit from the beach, ran across these makeshift bridges, and dove down the steep drop on the far side. Somewhat protected in a small fold in the ground, they took a moment to re-group before proceeding inland.

No more than twenty of the two hundred in their party had fallen on the beach, fewer than predicted. A medic volunteered to stay behind, but the rest had strict orders: leave the dead and injured. Get off the beach, regardless of casualties, "regardless of anything." They moved inland.

Group I's gunfire from the front distracted the soldiers in the battery as hoped, while F Troop and the rest of Group II wound their way around, single file, silent, through a small passage in the hefty barbed wire guarding the back of the concrete monstrosity. Porteous and his troops lobbed explosives inside, killing the German soldiers and silencing their guns. Porteous was shot twice, once through the hand and once in the thigh, but he survived, as did most of Number 4. Their mission was a success.

However, theirs would be the only victory the Allies could claim that day in 1942. By the time Number 4 and their freshly taken German pris-oners evacuated from the base of the cliff at 0815, with the wounded Captain Porteous in tow, they were faced with new horror. Thick layers of smoke obscured their view, but the raging thunder of machine guns and mortars at the other landing locations forced a brutal realization. Their compatriots were caught in the worst hell imaginable. Everyone else was stuck on their beaches.

Every snowball begins first as a harmless flake, a diminutive nucleus of crystallized water that, given a downhill nudge, quickly accumulates power. At Dieppe, one nudge was a case of mistaken identity.

In the dark, Gunboat 315 looked an awful lot like Gunboat 316. And by the time the Royal Regiment of Canada, headed for landing site Blue Beach, recognized the error and realigned behind the correct vessel, they were sixteen minutes behind schedule. The Royal Air Force had dropped a thick layer of smoke to obscure the aim of any Germans who happened to be awake, but by the time the tardy Royal Regiment arrived, the punctual smoke had begun to clear.

Aerial planning photographs taken months earlier had revealed the batteries gracing the tops of the cliffs, but the top-down views hadn't hinted at the surprises in their vertical faces. The Royal Regiment landed on Blue amid the parting smoke and looked up to a nasty revelation. The shelling that had started on time at the other beaches had roused the Germans, and the Germans were not only awake; they had dug into the very cliffs themselves, creating machine-gun-filled nests burrowed into the sides of the steep white walls. They unleashed a furious thunderstorm of metal onto the Allies below.

The few Royals who survived the initial mad dash across the impossibly broad shore hunkered into small shapes, curled helpless against a meager seawall, hoping that the ships offshore and pilots in the air could take aim at the machine guns and mortars mowing them down. But it was hopeless. The officers on the ships couldn't see through their own protective smoke screen. The pilots in the air, who were not equipped with rockets, struggled to hit the tiny targets dug into the cliff face with guns. Those down on the beach couldn't fire upward and around corners into the machine-gun grottoes. Bullets and grenades peppered their sky.

Not a single soldier made it off Blue Beach. Almost half of the five hundred in the charge died. The unit had a casualty rate of 97 percent. But, since the Royal Regiment never got off the beach, they also never accomplished their goal: destroy the battery at the top of their cliff.

Other batteries in front of the main landing approach resisted Allied attack, too. In order to get to the town, the rest of the troops would have to run south across broad, open, unprotected beaches Red and White, with active guns crisscrossing their path. The deaths of the warriors at Blue, and the intact battery above them, were a snowball, and they triggered a crossfire avalanche. About a thousand of the Allied landing troops were British, but more than four-fifths of the 6,086 total were Canadians, with only about fifty Americans. What happened next is considered one of the most horrifying disasters in Canadian military history.

The inexperienced Canadian troops headed for Red and White were convinced that tanks alone would provide the power needed to break through the German defenses. Everyone would jump out of the ships

An aerial photograph of the cliffs above Blue Beach, taken a few years after the raid. The gun caves are still visible in addition to the batteries along the top. *(NARA photograph)*

together—the engineers with their packs full of explosives, the soldiers on foot, and the fearsome new tanks nicknamed "Churchills." The soldiers would face down the guns while engineers blasted paths through the barriers blocking the exits from the beach; then the Churchill tanks would proceed smoothly and easily through those paths into the town. Their commander, General John Hamilton "Ham" Roberts, had even promised in his prelaunch pep talk that it would be "a piece of cake."

With this plan fixed in his mind, Dutch-born noncommissioned officer Laurens Pals of the 2nd Canadian, on board an impressive tank-carrying ship, confidently prepared to land on beaches Red and White. He did not know that the flanking batteries remained intact.

The barrage was instant, and staggering. The tanks had been waterproofed with a thin material, but their carrying vessels were not bulletproof. Both got riddled with projectiles. The vessels began to sink; the tanks began to swamp. Some coasted to land too crippled to function; others burbled to the bottom of the sea, drowning the people inside.

The process to unload the tanks was slow. The beasts were cumbersome, and soldiers on foot were stalled by tending to the monsters as they rolled out. The struggling clusters of troops provided dense stationary targets for hungry German gunners.

The combat engineers carried explosives to clear paths, but their ordnance caught fire when hit by bullets. Some engineers were forced to throw their flaming piles of munitions overboard, while others pressed on, knowing they were human bombs. All of these specialized soldiers were in the first wave of the assault. All were mowed down. As a result, paths were never cleared. An official report described the problem bluntly: "The road blocks were never destroyed because of the failure of Engr [engineering] personnel to reach these blocks alive."

With the deaths of the engineers, no matter what, the tanks were blocked. However, the Churchills had never been used in combat before, or even thoroughly tested, and the beach itself proved another unexpected obstacle. The waterfront was made of pebbles, billions of smallish rocks polished smooth by the eternal grinding of the waves, and the rocks got stuck in the tanks' exposed gear mechanisms. The tanks threw off their tracks. Almost all got stuck, at least one flipped over, and several somehow caught fire. Every single tank was destroyed. Some of the troops had been equipped with bicycles so they could travel farther into the town, and these, too, sank feebly into the pebbles. Almost nobody was able to clear the broad, steep shores.

When Laurens Pals landed, he took what cover he could in a small divot in the sloping ground. When ordered to advance, he dashed to a second small hole. From there he could see the street where he was supposed to start his specific mission during the raid. He would never reach it; the opposition was too great. A haze of death filled the space between. Dodging mortars and machine guns, he dove behind a landing ship that had tipped over, joining dozens of troops who had been shot. He helped establish a sort of field hospital in the cover of the sideways vessel. Pals and the others hunkered down and tried their best to survive the ongoing hailstorm of bullets, periodically darting out to drag more wounded and dead behind their improvised protection.

Laurens Pals was charged with a secret mission. He had been ordered to sneak into town, aided by a select handful of explosives-carrying engineers—all now dead—and steal a big yellow code book out of a hidden safe. There were others also charged with intelligence-gathering missions, and these missions were the true purpose of the entire raid. The Allies hadn't wanted to bomb the town of Dieppe beforehand and risk destroying the fragile information they sought. But the official explanation for the lack of bombing was that it was hard to teach pilots to aim properly in such a short amount of training time.

So, because of the lack of bombing, the intact, beautiful residences along the waterfront stayed packed with machine guns and machine gunners. These gunners, too, fired at Laurens Pals every time he ran out to retrieve more wounded and dead from the bloodstained waves.

Desperate to accomplish the secret missions and gather intelligence from the town, the commanders offshore continued to send wave after wave of new troops to Red and White in the hopes that a few individuals would manage to break through and maybe find a code book or an encryption device known as an Enigma machine. The new model of Enigma was being used by German submarines to receive encoded messages and wreak havoc in the Atlantic. However, the reinforcements served only to add to the growing piles of Allied bodies strewn about the rocks. At 1100, after hours of abuse, the exhausted and battered survivors finally received a radio transmission ordering them to surrender.

The convoy offshore never sent ships for an evacuation. One group of sailors made an abortive attempt, but their lone vessel got overwhelmed by enemy fire and too many troops trying to scramble on board. They managed to evacuate only a small fraction. The rest were abandoned. Slowly, the screams went quiet.

Under pressure from the Russians to open a second front in the west to divide and distract the Germans, and under pressure from the Americans to try an invasion of the European mainland, Winston Churchill had reluctantly permitted the raid on Dieppe. Churchill had already been part of one beach debacle, when during World War I he had been responsible for a failed beach-landing attempt on the Gallipoli peninsula in modern Turkey. The travesty had resulted in 250,000 casualties stretched out over nearly a year of misery without resolution. He knew the Allies were not yet ready to assault beaches, but when pressed, the infamously fractious leader had been willing to prove his point in blood.

Ham Roberts, the commanding general of the 2nd Canadian who had made the quip about cake and kept sending in reinforcements to die, fled his home country. He would allegedly receive an anonymous parcel on his doorstep containing a single stale piece of cake on the anniversary of the raid every year for the rest of his life.

In the weeks following the massacre, when the tides had finally washed the blood off the pebbly shores and the Allied news media lied to pretend it had been a victory, the Allies would title the key section of their official report simply, with the most optimistic veneer they could muster: "Lessons Learnt." The Dieppe "Lessons" report is unequivocal in its language: "As soon as it is known that a project involving a combined operation is under consideration"—meaning an operation like an amphibious invasion that combines sea and land—"the question of beach reconnaissance in all its aspects must be investigated."

At Dieppe, landing ships had encountered underwater obstacles for the first time. The obstacles were fearsome-looking constructions of steel and explosives designed to tear and blow apart the ships from beneath. Landing parties needed to know if such obstacles waited beneath the waves. Perhaps most important, the report emphasized that they needed to do the scouting stealthily so they could keep the element of surprise: "Naval reconnaissance methods should also be used but care must be taken that they remain undetected."

However, the report does not mention what kind of undetectable naval scouting methods should be used. Possibly because, in the fall of 1942, most were unaware if any such methods existed. The militaries' only knowledge of the composition of the beaches at Dieppe had come from a photograph that someone had taken during a vacation trip years before the war—a photograph that showed finely clad ladies and gentlemen enjoying a day of relaxation onshore. The Allied officers had simply sent the tanks in and hoped for the best.

The Allies learned that to achieve a successful large-scale beach invasion, to have a chance at bringing the fight to Hitler on the European mainland, they would need real intelligence: solid and up-to-date information about the positions of the guns, the slope of the shore, the type of sand or rocks, and the locations of obstacles or land mines. They needed a plan for navigation that didn't rely on the luck of German-held lighthouses remaining lit—they needed detailed maps and the ability to

One of only three scouting photographs of White Beach, all of which were taken before the war.

send undetectable scouts who could place markers and beacons in advance. They were willing to throw bodies at the problem, but to succeed, they would need both bodies and cunning. They would need to send reconnaissance troops ashore. And the best way to do that, they realized, was to hide them under the ocean waves. To make a beach landing possible, they needed warriors equipped to survive underwater.

Fortunately, a small group of scientists was already on the case.

CHAPTER 1

Professor Haldane and
Dr. Spurway's Problem

A few days later in London, a resonant hiss vibrated through a narrow metal chamber as pressurized gas gushed in. The enclosed steel tube was a cramped, heavily riveted cylinder with thick walls and rounded ends, a mere four feet in diameter and laid down sideways on a platform in the corner of an industrial warehouse. Pneumatic pipes sprouted from the top like mechanical antennae. Inside, wood planking formed a humble, flat floor, and even when curled up on that planking in a sitting position, Professor John Burdon Sanderson Haldane took up most of the chamber's white-painted internal volume. Six foot one and unintentionally, awkwardly, massively imposing, the blunt yet mostly affable middle-aged geneticist had broad shoulders, a hefty mustache, and a prominent bald forehead that forced his bushy eyebrows downward at the seeming expense of his eyes. He sat staring across the narrow space at today's test subject: Dr. Helen Spurway. The gas roared around them. The chamber began to heat.

Twenty-seven-year-old Spurway, herself a PhD geneticist, was as lanky as Haldane was robust. She perched on a small stool opposite the prof, her slim shoulders forced into a curve by the tubular white walls. She had short dark hair with a subtle natural wave in it, cut into a practical bob that would have been easy to dry after getting in and out of the water repeatedly for diving experiments. Her smooth, heavy brows formed a sharp frame within her pale face to offset deep brown eyes, and gave her the animated look of someone who not only listened but scrutinized. For today's test, her nose was pinched shut by a spring-loaded clip. Her lips were sealed around a rubber mouthpiece that was connected through two large corrugated hoses to a square-shaped, flexible tan bag strapped to her chest. It was an apparatus designed to deliver pure oxygen to her lungs.

Haldane and Spurway's goal was to see how long she could breathe the oxygen before it began to poison her. She was an eager test subject. The poisonous effects of oxygen become far worse under high levels of pressure, and she wanted to use her own body to puzzle out how bad they could get.

Roaring air continued to force its way into the small steel chamber, which was dimly lit on the inside by the paltry warehouse lights shining through its one pressure-reinforced porthole. The intruding air drove the internal pressure levels up to those felt by divers swimming deep in the ocean. The temperature climbed as more gas pushed and compressed itself into the small space, and the heat became unmerciful, exacerbated by the thick, syrupy sensation of the newly densified atmosphere. Haldane and Spurway, pressed together by the unyielding confines of the chamber walls, sat curled close enough to watch each other sweat.

The hiss of the gas, deafening at first, began to taper slowly as they reached their target pressure, but even as it slowed the reverberations caused by its flow remained the dominant sound inside the white metal tube. At increased gas densities, such as the pressures inside a hyperbaric chamber, houseflies cannot take flight. People cannot whistle. Voices become cartoonishly high-pitched, increasing in frequency alongside the temperature and density. Breathing and movement acquire a conscious feeling as the thickness of the air deepens.

Haldane and Spurway reached a pressure level equal to ninety feet beneath the waves, as well as a scorching heat somewhere in the triple digits Fahrenheit, and an operator outside the chamber spun a valve to halt the inward rush of gas. New sounds pinged through the cooling metal tube as the chamber operators manipulated the gases to maintain the depth, while the temperature inside began to drop. Haldane noted the time.

Five days after Dieppe, not yet knowing of its horror, Haldane and Spurway were already unwittingly working on the next amphibious assault plan. The tan leather breathing apparatus Spurway wore on her chest was called a Salvus, and in a world before the invention of the now ubiquitous SCUBA, she was trying to figure out how long it could be used safely by swimmers and submariners beneath the surface of the ocean.

After precisely thirty-three minutes of sitting in the chamber calmly, patiently sucking down oxygen, Spurway yanked the rubber mouthpiece from her lips. She vomited. She vomited repeatedly. Later, she reported that she had been having visual disturbances beforehand: brilliant flashes of dancing purple lights she referred to as "dazzle." She gulped down chamber air, no longer on oxygen, and recovered slowly. Thankfully, today's symptoms were mild. Watching her gulping, gasping face, Haldane again noted the time.

Professor Haldane, Dr. Spurway, and the other members of their small clan of scientists had spent the last three years grinding away within the close confines of these hyperbaric metal tubes at the myriad questions surrounding underwater survival, with the initial goal of enabling sailors to escape from submarines. Their leader, Haldane, believed firmly that human beings provided better feedback than research animals.

Shortly before the disaster at Dieppe, but not in time to stop it, the scientists had been asked to pivot this work to focus on a new, more specific goal. The Admiralty had called. To help their countrymen and the Allies defeat Hitler, to help end the war, the Allies needed the scientists to use this same work to prepare for beach-scouting missions. There would be another beach landing. And the next one could not fail.

Decades before the experiment, a thirteen-year-old boy peered out of the tiny reinforced window cut into the side of a metal chamber, gazing at the small crowd of adults watching him back through the clear, hefty panel. This small hyperbaric chamber sat on the deck of a Royal Navy ship rolling with the waves off the coast of Scotland. As the hiss of gas filled his chamber, the towheaded boy practiced moving his jaw and swallowing to allow the air to enter the spaces inside his ears. He succeeded, proving he could clear his ears. He reached the maximum pressure safely, and was returned to the surface pressure without harm. After the circular chamber door swung open, the boy clambered carefully out onto the deck in the hot, late-summer weather. His father began to strap him into a sturdy canvas suit sized for an adult while the sailors and other researchers watched. It was time for the boy to have his first dive.

The boy, John Burdon Sanderson Haldane, was born in 1892 into the sort of Scottish family whose summer homes have turrets. Stately portraits of ancestors with facial-hair topiaries or dresses with miles of pleated fabric grimaced down from the high walls of their multiple estates, and manicured lawns and gardens buttressed the elaborate stonework buildings. But John Burdon Sanderson Haldane, called "Jack" in his youth and later "JBS" as an adult, had no patience for such pomp. He insisted on keeping an old bathtub full of tadpoles beneath the arcing branches of one majestic apple tree, within which he was determined to breed water spiders.

His immediate Haldane family's little team of four—himself; younger sister, Naomi; mother, Louisa Kathleen; and father, John Scott—formed one branch of a venerable lineage that stretched its documented pedigree back to the 1200s. As children, Naomi and Jack were bred into science the way some are bred into royalty.

Their parents, Louisa Kathleen and John Scott, seem to have gravitated toward each other because of the same fiercely independent, socially irreverent genius they would pass on to their children. She was a brilliant young woman with golden hair, classical beauty, an affinity for small dogs, and an outspoken confidence that, along with her propensity for the occasional cigarette, marked her as a rebel within the prim upper crust of 1800s Britain.

She stumbled across John Scott for the first time when he was asleep on the floor. He was a lanky young man with deep, kind eyes and a dark, full mustache, and she walked in on him sprawled on a rug in front of a fireplace, drooling into a scattered mountain of books. She and her mother were thinking about renting the Edinburgh home from the Haldanes for the winter, and John Scott was supposed to show them around.

He couldn't convince her that he had been awake, but he did convince her to marry him. Born stubborn enough to be independent and wealthy enough to stay independent, she relented only after negotiations. After rejecting him twice, Kathleen finally accepted Haldane's third proposal on the condition that he would not try to dictate her personal political views.

It stormed on their wedding day, and John Scott showed up late to his own ceremony in her parents' Edinburgh drawing room because he had gone back home to grab a forgotten umbrella. The bride, however, being equally utilitarian, did not mind. The ceremony was already planned to be short because Kathleen had tracked a blue pencil line through most of the dumbfounded minister's boilerplate script. She and John Scott were both firm agnostics and unafraid to buck convention, so she refused to wear lace, and she "strongly objected to the farce of being 'given away.'"

After the wedding Kathleen promptly moved in to John Scott's home in Oxford, England. He was a physician and professor of physiology at Oxford University, and the infamously eccentric researcher had already converted the basement and attic of the house into makeshift laboratories. There he could play with fire and air currents and gas mixtures. So could his children.

Eleven months later, it was into this kind of world that infant Jack was born. At one point as a baby, he screamed so forcefully that he gave himself a hernia, which unofficial scientist Kathleen calmly reduced herself. Sister Naomi was born five years later. A few years after Naomi's birth, the finished quartet moved along with Kathleen's mother to a larger, "comfortable and ugly" custom-built thirty-room mansion christened Cherwell. This building's entire design revolved around the pursuit of science, no longer limiting it to the attic and basement.

John Scott insisted that the bathtubs be made of lead for better thermal insulation. He refused to allow the installation of gas lines because he considered gas too chemically explosive. He and Kathleen packed every room with cozy chairs for reading and intellectual curios such as scarab beetles or painted Chinese bowls stocked with goldfish. John Scott's large office extended off the back of the house, and it was perpetually encrusted in a mosslike coating of books and papers. It held a room-length wooden table big enough to do the glassblowing required to make custom chemistry equipment, and a small step led down to their in-home laboratory. The lab had large windows looking out onto flower-filled gardens, and the walls were braced by broad shelves full of chemicals and supplies. The lab always featured at least one airtight gas chamber, just large enough to hold a person, nicknamed "the coffin." It could be flooded with any gas desired. It often was.

Even the family cat slept curled, tightly but voluntarily, inside a beaker in the lab, and Naomi's dollhouse was decorated by famous scientists. Nobel Prize–winning physicist Niels Bohr was among those who had contributed, bringing a little toy jug for her miniature estate.

Carved into the entryway stonework of the house was the Haldane family motto, one hovering word that would become an ominous portent. To enter the Haldane home, as if crossing the threshold of a new, far more foreboding world, all visitors and guests were required to pass under the simple declaration:

SUFFER

The two children were inseparable. Naomi and Jack chased globules of mercury across the laboratory floor, huffed chloroform and giggled at its effects, sucked in a miscellany of strange gases from storage bottles to test the effects on their voices, or simply whirled together dancing Viennese waltzes until the giddy spinning rendered them too dizzy to continue. He called her his "Nou." She addressed him as "Boy"; "Boydie," a contraction of "boy dear"; or sometimes just "Dearest."

"As children we were both in and out of the lab all the time," Naomi once wrote. They didn't need to sneak in; they and their father were the only three allowed. Naomi's job as a child was to monitor the test subjects through the observation window in the gas chamber and, should they fall unconscious, to drag them out and resuscitate them. By age three, golden-haired, chubby-cheeked toddler Jack was a blood donor for John Scott's investigations, and by age four, he was riding along with his "uffer" (father) in the London Underground while John Scott dangled a jar out the window of the train to collect air samples. The duo found levels of carbon monoxide so alarmingly high, the city decided to electrify the rail lines. Already, as a toddler, the younger Haldane was learning how to keep people alive and breathing in worlds where they should not survive.

By age four, Jack was also exploring coal mines with his father to figure out how people breathed in those cramped, dangerous spaces. Frequent explosions and gas leaks made mining one of the most lethal jobs in the world, and gentleman physician John Scott Haldane became known among the miners of the country for his willingness to clamber into the narrow, dark, coal-filled passageways on his mission to make the air supplies safer. "Canary in the coal mine" is still a common expression to describe early detection of any threatening situation, and using the small, chipper birds to detect gas leaks was John Scott Haldane's idea.

Every time a mine exploded, John Scott—and often his son, Jack—hopped on the next train to sift through the corpses and wreckage to find the cause. With every trip, young Jack watched his father practice firsthand what seemed to be the elder professor's most important lesson: volunteer yourself first; then, after you went, exclusively test on "other human beings who were sufficiently interested in the work to ignore pain or fear."

They chose to make informed consent their policy, in a time period so ignorant of the concept that physicians joked openly that it was easier to find volunteers for experimental surgeries than for experimental medical treatments because the unconscious surgical patients could not decline. Testing on animals was the absolute last resort, John Scott lectured, and he even designed a canary cage that would seal itself and resuscitate the coal mine birds with fresh oxygen the moment they fell off their perch.

From his mine trips, John Scott would send home periodic telegrams to reassure his family, but often the carbon monoxide inside postexplosion mines would temporarily addle him so badly, he would write the same words multiple times in one message, forget he'd already sent a message and send multiple repeats, or write entire messages of gibberish. (Kathleen did not find these telegrams reassuring.) To study the effects in a slightly more controlled manner, he sealed up his home gas chambers and willingly breathed levels of carbon monoxide that pushed him deliriously close to death.

Naomi wasn't invited on the field trips, mostly. She had a heart-shaped jaw, broad cheeks, a button nose, and signature Haldane golden ringlets spilling down her back. But Naomi was born a daughter in 1897, and so she was born a disappointment to her patriotic mother, because girls were considered less useful to the British Empire at a time when the Empire was the loftiest of concepts for the aristocrats of England. Kathleen and John Scott were feminists by any definition, but even the most ardent advocates for equality recognize that their ability to raise their children on an even footing within their own home far exceeds the level to which they can demolish the bias of the outer world at large. It is possible to see the value of a young woman yet understand the arbitrary and brutal shackles the world will force upon her. The world that Kathleen knew in 1897 would look straight through Naomi's genius or find it inappropriate. Indeed, despite the fact that Naomi grew up to be an extraordinary and prolific author as well as a vital key to adult JBS's future science, most biographers omit or minimize her. Even Kathleen's own memoir contains a beefy chapter titled "My Son" while only sparsely describing Naomi. Naomi summarized the end effect: "Certain avenues of understanding were closed to me by what was considered suitable or unsuitable for a little girl." Young Naomi was denied access to the paradise where Jack got to play.

However, she crept her way in. Naomi might not have been welcome in the mines or on the boats, but within the walls of their homes, she was mostly an equal. From her, Jack would learn firsthand the intellectual equality of women—a principle that he would absorb and apply with fervor, and one that would allow him to build a spectacular future lab during a war-driven shortage of eligible men.

Naomi demanded that other children in her preparatory school participate in the family science, giving brusque orders like "You come in here. My father wants your blood." A diary kept by Naomi at age six has colorful, meticulous illustrations of begonias with details about counting the petals to tell the sex of the blooms (a "she-begonia" versus a "he-begonia"), all written in the large, looping calligraphy of a child.

When Jack was eight years old, John Scott brought him to an evening lecture about Mendelian genetics, a set of mathematical explanations of how physical traits get passed down between generations—why some siblings turn out blond but others brunette; why some pea plants have violet flowers but others white. Science had not yet pinpointed the primary culprit as DNA. Jack became obsessed with the cutting-edge idea. Naomi had begun keeping guinea pigs after a nasty fall left her afraid of her pet horse, so they started using her carefully trained fuzzy "pigs of guinea" to test the theories of gene propagation themselves, along with mice, lizards, birds, and whatever other fast-breeding animal they could procure.

Soon the front lawn of their home teemed with a roiling carpet of three hundred guinea pigs, all carefully labeled, numbered, and partitioned behind wire fencing. The squeaking fuzz balls had been deliberately bred, observed, and documented so that the young scientists could compare the patterns of their whorls and colorful splotches against Mendel's math. They executed the calculus to describe the patterns of inheritance in their guinea pigs at a time when most adults were unaware that such fields of research even existed. Through this exploration, Jack Haldane learned statistics. He learned probability.

The two developed a perspective on the world that was based on an innate need to understand how it worked, a need to tease apart its inner mechanisms and find what drove the microcosmos, the animals, the planets, and the stars. Naomi said her favorite part of the day was opening the mouse cages and letting "the darling silky mice" wander over her in her specially designated blue mouse-tending frock. Years later, during WWI, when adult JBS saw some odd, hairy-tailed rats at an undisclosed forward combat location in Mesopotamia, he gushed in a letter to his sister about how he "would like to Mendel them."

As a child, Jack Haldane played with strange gases and saw their effects on test subjects in the sedate, controlled environment of the home gas chamber. He began to learn how breathing gases affected people under the menacing pressures of water.

Jack—soon to be JBS—Haldane was the next heir in a line of respiratory researchers trying to puzzle out the mysteries of the human body under pressure. People had been trying, and failing, to figure it out for decades. One of the first puzzles of life underwater was the mystery of decompression sickness, and this mystery was what had landed thirteen-year-old Jack inside a small hyperbaric chamber, trying to clear his ears against changing pressures, rocking on a Royal Navy ship near Scotland. He wanted to be part of the diving. The Haldanes were on this ship to prove that John Scott had answered the first major question of survival in the deep. However, long before the younger Professor Haldane and Dr. Spurway sweated in their chamber in 1942, even long before the elder Haldane's experiment in 1906, the question of decompression sickness had been forced into consideration.

CHAPTER 2

Brooklyn Bridge of Death

On February 28, 1872, two decades before J. B. S. Haldane's birth, Brooklyn Bridge laborer Joseph Brown was submerged well below the murky waters of New York City's East River...technically. The twenty-eight-year-old American-born foreman sloshed through a shallow layer of dark, frigid water in his work-issued, thigh-high black rubber Goodyear boots, but the rest of him stayed dry, free to hoist a pickax or a sledge-hammer beneath the low ceiling and bring it down with enough force to remove yet another hunk from the rocky sand of the moist river floor.

Caissons are massive constructions that look a bit like large rectangular wooden bowls. They're turned so the open mouth faces downward, and they get plunked into the water beneath a layer of hefty building material thick enough to ensure that they plummet firmly to the bottom. Once they have settled onto the topmost, waterlogged layer of unstable earth on the river floor, workers use gas pumps to push air down into them from the surface, and that compressed air blows out the water from inside the inverted bowls. The workers are then free to travel down into the wooden structures and dig inside, working the edges of the caisson downward inch by painful inch into the muck and soil as they remove bucketfuls of stone and earth from below. Each caisson used for the Brooklyn Bridge was thirty-one by fifty-two meters in size, divided into six long chambers by internal walls. They were sizable hunks of a city block beneath the waves, but with no sun, and with impenetrably

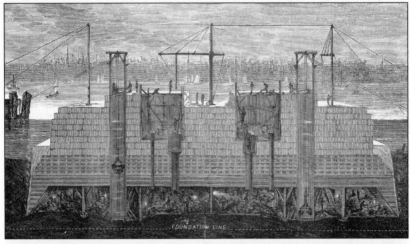

Harper's Weekly illustrations of caisson work for the Brooklyn Bridge.

thick, flat, iron-reinforced wooden skies looming only three meters overhead. Brown, like so many others, sloshed inside.

    The dark, soggy, air-filled crawl spaces had floors of mud and sand
and were peopled by "sand hogs"—workers wielding pickaxes, chisels,
sledgehammers, and shovels. As the sand hogs dug, the knife-edge perimeter of the bowl crept downward, the crushing weight of the tower
above held at bay almost exclusively by the upward push of the pressurized air within. More layers of stone were arranged carefully on top of
each caisson as it sank, and these layers became the pillars of the bridge,
creeping their way from the river floor downward to the bedrock like the

Inside a caisson beneath the Missouri River, 1888.

probing roots of a massive stone tree. When the workers hit bedrock, they would evacuate, the empty space inside the caisson would be filled with cement, and the entire monolith would be left in place as a support for the stable thoroughfare above.

The humidity of the river mingled with Joseph Brown's sweat and the evaporated stench of the laboring crew around him. Most of the workers had long ago peeled off their shirts, or they had gone down without them in the first place. The air temperature around them rose to tropical levels with each swing of their tools, even as their feet froze from the chill water, and they could feel their headaches slowly build as they breathed and rebreathed the same air inside the enclosed space. Brown and the rest of the roughly sixty-person working crew were at the end of their third hour inside the humid, dimly lit, wooden-walled caisson. For foreman Brown, it was the end of his watch. It was hot. It was humid. It was time to come up. He headed for one of the ladders.

A particularly brutal winter in 1867 had iced the East River, the obstacle between Brooklyn and Manhattan, and left the ferryboats locked at their piers. After that difficult winter, the people of Brooklyn had decided they wanted a bridge to Manhattan. But not just any bridge. With the East River more of a broad, active tidal basin than a river, stretching an immodest five hundred meters across, it would need to be the longest suspension bridge in the world. So it would need burly supports on either end to keep it strong. And in order to let ships pass under the bridge freely, each of those supports would need to be the tallest construction in America at that time—excepting a few stray church spires and the *Statue of Freedom*, which nudged the Capitol Building in Washington, DC, to a few feet of advantage.

To keep out the river, the air inside a caisson needs to be at least the same pressure as the water outside. So, as the depth of the roots increases, the pressure in the air increases, too. Joseph Brown's caisson was on the deeper, Manhattan side of the bridge. Far beneath the river, just like inside a pressurized chamber, the compressed air was thick. Because of the increased density, voices rose in pitch. Here, again like in the chamber, it became impossible to whistle. And it became dangerous. On February 28, 1872, Joseph Brown had been unknowingly working just beyond the cusp of where that pressure could become a deadly problem.

Brown and his crew, exhausted, shirtless, sweaty, filthy, and covered in river mud, climbed up the rungs of the stubby ladder and into the lock embedded in the roof of their caisson. The lock was a sturdy chamber in the shape of an upright cylinder, and it could hold up to thirty workers when densely packed. Once the last worker clambered up through the hatch in the floor of the lock, they closed and secured it behind them.

Finally, after hours at work beneath the suffocating body of the East River, the workers could see the sky again through the small reinforced window inside the opposing hatch in the lock ceiling. Only after the gas space of the lock was depressurized to the same pressure as the air outside could that upper door swing open, but then the sand hogs would be able to climb up the winding spiral staircase to fresh air, hot coffee, and freedom.

First, a crew member turned on the heating mechanism, which would send scalding steam flowing through the metal coils attached to the interior walls of the lock, in anticipation of the humid, bone-penetrating sort of cold that fills any space containing rapidly expanding air. The coils began to warm. Then the lock operator slowly and ever so slightly cracked open the brass valve to vent the pressurized air. The hissing scream of traveling gas dominated while the pressure dropped.

Steam condensed out of the very air itself, an eerie and frigid indoor fog. The lock approached the lower pressure level of the outside world. Then the trouble started for Joseph Brown.

At first he thought that he was fine. It took about ten minutes to vent enough gas to reach surface pressure, and after that, he climbed out of the lock and up the long, twisting route toward daylight. Little is known about what he did next, but perhaps he relaxed at the top, or drank a cup of hot broth offered to him by the on-site doctor. Perhaps he jumped into a carriage to head straight home. But whatever choices he made, almost precisely one hour after leaving the caisson, a sudden stabbing bolt of agony rippled through Joseph Brown's left shoulder and arm. Pain ran through his joints and along his nerves like lightning, "like the thrust of a knife."

Nothing relieved the vicious pain, and nothing would. Not twisting his hand, not massaging the joint, not a "drachm" of ergot—a small dose of a medically unhelpful hallucinogenic fungus that grows on wheat— that he might have been prescribed by the bridge physician Andrew Smith. Brown simply lived with the hot, throbbing, jagged, penetrating neurological sensation of stabbing fire until the next afternoon, when he needed yet again to go down into the damp, murky depths in order to make a living.

Immediately when he reached the bottom of the river, the pain disappeared. His arm was healed; it worked as it should; he had no discomfort. He blissfully completed his shift, and he came back to the surface again, in the lock, slowly. This time, after the ascent, he had no issues.

The next day Joseph Brown's arm still remained mostly pain-free, but it was covered with tiny red speckles. Freckles of blood had popped up like confetti on the parts of his arm and shoulder where yesterday's pain had screamed the loudest. He reported to the bridge doctor Andrew Smith that those areas were slightly sore and slightly numb, and Dr. Smith was perplexed. Smith logged Brown's strange symptoms in his notes. For months Smith would continue to wonder about their cause.

As the project progressed, the medical cases seen by Dr. Smith began to increase in tempo. More workers from each shift emerged from the caisson to become bent and contorted with often crippling pain. Each surfacing of the lock became the beginning of an ominous period of waiting for catastrophe to strike. Dr. Smith began recording everything obsessively, trying to understand, trying to treat, trying to figure out the pattern, trying—but failing—to save the workers from the mysterious ailment.

John Roland took a hit on March 8. While in a horse-drawn car and going home, he felt suddenly dizzy, lost all his strength, and could not stand. His feet were cold but his head was hot and he began sweating the disturbing, cold perspiration of an immune system fighting a demon. He recovered on his own within a few hours.

Hugh Rourke was next. On March 12, he, too, was going home when his right knee felt the telltale stabbing of knives deep inside the joint that left him suffering through the night. Rourke turned to morphine for relief. It became an addiction.

The deaths began in April, when the caisson roots stretched beyond sixty feet deep. German immigrant John Myers was the first on record. He dropped dead without warning after leaving a shift, having mentioned only that he was "feeling oppressed by the impurity of the air." Patrick McKay collapsed the next day, unconscious, while walking down Roosevelt Street, having just left work. He was carted to a local hospital where he died in agony. His death was labeled "congestion of the brain" by perplexed physicians.

By May, a mere month before the pressurized work was done, when the caisson was nearing its deepest depths, the trickle of cases bloomed to a flood. On May 2, James Hefferner had pain, projectile vomiting, and extreme constipation. On May 4, Thomas Kirby had severe pain and swelling of the right arm. The sand hogs began to strike, demanding three dollars for every four hours of the dangerous work, instead of their previous rate of $2.43. On May 8, German immigrant Bruno Wieland was hit with pain, vomiting, loss of bladder control, temporary paralysis, and inability to stand or walk. They got their wage increase.

Then Dr. Smith wrote one medical case report that would provide more information and more useful clues than all the others. Yet ironically he did not list the victim's full name. He wrote: "Reardon, England, 38." His report is simple but brutal: "Began work on the morning of May

17; was advised to work only one watch the first day, but nevertheless, feeling perfectly well after the first watch, went down again in the afternoon." Determined to work a second shift, Mr. Reardon refused to give his body the acclimation and trial period the doctors recommended. He refused to wait to see how his body responded to its first day of pressure. And so, after a morning of 2.5 hours hacking away at the river bottom, seventy-eight feet beneath the water, the stubborn Mr. Daniel Reardon, his full name known only through his death certificate, ate his lunch and climbed back into the shaft for a second shift.

It was his undoing. Immediately after his second depressurization, Reardon "was taken with very severe pain in the stomach, followed by vomiting," and that pain quickly escalated. It spread to his legs. Within minutes, he was paralyzed from the waist down, but still feeling the agonizing fire. Unable to move his limbs, he could feel rivulets of pain coursing through them, excruciating stabbing needles from the wrath of increased pressure released too quickly. It could not be controlled by morphine. Daniel Reardon died the next day after having vomited through the night, having felt no release from the chronic lightning bolts of agony that made him twist and bend his body into strange and terrifying shapes.

Reardon, unlike the others, had an autopsy. When the doctors cut him open, they saw that his lungs were engorged, filled with the fluid and phlegm produced by a body struggling for breath. His spinal cord was "intensely congested," meaning that it showed a telltale eruption of blood just above the nerves that, in life, should have controlled the movement of his legs.

Still, he was not the last. By June the newspapers were reporting the deaths with an air of fatigue. On June 18, laborer James Reilly's paralysis and critical medical state were announced under the worn-out-sounding headline "The Caisson Again."

The average sand hog on the Brooklyn Bridge lasted only three weeks on the job. Most wandered off without notice, either too hurt or too scared to continue. Although they were referring to other construction accidents, too, the newspapers called it "the Bridge of Death."

CHAPTER 3

Devil's Bubbles

The mysterious disease is now called decompression sickness, abbreviated DCS and more commonly known as "the bends." While the disease in its severest form can indeed force victims to contort their limbs and spine in agony until they look "completely bent up," its name originally started as a joke. The earliest known victims had milder cases from working in the lower pressures of shallower, less impressive bridge supports. These first workers started with minor pains and twinges, and to alleviate the issues, they sometimes walked with an odd, warped posture. The posture looked similar to the way rich women would warp their backs to bend forward and accentuate the bustles of their fancy dresses, a posture known as "the Grecian bend." So, to mock them, the affliction became known as the bends. The joke became less funny when the caissons reached depths where the pressures began to cause paralysis and death, but the name stuck.

The devil of DCS is nitrogen. The odorless, colorless, otherwise harmless gas with its large, lazy molecule forms 79 percent of the air around us, and because the insides of the caissons were filled with nothing but pressurized air, nitrogen formed 79 percent of that, too.

Normally nitrogen is a bit of a good-for-nothing gas. Made up of two tiny nitrogen atoms tied to each other in triplicate, it's a generally stable chemical, happy, almost boring. It's too heavy to use to fill a good dirigible, too stable to burn and give off pretty colors. It's not even interesting enough to make a good poison like carbon dioxide, or provide a funny smell like methane. Nitrogen is the boring neighbor on the cul-de-sac, smiling to everyone from its lawn, always ready to loan its garden tools to carbon when carbon wants to build a tree or a flower. Through its quiet existence and predictable early bedtime, nitrogen makes its neighbors oxygen, helium, and hydrogen look like wild partiers in comparison, with their balloons and their explosions, because—mostly—nitrogen just sort of exists.

Sometimes, however, it's the quiet neighbor who turns out to be the serial killer. And in the caissons, once the nitrogen reached high enough pressures, the depths became John Wayne Gacy's murder basement. The nitrogen crept in quietly while the sand hogs were at work. It oozed its way from their lungs into their bloodstreams with every normal-feeling breath, then from their blood into the cells of all their other tissues. Muscles. Fat. Joints. Nerves. As the workers chipped away at the rocks and the mud and the sand, silently their tissues grew soggy with dissolved nitrogen.

By the end of the unprecedented bridge construction and the unprecedented pile of dead bodies it produced, bridge physician Andrew Smith had deduced that the problem was nitrogen. Nobody knew exactly how the gas was getting into the workers' cells. Still, as of this writing, nobody knows how, exactly, nitrogen behaves as it does inside the body. But gases inarguably have some way of diffusing inward, of creeping across barriers, of leaking into and out of all containers, including the cells of the human body. Somehow, while Joseph Brown and the others were hacking away at the mud and the boulders of the river bottom, the nitrogen crept into them.

After the workers let the nitrogen ooze for three hours, spending a scant few minutes decompressing was not enough to let it back out without a fuss. With every cracking of the brass valve to release the pressure in the lock, the pressure within the workers' cells was also released. Suddenly loosed from the pressure dam keeping it in its prison, the nitrogen roared back as a flood. Tiny bubbles cascaded out and got lodged in the smaller passageways of the body, wedged themselves in, refused to move, refused to reabsorb, and, most important, refused to let any fresh oxygen-carrying blood pass by. DCS did not happen to every worker, and it did not happen on every ascent. But for those it struck, it caused mayhem.

The tissues, deprived of oxygen, would start to die, and in their scream of death, they would throw out jagged daggers of pain. Some, like survivor Joseph Brown, were lucky, because their bodies managed to carry on, to absorb the bubbles over time, to return to normal function. Others, like Daniel Reardon, were not. The bubbles lodged against their lungs, in their hearts and kidneys, and sometimes in the delicate nerves of their spinal cords. They caused difficulty breathing, cardiac arrest, paralysis, and death. The bodies, when opened up, were filled with bubbles and congestion, visual proof that the pain was caused by nitrogen-induced clogging and chaos.

The doctors knew that the risk and severity of the "caisson disease" increased with increasing pressure and increasing time. Some suspected the root cause was nitrogen, but nobody could prove it, and worse still, nobody could prevent or treat it. That is, of course, until the Haldanes came along. The Haldanes, who spent their lives breathing weird gases.

———

For young Jack, the home gas chamber was a sanctuary, and science time with his mustachioed father, John Scott, was his favorite escape. The tall, strong, self-described "fat" boy was relentlessly bullied at boarding school. The older boys physically beat him, sometimes every day for weeks, and they also found endless other ways to torment the socially odd, atypically brilliant young man. They chanted mocking rhymes about a small patch of white hair he had on the back of his head. When his uncle and namesake, John Burdon-Sanderson, visited, Jack missed seeing him because the other boys had pinned him beneath a flipped table loaded with bags of sand.

Sometimes he would respond to the bullying with dramatic losses of temper, which further encouraged the sadistic bullies to treat him like a spectacle who could be needled into hilarious explosion. On one occasion he tore an entire sapling out of the ground, including the roots, "and attacked his tormentor with it." After flinging down the uprooted tree,

Jack stormed into the classroom, fuming out loud, "I wish I could kill. I wish I could KILL." He would never explain to his parents why he hated the school; his mother knew only that "he just cried and said that he 'couldn't bear the place.'"

Because of such stories of unusual behavior, people have historically fixated on both JBS and John Scott Haldane from a sensationalized perspective. Records about both men focus on the anecdotes of oddities and atypical traits as if the two were zoo animals, a special species of nerd to be marveled at rather than treated as strictly human. Many records have spun and whipped true stories until they include overt fiction to augment the jokes at the scientists' expense. The take-home message written about both Haldane men is often the same: a patronizing version of "Let's marvel at how odd these people were." People who continue to propagate this message about the Haldanes fail or refuse to accept that some minds work differently, yet still beautifully. That some people communicate using less expected patterns, but they still communicate. People who propagate this message immortalize the bullying and torment after death.

When they were with each other testing gases, John Scott and his son were no longer social oddities. They were peers. They spoke the same languages of science and chemistry, and they both found science to be more delirious, more joyful, than any drug. They did not need to explain the passion to each other. Math always made sense; people were unreliable. So, when it was time for John Scott Haldane to test his latest theory—his revolutionary, mathematically based idea of how to prevent decompression sickness—he invited his son, Jack, along. Jack got an education about DCS thirty years before he would turn out to need it.

Decompression sickness happens not just because nitrogen is leaving the tissues, but because nitrogen is leaving too quickly. The body can process nitrogen harmlessly if it is given enough time. If the external pressure is removed slowly, blood vessels can do their job, pushing the nitrogen-filled blood along the surface of the lungs like a river grazing the sky, and the gas is harmlessly exhaled with the next breath. The most common metaphor used is a seltzer bottle either shaken up and released in a fury or cracked and decompressed one cap twist at a time.

Prior to Haldane, scientists and physicians tried to prevent the disease by advising workers to travel at a steady, constant pace from the maximum depth up to the surface, but some people also thought they could prevent it by installing fans in the caissons, showing how poorly most people understood why the affliction occurred. John Scott's idea was to ascend by ratios instead. He figured that if a body was still absorbing gas when it was at eighty feet, then, logically speaking, traveling slowly and lingering at seventy-nine feet, then at seventy-eight feet, then at seventy-seven feet, probably was not doing much to release that nitrogen. In fact, assuming the person had not completely filled their nitrogen stores to the brim by the time they left eighty feet, he reasoned that staying at those marginally shallower depths was actually shoving more nitrogen in.

So he decided to try removing the pressure by half. Nobody seemed to get caisson disease when they came to the surface from depths shallower than thirty-five feet, he observed. Every thirty-three feet of salt water provides one atmosphere's worth of pressure, so going from thirty-three feet to the surface roughly divides the pressure in half, from two atmospheres of pressure to one. So, if it's safe to cut the pressure in half, why bother going slowly from eighty to seventy-nine feet? Why not skip that long, malingering, cold, deep drama and instead cut the pressure in half immediately, then stay at the shallower depth to let the nitrogen bleed out faster?

According to the official records, for these potentially deadly experiments, goats went first. Stamping and clanking their hooves against the curved metal arc of a hyperbaric chamber inside a wooden shed at the Lister Institute in London, England, the patchy-colored furry conscripts shook their heads to clear their ears during the pressure change as the air grew hot around them during the distinctly ungoatlike experience of surviving the pressure of the ocean. But survive they did. So the humans went next, including Jack.

Around dinnertime on the warm summer evening of Monday, August 20, 1906, the HMS *Spanker* of the Royal Navy puttered into the harbor of

Rothesay, on the west coast of Scotland. Rothesay is a small town with rocky beaches and wild forests nestled safely alongside the still blue waters of the Isle of Bute, whose waterways eventually connect with the Atlantic. A circular, rock-hewn, surprisingly intact medieval castle and its moat perch on a high point of the town overlooking the isle, but from there the land drops sharply into the water. The cliffs plummet to hundreds of feet of depth without requiring much traverse out to sea, making them perfect for the extreme experiments that the Haldanes and their research partners had in mind.

The first full day of work, Tuesday, was spent checking equipment. John Scott Haldane and the other researchers—physician Arthur Edwin Boycott and Royal Navy Lieutenant Guybon "Guy" Damant and Gunner Andrew Catto—test-ran the gas pumps that would supply the divers with air, and made sure their hoses could flood a proper quantity of gas down to the depths without a struggle. Perched on top of the torpedo charging station was the small metal chamber in which young Jack proved he knew how to clear his ears. Without people inside, they also tested the limit of how fast they could smash the gas into the chamber, how quickly they could pressurize it should the worst occur and they needed to return a screaming, wretched diver back to the lifesaving depths to force the nitrogen bubbles smaller.

Divers Damant and Catto checked the underwater equipment. They would dive one at a time and stay warm inside a watertight canvas suit, which guarded against the life-sucking chill of the Irish Sea by enveloping them in thin layers of air. Their hands poked out through rubber seals at the wrists of the suit, and a thick-walled copper helmet screwed onto a metal ring around the neck. Each metal boot weighed twenty pounds, and the entire setup totaled 155 pounds, intentionally, to keep the divers lodged against the muddy bottom. An umbilical hose led from the helmet to the air pumps on the ship above.

Damant and Catto spent nine days repeatedly lowering themselves in this bulky diving dress down a ladder into the frigid sea, each carefully balancing the top-heavy copper helmet on his shoulders. Once they reached low enough to sink into the water, the buoyancy of the life-giving air inside the hard hat supported the weight of the helmet for them. They slowly increased the depth of their dive each day, taking little time off during a nonstop marathon of risky research. They started shallow, then inched the *Spanker* into deeper and deeper waters, lowering the divers down into the depths to test Haldane's new decompression procedures at extreme limits.

In 1872, caisson workers died after working deeper than sixty feet. In 1906, Damant and Catto revolutionized the industry by diving to two hundred ten feet. They set a new world record for the deepest, longest, most dangerous dive made up until that date, and, more impressively, they did it without a single injury. John Scott Haldane wrote the schedule for their return to the surface, and his pattern of ascent, called "staged decompression," is still the standard used by divers today.

During the sea trials in Rothesay, Jack and Naomi had started out onshore with their mother in an inn, playing among the forests and walking along streams to find their sources, pretending in the blazing summer heat that they were explorers trekking through the wilds of the Nile River. But, as Naomi looked on with jealousy, Jack earned an adventure on the *Spanker* by calculating for his father a handwritten table of logarithms to aid in the diving research. He was allowed on board to sleep in a cot that held gear, like he was a piece of driftwood tossed among the sailors' rifles and small arms. He got to be there while his father and others disrupted the world of diving in a way that permanently and irrevocably transfigured humanity's ability to explore the ocean. Naomi wrote in her diary, "I wish I were a boy."

On the tenth day, Jack received his real reward: He got to dive. He went for his pressure test in the chamber. He was fitted in the canvas suit. It was oversized on him, billowing, because even though he was tall and large for his age, he was still a child putting on gear made for adults. The wrist seals were loose around his teenage arms, but he did not complain because he did not want to jeopardize his chance to go.

He drifted slowly down into the ocean, plunking the heavy, oversized metal boots into the soft mud at thirty-six feet. Billowing clouds of silt puffed into dark whorls and consumed his legs in response to the brazen intrusion of his feet, rising higher and growing more opaque as he moved clumsily through the chill waters. The sea looked crystal green from his vantage, as if lit by an iridescent emerald glow. He held on tightly to one end of a rope tethered to a large weight on the bottom so that he would not get lost. The frigid water trickled slowly in through his ill-fitting wrist seals as he plodded along the soft floor, his hefty boots and helmet making more sense in the water, where the forces of buoyancy helped to lift them. He bounced along the ocean bottom like astronauts would sixty years later on the moon.

Jack Haldane returned to the surface hypothermic. It was the middle of an August heat wave, but the slow trickle of water in through Jack's wrist seals meant that he, unlike the larger, adult divers, was exposed to the cold temperature of the ocean. He had done his best to use the pressure valves to keep the water out, but it had made headway, creeping its way slowly up his legs and chest. Because he had refused to end his grand adventure early for something as irrelevant as hypothermia, he crawled out of the water wet, shivering, bluish, and happy. His laughing, delighted father fed him a shot of whisky to warm him; then Guy Damant helped the boy onto his own bunk and heaped it high with blankets.

To avoid injury, the caisson workers of the East River should have spent eighty-seven minutes decompressing from their deepest depths, rather than the mere nine they had given the process. Instead of steadily letting air bleed through the brass lock valve at a constant rate, they should have repeatedly divided their pressure levels by half until they reached the surface. The Haldanes had proved it, and in so doing, they had taken the first giant stride toward letting humanity exist underwater and—more important—return to the surface. In those ten days of science, the family established themselves as legends in undersea survival. John Scott was offered a knighthood for the work, but seeing no practical utility in a ceremonial title, he politely declined.

They took this large, bouncing, underwater step in 1906, but the rest of the leap for humanity would have to wait until one key, tragic day in 1939. June 1, 1939, to be precise.

By 1939, Helen Spurway was a fully credentialed PhD researcher in genetics and a grand master of the Mediterranean fruit fly. She prowled the halls and laboratories of the genetics department of University College London as part of J. B. S. Haldane's growing lab, where she developed her inherent talent for sustaining, nourishing, and relaxing into a willingness to breed all the myriad animal species she studied. In the year since she had finished her doctoral degree, she had coaxed into existence, like magic, innumerable new generations of creepy crawlies so she could examine the math of their special recessive traits like red eyes or curly wings. She feuded with lab mate and fellow young geneticist Cecil Gordon, but the two also pushed each other to complete an extraordinary paper that they would coauthor that year. It was fortunate that she was exceptional in her field because, five years earlier, at the age of nineteen, she had alienated many of her fellow students when she announced on the first day of class that she planned to marry the professor.

CHAPTER 4

Subsunk

When duty officer Commander George Fawkes peered in through the office doorway on June 1, 1939, Commander Ronald Bayne was already in a meeting. There had been no word from the submarine HMS *Thetis* on its sea trials in Liverpool Bay, so Fawkes interrupted to share the news of the sub's continued silence.

Bayne finished his talk but did so quickly, sending his visitor back outside into the crisp British summer breeze blowing off the English Channel that late afternoon. He hustled over to the Operations Staff Office within the stately, boxy government gray Royal Navy building, near a military shipyard on the southern shore of England. Fawkes had beaten him there and was already on the phone. Grabbing the handset, Bayne began to listen to the envoy of a rear admiral who was parroting official guidance into the other end of the line.

Nothing to worry about yet, the envoy attempted to placate Bayne. At 1356 that afternoon, the brand-new submarine HMS *Thetis* had signaled the start of its first-ever probing descent beneath the waves. The sub was overloaded with 103 crew members and civilian submarine engineers on board, and they were on a jovial trip to celebrate the completion of its construction and assess its function in the blue slip of the Irish Sea a few hundred kilometers south of where the Haldanes had emerged victorious decades before. Unlike those divers, however, the submarine had yet to return, even though it was nearly four hours later. But, the admiral insisted through his messenger, it was not yet time to worry. Or to take action.

Commander Bayne laid down the receiver dissatisfied. He had heard the Admiralty signaling to *Thetis* by wireless earlier in the afternoon, and he knew that the crew had given no response since their descent. He disagreed with the Admiralty's inaction. Bayne grabbed his maps and his lists of all the signals to and from the Royal Navy ships currently at sea, and he began to plot where and when those signals had been transmitted. He was plotting which ships were close enough to help, if help should prove to be needed.

HMS *Brazen* might be the submariners' best hope, he concluded, assuming his worst fears were true. The *Brazen* had recently steamed past where the *Thetis* should have been, so it wasn't far. However, it was traveling farther away, so every minute of delay now was worth two minutes later if the ship had to double back. Bayne drafted the order to command the *Brazen* to divert, to send its crew back for the *Thetis*. To begin a search.

He did not send it. He waited. The rear admiral had been concerned that any signal crackling across the open and easily intercepted airwaves would leak the news of the missing submarine to the public. But the clock had already ticked forward to 1755, so in one more hour, according to Bayne's math, it would be too late for HMS *Brazen* to make it back to the submarine's dive site before sundown. It would be too late for anyone to be able to find any escaped and drifting survivors who might have been floating unprotected in the frigid, dark chop of the North Atlantic, or to coordinate a rescue. Bayne kept his signal to himself. For now. Instead, he watched the clock. He waited.

By 1822 he had waited long enough. He never got the order to send the signal, but he was never expressly forbidden from sending it either, so he made a choice. He diverted the *Brazen*. Bayne and his staff buzzed into action, laying out charts, pulling protocol manuals, trying to find the most effective way to launch a rescue should the *Brazen* arrive on-site to find nothing or, worse than nothing, to find a telltale slick of oil on the surface of the water, which might have been the only trail left by a sinking submarine.

The insistent ringing scream of the office phone sliced through the noisy hum of the planning. It was the rear admiral's envoy again. They had gotten a message from the tugboat that was attending to the HMS *Thetis* during its diving trial. The tugboat had not seen it. The tug crew had called asking when *Thetis* was supposed to surface. The tug crew was alarmed, and so the situation was now, even by the admiral, officially "regarded as serious."

Thanks to Bayne's preparations, HMS *Brazen* was the first on-site, followed quickly by everybody in the Royal Navy who thought they could help. As the hours passed, more ships and more sailors arrived, adding to the burgeoning impromptu fleet surrounding the former dive site, now suspected to be the submarine's wreck location.

Telegraphist William Harold Bradshaw of HMS *Bedouin* was in a pub five hundred kilometers south in Weymouth when a barmaid announced what had happened. She was part of the informal network of civilian messengers who helped ripple the news through the harbor town, telling the sailors to hurry back to their ships. Bradshaw and the others ran, wild to help. Six Tribal-class destroyers, including the *Bedouin*, left immediately. The captains sailed so quickly, they left without huge chunks of their still gathering crews, instead designating two vessels to stay behind to scrape up the stragglers and bring them later.

However, once they reached Liverpool Bay, the peacoated mariners on the decks of the destroyers floated uselessly in the chill and otherwise peaceful salty breeze. They knew where the submarine was because the waters were shallow and they could see pieces of the stern poking above the low tide, but they could do nothing. Many broke down in tears at their own helplessness.

For a while the stern and propeller of the mighty *Thetis* poked above the surface of the ocean chop, providing a tantalizing visible lure offering the false promise of an easy exit for the 103 people trapped inside. Helpers on board a tugboat tried to get a line around the stern while it was visible, to stabilize it and raise it farther so the hull could be cut open, but the line slipped, the tide changed, and the water once again swallowed the *Thetis* whole. On the cusp of June 2, at 2300 hours, the Admiralty finally made the news public with simple words: "The

Admiralty regrets that His Majesty's Submarine *Thetis* . . . has failed to surface."

Less than four hours after the announcement, in the early darkness of the morning, among the ships circling the wreck site was the specialized diving ship HMS *Tedworth*, carrying now retired Captain Guybon Damant. The same Guy Damant who thirty years earlier as a fresh lieutenant had wrapped young, dripping-wet Jack Haldane in a blanket and nestled him into his own bunk after a shot of whisky. But now Damant was in charge. He immediately began feeding information back onshore, where it trickled its way, perhaps deliberately, to the physiology-savvy adult that Jack, now known as JBS, had become.

The circling surface ships took turns returning to Liverpool for supplies. The sailors knew little, but had permission to share nothing. Each time they returned to port, they had to disembark and walk silently past the waiting, worried crowds of families searching for information about their missing loved ones.

In the end, four *Thetis* submariners managed to escape. They were scooped alive, but barely so, out of the frigid and brutal ocean. The rest died inside. For most of the sailors, the deaths of those inside the submarine would have been only the first of many familiar lives lost in the coming war, but still they were the first. For telegraphist William Bradshaw of the HMS *Bedouin*, at least, June 1, 1939, "was the day the war started."

It was not immediately clear how the remaining ninety-nine had died or why so few had been able to escape despite having the breathing apparatuses and tools theoretically required to do so. Yet again the ocean had presented a fatal underwater mystery.

In the early 1900s, the British had assumed their submarines would be able to come up for fresh air at their leisure. They focused their submarine development on designing patrol-ready boats for defensive purposes only, to be used in territories where they had full control. The Germans, however, did not. The Germans built swift and silent killing machines. As a result, the first nine deadly submarine encounters of World War I were unequivocal clean sweeps by German U-boats, including one dramatic event in which a single German sub sank three armored British cruisers in less than one hour. The British would spend the next few decades trying to catch up with those mechanical designs, and they arguably did, but they seemed to have ignored most of the physiology.

Some British subjects were more prepared to address toxic gas problems than others. During World War I, poisonous gases caught the soldiers in the trenches by surprise. The toxic gases filled their lungs and choked them out. One of those soldiers was twentysomething J. B. S. Haldane.

JBS had signed up early for the vaunted Scottish Black Watch. By February 1915 he was an officer "teaching bombing to young ossifers." One field marshal described him as "the bravest and dirtiest officer" in the army, consistent with Haldane's statement that "the dirt is too far into some parts of me to wash out," so he did not bother with baths for months. J. B. S. Haldane, indoctrinated into a tolerance for pain by science, described a month of nonstop bombing as "one of the happiest months of my life" and WWI overall as "a very enjoyable experience," despite the fact that he was twice physically torn apart by shrapnel from major explosions.

His father, John Scott Haldane, might have been too old to be drafted into combat by the time of the Great War, but he found a way to contribute to the war effort through his knowledge of respiration. He locked himself inside a home gas chamber again, this time with material designed to absorb carbon dioxide inside enclosed spaces like submarines to test its performance. During one experiment, he stayed inside the chamber, breathing, for fifty-six hours.

John Scott also sealed off portions of the house and asked for samples of the poisonous gases wreaking terror across the front lines. He first helped to identify the mystery gas wafting across the trenches as chlorine, then decided to find out how much of it the human body could tolerate. He used his contacts to have his son pulled from the front lines, not to keep him safe, but so that he could put JBS in the box, too, and see how he responded to inhaling deliberate clouds of the horrifying weapon being used to blind and choke soldiers in the field. More data points are always better in the world of science, and his son could be counted on to head into the box of poison willingly, aware of the dangers and ready to report back. Meanwhile, JBS learned more and more about how to examine the fragilities of human lungs and about breathing with limited gas supplies.

During these experiments, Naomi and Kathleen were determined to design better gas masks, so they shredded their dresses to test the gas-absorbent properties of different types of materials. They stuffed the materials into prototype models, working as a team with John Scott, JBS, and author Aldous Huxley, a family friend.

In the midst of the war and the chamber experiments, Naomi, at age seventeen, also wrote an academic manuscript to publish the results of some of the earliest genetics experiments that she, JBS, and collaborator A. D. Sprunt had started with mice prior to JBS's deployment. Although later biographers would downplay and sometimes omit her contribution to the work completely, Haldane's own letters home tell the truth. When their coauthor Sprunt was killed in combat, Haldane asked famed geneticist William Bateson to help Naomi finish the paper if Haldane died, too, but only, he specified, "if she wants it." No matter how all-encompassing the war, the Haldanes, and everyone they had sucked into their orbit, looked at it through a lens of science.

For JBS, during the peacetime stretch between the two world wars, while HMS _Thetis_ remained a pile of metal sticks and a forming idea, that orbit grew to include a group of like-minded scientists in a proper academic lab at University College London (UCL). All of the lab members would shortly find themselves learning about the science-meets-magic of surviving underwater when they decided to solve the question of why so few escaped from HMS _Thetis_. Many of the members were in the UCL lab for the same reason the Admiralty had built the _Thetis_: a mustachioed madman named Adolf Hitler.

———

Haldane and Hitler both had thoughts about biology, both strong, but different. By 1933, six years before the sinking of the _Thetis_, Haldane and a slowly growing crew of misfit nerds had taken over a few small rooms in the corner of University College London in the bustling downtown center of the city. They were part of a group who did research in the

Galton Laboratory in the zoology department, where they could breed flies, ducks, newts, mice, rats, and any other small animals on which they could test mathematical theories.

The flies were special favorites. They bred the fastest, which let the group test their theories more quickly. The vibrations of wings humming inside bottles and mesh cages provided an ever-present soundtrack that filled the musty, mouse-scented rooms. Inside UCL's white stone buildings in the middle of the lively interwar city, the scientists with their flies searched for answers to the mystery of how genetics rippled physical traits down through generations.

The 1859 suggestion by Charles Darwin that mutations gave some organisms a survival advantage had sparked immediate furor in the scientific world, but in a time before the identification of DNA, it also divided the scientists firmly into two camps. The biometricians, like Haldane, wanted to study genetics by looking for observable differences and using math to describe how those differences propagated down through generations. The eugenicists, in contrast, believed the theories of genetics should be used to inform deliberate breeding choices for the betterment of species.

The idea of breeding burlier pack animals or meatier crabs was globally popular, as it was simply what farmers had been doing for thousands of years with less math. But when the science was applied to people, things grew controversial. The idea of eugenics being applied to promote the nebulous concept of "good breeding" in humans was often an implied or even overt facet of the field. It was a classist, ableist, racist, but disturbingly commonly proposed application of the science, fueled by layers of bias over the interpretation of what "good" meant. Usually, that definition meant white, wealthy, abled, and European. By 1933, the scholars of UCL were firmly divided into the two camps of biometrics versus eugenics, like they were at most academic institutions of the era. The biometricians of the Galton Laboratory had already been fighting eugenicists for decades in the academic sense, but soon the battle against eugenics would get more literal.

With the rise to power of the eugenicist zealot Adolf Hitler in early 1933, the higher-ups at University College London swiftly realized how many ingenious Jewish minds would need refuge from Hitler's planned anti-Semitic application of the theory. On April 7, 1933, Hitler banned all Jewish people from government employment, and as a result, German public universities fired all their Jewish scientists and professors. Quickly thereafter, all Jewish students were expelled, too.

Hitler made no secret of the fact that the mass firing was merely a start and that he planned to use the theories of eugenics to "improve" the human species as a whole. So organizations like the Rockefeller Foundation and the Academic Assistance Council sprang into action and provided scientific funding to pay the salaries of Jews who wanted to flee. By November, sixteen exiled Jewish academics were working on their groundbreaking science at UCL instead of in Germany. Even though anti-Semitism was prevalent in England as well, as it was in most societies of the time, other British universities quickly joined the effort.

The often bullied Haldane was happy to accept scientists fleeing tyranny. He was notorious for being prickly, bordering on vicious, toward individuals who crossed him. His fights with a eugenicist named Ronald Fisher have become the stuff of legend. However, he was also known for optimistically kind delusions like believing he could eliminate global racism by writing a sufficiently clear essay about how skin color is nothing more than genetic variations in pigmentation. In the 1930s he wrote that he could see no genetic issue with interracial marriage, which, to a genetics obsessive like him, meant that he could see no justifiable problem with it.

His temper aside, he seems to have genuinely believed that all categories of humans held equal potential for greatness, regardless of race or religion. By the late 1930s he had married, so, possibly encouraged by his wife, Charlotte, who was both Jewish and the daughter of immigrant German Jews, he became an active lobbyist for hosting as many of the refugees as possible. He would later jokingly declare that he had two men to thank for his lab: Adolf Hitler for firing the scientists, and John D. Rockefeller for funding their flights from Germany.

Two of those refugees were Ursula Philip and Hans Grüneberg, both stunningly brilliant PhD scientists who happened to understand genetics, mathematics, and the noble fruit fly. German scientists who happened to be Jewish. And so, in 1933, they came to the entomophilic lab at UCL. By 1939, in time for war, they were settled members.

––––––––

Hans Grüneberg said that when Adolf Hitler spoke about his plans, Hans believed him. In contrast, many of his friends and family members murmured general disbelief in the early years of the 1930s. They lived with hopeful denial and a tendency to shake their heads at the madman's proclamations and anti-Semitic ramblings before he came to power, declaring "Well, this is so absurd, this can't happen" in response. Grüneberg, however, heard the proclamations as promises. And so, when the Nazis came to power, Hans said he knew he "would have to leave Germany as quickly as possible." Within the year, at the latest. With limited planning, if necessary. "My main plan," he stated, "was to try and survive."

Grüneberg was a medical student with a high forehead, an oval face, and full lips at the age of twenty-five when Hitler expelled all the Jewish students from the universities. He was roughly five foot six in height with a slight tendency toward being overweight. Clean-shaven, with meticulous grooming habits, he carefully slicked back his opulent head of jet-black hair each morning.

Grüneberg's first action after the expulsion was to ask for help from a friend at a local, privately owned hospital near his home in Freiburg. If he could make it through the final few months of his degree program, he figured, he could take his newly minted medical degree and run with it to one of the three countries that would honor the German credentials. Those options were Ethiopia, Siam (now Thailand), and Iran. He knew nobody in and nothing about any of the three countries, but he thought Ethiopia might have the best climate. To execute his strategy for survival, he began pestering Ethiopian embassy personnel in Berlin for tips about connections, and sending letters out blindly.

One day, out of the blue, a letter arrived for him. Hans Grüneberg had never heard of J. B. S. Haldane, but J. B. S. Haldane had apparently heard of Hans Grüneberg. When the Academic Assistance Council started funding the exodus of well-established, already venerated Jewish scholars and academics from Germany, Haldane had reasoned correctly that the younger scientists who had less fame might have a harder time getting funds, so he began hunting down and recruiting new graduates among the applicants. Grüneberg said that even though his father initially thought "the whole thing was a hoax" because it was too good to be true, the receipt of Haldane's invitation "disposed of the hare-brained scheme to go to Ethiopia."

By a Sunday evening in mid-August 1933, Hans Grüneberg found himself safe in a hotel in London, England, just around the corner from the UCL campus. He had been single when he sent out his blind application letter, but he had married just before leaving Germany, and his wife, Elsbeth, planned to follow as soon as she finished her exams to become a dental surgeon. He had told nobody in England about the recent marriage for fear it would ruin their chances for flight.

Grüneberg arrived on the steps of UCL with stereotypical German punctuality at eight a.m. the Monday following his arrival. He found himself alone, with all the building doors locked. By nine a.m., a lone beetle crawled by. It was the first break in his solitude.

It was not until ten fifteen that Haldane manifested, and when he did, he appeared as a whirlwind. Haldane greeted Grüneberg with hectic enthusiasm and insisted that for his first day they should take a trip to the London Zoo to pick up Grüneberg's spirits. The zoo trip was followed by a family dinner at home with Haldane and his wife, Charlotte.

Haldane was a brand-new department head with a hands-off mentorship style, and Grüneberg was a young but self-piloting scientist brimming with questions of his own, so they got along immediately. Hans Grüneberg "got the feeling that [he] had got into a world of freedom" in England and at UCL, "an environment where [he] could develop as a scientist unfettered." Shortly, he would start rushing with regularity into Haldane's office to announce his new discoveries, feeling extra schadenfreude because he had gleaned the knowledge from a line of rats he had gotten from Germany, where the scientists of the country that had rejected him had "failed to trace down the cause [of the mutations] or to work out the genetics."

Grüneberg was alone for only a few weeks. Soon, Haldane would scrape together funding for another scientist.

Ursula Philip.

Dr. Anna Ursula Philip went by Ursula, or Ulla to those she considered friends. At five foot four, she was a sturdy, large-chested woman whose natural posture showed a slight stoop in her shoulders. She had extraordinarily clear, peach-tinged porcelain skin and high elegant cheekbones that were part of her strong, angular facial bone structure. Her distinctive black eyebrows formed heavy, dark punctuation marks to accentuate her bright, kind brown eyes, and the long, tightly coiled black curls of her hair often swirled with a mind of their own like a dark halo around her pale features and bright smile.

She had been raised wealthy. Her father had been a banker and her mother had descended from the Lithuanian dynasty of a successful kosher sausage business. The elegant couple had welcomed Ursula into the world in Charlottenburg, Germany, on September 6, 1908. Her first few years were happy, full of card games and joy, but with the advent of the First World War, everything changed.

Ursula's father, Dagobert, was a physically large man, full of strength and power, but time in the trenches and the mud and the shelling and the rampant disease changed him as it changed so many others. He became quiet, sullen. Dagobert withdrew from his family, his life, his children. There is no record of an official diagnosis, but his family described his condition as "shell shock."

In Dagobert's mental absence, Ursula's mother, Hedwig, was forced to bloom. She was the immaculately dressed picture of grace and style, and her smarts and confidence served her well as she learned to navigate the new trials forced upon their little family by the horrors of the trenches. Having lived firsthand the need for women to be able to function autonomously, she imposed the same strong, anachronistic feminist idea of women using their intelligence, getting an education, and being independent on her eldest child, Ursula, whom she encouraged and supported in school and in science.

As a result, on August 27, 1933, the independent and newly graduated Dr. Ursula Philip could submit her application to escape. She spoke English, French, and Italian in addition to her native German, and she had earned her way through degrees in mathematics and biology, followed by a PhD in genetics. Her immigration application humbly listed her pioneering work researching the way chromosomes interact with one another during meiosis, the process through which cells divide to become sperm or eggs. After Hitler banned Jewish employees and most Jewish university students, she barely managed to cram her doctoral defense onto the calendar on July 20, 1933, before he shoved her out. When she departed, she left behind her parents and her two younger brothers, who trusted a friendly police officer to tip them off when they needed to run.

Rumors about Ursula's flight from Germany suggest she had to pack in such a hurry that she had to abandon the physical piece of paper proving her PhD and have a friend mail it to her later, for fear of being trapped in the coming tidal wave of hatred. However, her daughter dismissed that suggestion with a loving chuckle. Ursula was strong, kind, and ebullient in a way that would make her a unilateral favorite with friends and students for decades to come, but she never quite fit in with the elegant, refined, polished world of her mother. Ursula, in contrast, was always a bit scattered, always a bit uncomfortable in her fancy clothes, always a bit slouched, always a bit likely to spill food on herself at a nice party. According to her family, she simply forgot her degree paper and had to ask for it to be shipped once she remembered.

Ursula's failure to fit in with prescribed but functionally useless social mores might have been why the outcast Haldane immediately admired and liked her. By September 1933 she was firmly ensconced in his lab, studying the underlying calculus of fruit flies and greenbottle flies and applying the elegant but wildly complex language of statistics to determine how they passed traits to their children. In an era when women were often barred outright from science, and the few other researchers at UCL who were willing to hire women hired them only because they could pay women less than men, Haldane referred to Philip in letters as "invaluable" and "of the greatest value." But perhaps more important, he was willing to pay Philip and Grüneberg the same salary.

Despite her being inside the relative sanctuary of the lab, Ursula, like anybody living in London in the 1930s, was not immune to the struggles and conflicts of the rest of the world, which swirled through the streets well before the war. Unemployment in the interwar period reached astronomical levels worldwide, and in England the Great Depression of the 1930s was characterized by protests and hunger marches of the unemployed, who often passed the buildings of UCL on their way to the steps of Parliament.

Each political party screamed that they alone had the solution to the plague of economic depression, both globally and locally, and in an eerie forecast of the global war to come, some groups were willing to fight in the streets for their version. In one particular march, anti-Semitic fascist Blackshirts blamed the Jews for their problems and marched through London's heavily Jewish East End. The march turned into a brutal and now infamous urban battle with the fascists, along with their police escorts, clashing violently with Jewish, Communist, Socialist, and other anti-fascist protesters who had turned out to stop them.

Through all the conflicts ran whispers of the eugenics movement, murmuring not so subtly in the background: *These people are the source of our problems because they are "less than." They should be eliminated, either through selective breeding or more directly.*

Hans Grüneberg and Ursula Philip watched these conflicts through the windows of their lab in the 1930s. But despite it all, in another few years, they would prove willing to put their bodies and lives on the line in defense of their new country.

During the social unrest, Communism gained new and shiny appeal as one of the ways to oppose fascism, and so it would reach its tendrils for J. B. S. Haldane. Because many English citizens were inspired by Russia's 1917 Bolshevik Revolution and the falsely rosy descriptions being fed out to the world by the propaganda machines of Lenin's new government, England would become one country of many to host homegrown Communist parties demanding the same purported egalitarian utopia for themselves.

By 1937, after a year marked in London by protests, riots, and beatings in the name of employment and fairness, the Communist party's supporters were joined by the always underdog, always ready for a fight, outspoken J. B. S. Haldane. He was one of many scientists taken in by what would turn out to be false reports of "the remarkable advances of scientific theory and practice in the Soviet Union." And because he was willing to chat with Communist Russia, Britain's elite spy agency MI5 started watching him. Unbeknownst to Haldane, they began tapping his phones. Reading his mail. Attending his lectures and taking notes to be filed in hefty reports with his name in the title.

Once on board with Communism, Haldane submitted himself to it fully, as he did with everything else. He would take his extreme beliefs along with his expertise in gas masks into combat on the European continent during the Spanish Civil War when he volunteered to fight overseas with the International Brigade. He seems to have been universally beloved there for his lectures to the troops about science, and for his notorious jacket with its one remaining straggling button, which at one point he refused to change for nearly four months straight. So it wasn't entirely surprising that in 1937 he brought back with him a group of Communist friends and compatriots. Just two years later, this group would be willing to put their bodies on the line at his request to prove what had happened inside the British submarine HMS *Thetis*.

Four submariners from HMS *Thetis* escaped, but the boat was raised after the final ninety-nine had perished. It was brought to a beach where Royal Navy trainees were among those who stood and watched the removal of the corpses. The retrieval personnel used breathing apparatuses that had been designed to allow safe entry into mines filled with poisonous gases, and they climbed down into the submarine's hull to carry out the bodies of their lost brethren from within an atmosphere now filled with the fetid stench and myriad chemicals produced by rotting corpses. Those removing the bodies put up a massive eight-foot canvas sheet to block the morbid view, to hide from the public what the water and the submarine and the limited supply of air had done. The trainees onshore never saw the contorted ways that death inside a submarine caused the dying to warp themselves in pain.

Almost before the bodies were removed, the confusion and the debate began. The four who got out were in unexpectedly rough shape, surprisingly extreme physiological distress, which led the Admiralty to ask: Would it even have been possible for an entire submarine crew to enact their own escape? All of the Allies' plans for escapes from submarines had been based on the theory that the crew would be physically able to save themselves. Had those theories been based on a total lack of understanding about what being inside a downed submarine would actually be like?

The bodies inside revealed a protracted, painful, helpless death, and the Allies knew so little about survival underwater that they did not know how long it took for people to become unable to help in their own escape or rescue. The *Thetis* disaster forced the harsh revelation that, even though the Allies had built these new subs, once they disappeared beneath the surface of the ocean, how to stay alive inside them remained an understudied mystery. A boat surrendered to the ocean meant a boat surrendered to a realm they did not understand.

Except, of course, for J. B. S. Haldane. He had first stepped onto a submarine in 1906 at the age of fourteen. A year after his diving adventure, he went on board to help his famous father conduct scientific measurements of the breathing gases inside an early prototype, and even at that age, he had proved to have a superior understanding of the trials inside the boats by reciting from heart the chemical formula for carbon dioxide absorbent. So, while the country reeled from the tragedy of the lost *Thetis* and the Admiralty gathered its experts for a tribunal to discuss and wonder what on earth had happened, J. B. S. Haldane planned an experiment to find the real narrative of the submariners' final moments. As he had been taught since birth, he would try the experiment on himself first.

Haldane wrote to his contacts at Siebe Gorman, an English company known for its manufacture of diving hard hats, the mine breathing apparatuses used for the body recovery, and even the Davis Submarine Escape Apparatus (DSEA) used by the lucky four who had managed to squeak their way out of the doomed *Thetis* atmosphere. Siebe Gorman had a magnificent brick-walled factory in the heart of London, a stone's throw from the bustling Westminster train station and across the street from County Hall. Their engineers were the ones who had been receiving information about the sunken sub directly from Guy Damant while he was at the wreck site looking for survivors popping to the surface, because they knew any survivors would be wearing Siebe Gorman DSEA breathing devices. Inside their factory, they had all the equipment needed for designing and testing breathing equipment, including several large water tanks, plenty of bottles of compressed gas, chemical analysis laboratories, and, perhaps most important, three hyperbaric chambers.

Haldane chose the largest gas-tight chamber for his *Thetis* experiment. The large chamber was an upright cylinder with a flat floor and curving walls like a tiny grain silo. Stuck onto one side was a small air lock, which was a separate attached metal room that could be brought to pressure or released independently from the main space. This lock allowed people to enter or leave during the experiment, if needed, or send in supplies. At Haldane's height, he would have had to stoop to shuffle in through the air lock's heavy, rounded steel double doors. By his math, the *Thetis* contained 4,710 liters of available breathing gas per submariner inside, and this biggest chamber contained 5,800 liters for himself alone. Close enough to be his own personal faux-submarine.

At 10:10 p.m. on an otherwise average evening in July 1939, Haldane stooped. He shuffled through both doors. The inner door was closed and latched behind him, pulled snug and secured with metal bars to force its rubber gasket to form a seal. With a hiss, a new gas began to penetrate the space.

The goal was for Haldane to see what the submariners would have felt like after their first night spent accidentally beneath the surface of the Irish Sea. He started with 2.5 percent carbon dioxide mixed into the air. A slow gas leak out the door's seal meant that the chamber operator outside had to periodically twist a valve to replenish the pressure inside with the occasional reverberating hiss. Even still, JBS was asleep by 12:30 a.m. Overnight, he drifted in and out of dreams while the operators outside observed him through the chamber's tiny round window.

By 8:30 a.m., he was awake, and he was panting. The carbon dioxide had risen to 5.1 percent overnight, and his body was forcing him to breathe harder, automatically triggering his tendency to hyperventilate in a misguided effort to remove the toxic chemical from his bloodstream. Humans evolved to breathe harder in order to eliminate the carbon dioxide we generate from exercise, but we have not evolved to survive inside submarines. So, because the carbon dioxide was coming from outside rather than inside, his body's efforts to eliminate the poison were useless.

By 10:50 a.m., the headache set in. His eyes became sensitive to light. Waves of nausea began to come and go. His pulse was lower than normal. At 12:40 p.m. came symptomatic confusion. By then he was breathing a survivable level of 6.6 percent carbon dioxide, but he'd been breathing it for fourteen hours, so the chemical had built up in his body and his tissues to an alarming degree. At 12:40 p.m., he stopped. He left the chamber, and once outside, he put on the same model of Davis Submarine Escape Apparatus issued to any British submariner hoping to escape a waterlogged coffin. It was an oxygen-filled, airtight bag with hoses connecting to a mouthpiece, and a canister to remove exhaled carbon dioxide. It had been designed to supply desperate submariners with oxygen while they clawed their way through dark waters to the surface.

The results speak for themselves: "After 2–3 minutes [Haldane] removed the mouthpiece in order to vomit, which he did at intervals, probably bringing up about a pint of clear fluid. He had not eaten or drunk for the previous sixteen hours, in order to parallel conditions in the submarine. A violent headache developed, mainly front, but somewhat diffuse. This was severe for about an hour, and passed off about 5 pm."

A photograph of the five participants in the *Thetis* testing, which appeared in the Communist newspaper *The Daily Worker* on July 21, 1939 (*from left to right*): Bill Alexander, Don Renton, J. B. S. Haldane, Paddy Duff, George Ives.

These were the previously undescribed effects of a sudden reduction in carbon dioxide, and they were brutal.

But one data point was not enough for good science. This set of symptoms was how J. B. S. Haldane reacted, but was everyone the same? Or was it just him? He began to recruit from among his friends. He focused on those he thought of as his bravest friends—namely, other Spanish Civil War volunteers.

Four days later, fellow International Brigade veterans Bill Alexander, Don Renton, Paddy Duff, and George Ives performed their version of the stoop and shuffle into the same chamber with Haldane, just as they had walked with him into combat in Spain. This time, they started the trial with 6.1 percent carbon dioxide to move things along. By the end of the first hour, now with five bodies generating CO_2 instead of only one, "all were panting severely." Paddy Duff "was in considerable distress." He had five major shrapnel wounds from the conflict that had inspired Picasso's painting *Guernica*, and they exacerbated his anguish. One of his wounds was sufficiently severe that he had lost the use of his left arm, and other wounds still needed reparative surgery. They left the chamber and everyone but Haldane grabbed a DSEA mouthpiece to start sucking down pure oxygen. Alexander "removed the mouthpiece almost instantly, and vomited repeatedly, though he . . . is a good sailor and does not vomit readily in other circumstances." Duff and Renton were "temporarily incapacitated" by the severity of their ripping headaches, and Haldane and Ives, too, had migraines.

The take-home message of the experiment was clear: Whether they had used oxygen or breathed normal room air after the carbon dioxide exposure had not mattered. The Admiralty had at one point theorized it might have been the oxygen making the survivors vomit, but with either oxygen or air, they all got sick. Therefore it was not the oxygen that was the problem; it was the carbon dioxide. Carbon dioxide was why the crew of the *Thetis* had contorted themselves before death, huddling into the fetal position, often biting into their own tongues in agony. Nitrogen caused the bends, but carbon dioxide under pressure could be just as deadly.

Haldane took the stand at the Admiralty's formal inquiry over the submarine accident. He presented the results of the carbon dioxide experiment to the government officials and stated clearly that the odorless, colorless gas was the limiting factor in survival. His father, John Scott, had died and so was not able to speak for himself, but JBS revived his father's idea, the idea that JBS had learned as a teenager: that for anyone to survive inside any enclosed submarine for long periods of time, carbon dioxide management was key. By the end of his testimony, the Admiralty was convinced. They asked Haldane not *why* absorbing the carbon dioxide was key, but *how much* absorbent they should start carrying on their submarines. Haldane whipped out a pencil and, complaining that he still had a slight migraine from leaving the chamber twelve hours before, did the math on the stand with everyone watching. One hundred pounds of absorbent would have allowed many more to escape, he said, to survive, to stay vomit-free as they swam for sunshine and life.

At least one of the dead crew of the *Thetis* had removed the retaining strap and yanked the mouthpiece of his escape apparatus out of his own mouth before he died. He was unable to operate the escape chamber, and it might have been because of his need to vomit. When the crew climbed into the escape hatch and inhaled their first breaths of pure oxygen from their DSEAs, after having breathed CO_2 for days, they would have been immediately stricken by nausea, vomiting, migraine, and the various other symptoms that had hit the volunteer divers in the chamber on land. The official write-up of the trial put it bluntly: "It is suggested that some of the remainder were incapacitated by vomiting or other symptoms." Thus the crew failed to operate the escape hatch. The hatch was rendered useless by their failure. No more could crawl in.

In the newspapers, Haldane spared the families of the dead when he told the press that the submariners had felt little pain, and had gone quickly without agony. It was a lie.

———

Fresh on the heels of the carbon dioxide revelation and in the face of mounting political tensions, it did not take long for the Admiralty to reach out to Haldane and the UCL genetics lab with more questions. By the end of that same July, Royal Navy commander Martin Dunbar-Nasmith, recipient of the Victoria Cross during WWI, wrote asking for help. At first, the help he requested took the form of a simple memo formalizing the explanation of the *Thetis* chamber tests. The Royal Navy had repeated the same experiment and gotten similar results, and they wanted to know more about why. But their definition of "help" would soon grow.

The researchers began studying submarine escape after the accident of the *Thetis*, not yet knowing that by the end of the war, their work to allow underwater survival would become critical to divers and to scouts leaving tiny submarines that lurked on the bottom on purpose.

Siebe Gorman told Haldane the chamber "facilities are unreservedly at your disposal for any further tests on yourself or others that you may wish to make." MI5 quietly added a press clipping about the *Thetis* test to Haldane's espionage surveillance file.

———

Clearly, more work needed to be done. However, the veterans of the Spanish Civil War who had climbed into the chamber for the *Thetis* trials were not officially lab members. They had other jobs, and they were not able to dedicate the hours, days, weeks, months, or years of their time that would have been required to puzzle out the painstaking and elaborate science now being requested by the Admiralty. Alexander, Renton, Duff, and Ives went back to their normal lives. Citizens of London began to evacuate as international tensions heightened. Haldane's secretary was one of them, in her justified and accurate fear that soon German planes would start directly attacking the juicy target of the heavily populated city. So the lab would need new members to serve as guinea pigs for the newly assigned war work.

Haldane had already been warned by his sponsors at the Rockefeller Foundation about having too many Jewish people in his lab. One Rockefeller representative commented about the fact that Haldane's wife, Charlotte, was a "Jewess" and cautioned him about the optics of stacking his lab with too many refugees of that religious persuasion. Given that they were funding the Jewish lab members' salaries and therefore the organization must have been supportive of helping refugees, the reasons for the warning were not clear. Perhaps they were responding more to external social pressure than organizational beliefs. While their motives were not clear, their message was.

As Hitler's troops began to wreak their first glimmers of true mayhem across the farmlands and cities of continental Europe, more Jewish scientists submitted desperate applications, hoping to flee. With the end of the cold, intense, but precombat standoff period of the "Phoney War" and the looming horror of real, flesh-tearing conflict with the Axis, Haldane used the evacuation of London to escalate his efforts to save refugee lives.

Despite the caution he had received, Haldane accepted at least two more Jewish scientists, one of whom would stay with the lab through the war.* He also needed a new secretary. Two old scientific friends from the past would also send desperate letters to Haldane looking for work and get sucked into his whirlpool.

Once these four new lab members were added alongside Helen Spurway, Ursula Philip, and Hans Grüneberg, their wartime cohort would be complete. They would put themselves through hell together.

---

*Dr. Dora Bak, born in Austria in October 1889, was a hematologist who received refugee funding and worked with the lab for a short time in 1939 on her arrival in the UK. Unfortunately, she is not a part of this story because of lack of information about her, and because she did not stay with the lab through the experiments.

CHAPTER 5

Love, Cooking, and Communism

Clutching the flimsy pages of an evening-edition newspaper in his hand, geneticist Hans Kalmus settled his lanky, athletic frame down to wait. He was perched on the bleach-white cliffs above the port of Dover, England, where tourists typically flocked for views of the soaring, naturally white walls and the waters dividing England from mainland Europe. But Hans had no eyes for the view. Not today. He looked down, locking his anxious gaze on the exit doors of the immigration building below. Hans was worried about his family . . . and the headlines in his hands. He was Czechoslovakian, and Jewish enough to fit the Nazi definition, and according to his newspaper, Hitler had just invaded Prague.

Hans Kalmus did not know the name of the ship his wife and two sons were supposed to have taken to escape. He knew only that they were supposed to arrive today. He did not know if they had managed to slip between the dark gnarled talons of the Nazi army, now tightened around their formerly warm and loving home. He would have to wait. He might wait hours. Or he might wait an eternity. Hans, sitting on the windswept cliffs overlooking the vast, ocean-filled gap between him and his homeland, watched the doors.

———

Hans Kalmus had learned as a child that there was Jewish heritage blended into his bloodstream, but he had been raised Lutheran, attended a mixed-religion school, and made friends with children of all back-grounds. He had never identified as Jewish, or religious at all for that matter. Rather, the skier, mountaineer, and competitive track star had always felt more German than anything else, since he was part of the German-speaking minority population of Czechoslovakia. In the first few years of expanding anti-Semitism immediately following Hitler's rise, the young man had not yet been frightened by the cartoons being used to depict Jews, because he did not recognize the features of his own long, angular, granite face in the grotesque caricatures.

The first dire warning of his changing new world came on March 12, 1938, when Kalmus was fired from his university job. He had refused to start classes with *"Heil Hitler,"* and his wife, whose maiden name was Rosenberg, was inarguably Jewish. Therefore, so were his sons. A friend, who now had anti-Semitic fires burning behind his formerly welcoming eyes, strongly suggested Hans abandon his wife and children and find a way to show allegiance to the still-growing Nazi Party.

Hans stayed in bed for two days. But it wasn't to think about his decision. It was from shock. He already knew what he had to do. Once he recovered, he acted quickly. In his words, Hitler could "lick his ass."

Hans Kalmus had fallen in love with his wife, Anna, who went by Nussy (pronounced *NOO-shee*), the second he saw her descending from the steps of a ferry. Her hands were nestled inside a fur muff, and she navigated the steps with such bearing and beauty that Hans forgot he was there to meet a friend. Luckily for Hans, that friend turned out to be Nussy's brother.

Nussy's regal lifestyle included things like servants, bodyguards, and a villa; Hans was a humble scientist. He proposed quickly, and she accepted, even though the brash, irreverent young scientist was considered by her family to be "below" her. The two were silly together: she insisting that cloth dinner napkins be folded neatly so they formed two precise horns, and he plucking her fabric origami out of the water glasses to hold the horns behind his head and chase her wildly around the table. He would make braying animal noises, and she would laugh while hurling words at him in Hungarian that they would never teach their children.

In Prague, they made a home characterized by love and exceptional

Hans Kalmus and Anna "Nussy" Rosenberg Kalmus.

cooking, and by the time Hans Kalmus was fired, they had welcomed two sons named Peter and George. Hans used to take the boys for long walks along the river and explain to them the mysteries of the world through the language of science. Since his first sight of her wide smile at the steps of the ferry, Hans had loved Nussy with the kind of poignant ferocity that, when money was thin and they were apart, made him choose an expensive phone call to hear her voice over eating dinner.

He was determined to get her and the boys out of Czechoslovakia. He was going to save his family, not abandon them.

The first step was to send the children away. They sent Peter and George southeast, farther from the western Czech border, which nestled against Germany as against the shoulder of a volatile and controlling partner. By the fall of 1938, when Hitler managed to grab the fringe German-speaking regions of Czechoslovakia through foot stamping and bellicose threats of violence, the children were in relative safety at Nussy's mother's home in the small village of Beočin, near Belgrade, Yugoslavia. But Hans and Nussy stayed behind in Prague. At least, they stayed as long as they could.

The humming frenzy of the Nazi Party continued to grow until anti-Semitic and imperialistic tensions vibrated through the very air of Europe like an electric threat. The people of Prague knew they were living inside a high-value target for the German leaders. Their city had been one of the capitals of the Austro-Hungarian Empire until its forcible dissolution at the end of WWI, and some Germans wanted the city back. Among other things.

Countries beneath the storm clouds of this expanding fervor began to isolate themselves, preparing to cut off escape for the Jews who were their scapegoats for the global economic depression. Train service was stopped between Yugoslavia and Prague, and when those trains were stopped, Hans and Nussy, still in Prague, were separated from their sons by nearly a thousand kilometers full of mountains.

Through the dark hours of the night of November 9, 1938, angry anti-Jewish mobs broke out in Germany, Austria, and the German-speaking parts of Czechoslovakia, where Hans and Nussy had previously thought they belonged. Violent anti-Semites attacked and murdered Jewish citizens, destroyed synagogues, and smashed the windows of Jewish-owned businesses. The glittering shards of glass from the thousands of broken windows would spark an evocative name for the hate-fueled rampage: *Kristallnacht.*

Neither Hans nor Nussy left a record of what *Kristallnacht* was like for them, but afterward, Prague was clearly no longer their home. They took what trains were still running, when possible, but the wintry, snow-covered border they had to tackle on foot. So they hiked. Hans was an experienced mountaineer who had grown up winding his way in boots and on skis through the very lands that now formed the barrier between him and his sons, and he knew how to navigate a safe path. They managed to scrape across the border during the final days before Christmas, and before what would turn out to be an unusually dangerous new year. They reached Nussy's mother's Yugoslavian villa safely, after seeking sanctuary and companionship with what friends they could find along the way. Jewish friends. Friends they would never see again. Friends who would not survive the coming storm.

Reaching Yugoslavia provided temporary safety, but they still wanted to leave mainland Europe, and Hans Kalmus would find their salvation through science. He received a small amount of academic refugee funding to travel to University College London, but because of the way the travel permissions worked, he would first have to go to England alone. On the train, which wove across the continent through Italy and France before taking him to a northern shore where he would board the boat for England, the depressed young outdoorsman refused to look out the window for fear he would never see a towering mountain or crystal blue lake again. Once in London, he found a lab to host him, and he scraped together more funding to stay. But more important, his new lab sent back to mainland Europe a letter arguing for the governmental travel release of Nussy, Peter, and George.

They would be required to leave their temporary Yugoslavian sanctuary and return to Prague first, though, to get the necessary visas, rubber stamps, and other mandatory yet fundamentally meaningless paperwork trappings that signaled formal government approval. The trappings used as an excuse to detain so many others. The trappings that would have placed them in Prague when the city fell. Hans Kalmus waited on top of the cliff to see if they had made it out or if they had been caught by the need for paperwork.

Nussy, Peter, and George Kalmus emerged from the immigration building at the base of the historic white cliffs of Dover on the evening of March 15, 1939. Not only were they safe, but they had not even heard the news of the Nazi invasion of Prague. For all the stress that had carved new lines into Hans's face, they had had a lovely journey.

Technically, their paperwork had required them to return to Prague before they left for the escape that they were pretending was not an escape. But sensing the danger, they had not. Nussy brought a friend with her to the customs office in Belgrade, and this clever friend had used a dash of pushiness mixed with a heavy dose of flirtation to convince the Yugoslavian government agent to let the family travel to England without the trouble of going back home to Prague first. They had managed to leave directly from Belgrade instead, transporting and shipping silverware, nice linens, family heirlooms, and, most important, themselves out of the country with little fuss.

That friend had saved their lives. Because of her, they learned of the invasion from Hans's newspaper rather than living through it. Had they returned to Prague, they would have been trapped. Like Hans's parents now were.

Ursula Philip's two younger brothers and her parents were still in Germany. It was time for brother Georg, "George," to leave, according to the covert warning from their local policeman. Young George had crossed the threshold to the age of fourteen, and as his bones lengthened and his voice deepened, the growing Jewish teenager became ever more a perceived threat to those who wished his religion dead.

Ursula began pleading for a pay raise to fund his trip, and Haldane joined her in the cause. The Rockefeller Foundation stepped forward with the funds, UCL increased her salary, and Ursula bought George his life for the cost of a boat trip to the United States. Her parents began to make arrangements for themselves to travel the same way, and for their other boy, Manfred, too. Thus, four more safely escaped.

At the same time, rumors began to trickle across the broad, soot-stained paving stones of the streets of London, spreading like a river of whispers, that those streets, too, might soon be unsafe. For everyone.

During the First World War, intrepid German pilots had crossed the Channel using the then new technology of airplanes to drop bombs on the startled people of England. Since then, the airplanes had gotten better. Faster. The bombs bigger. More terrifying. People with the ability to do so increased their rate of exodus out of the city and into the country, abandoning the easy and appealing target that was densely populated downtown London. However, those with no choice, or with the stubbornness to stay, stayed.

Haldane's secretary evacuated, but J. B. S. "Bomb-o" Haldane was stubborn. His wife, Charlotte, would write in her memoir, penned after their divorce, that JBS stayed because he was terrified. She described him essentially quivering in place because of a fear so paralytic, he found himself unable to run or hide. However, their marriage had soured years before, and she had become known for "taking every possible opportunity to embarrass and even insult H[aldane] in public." Some implied that her petulant behavior was because she was "a Jewess," showing that not even the wife of a famous professor was exempt from anti-Semitism. Since her account is the only record to depict Haldane that way, and all other sources disagree, it seems more likely she was attempting to embarrass him. After all, by then JBS had been rumored to be having an affair with his young, gangly genetics protégé, Helen Spurway.

Spurway's prediction of being with the married professor had become reality, which could seem a scandal for the time period, but the Haldanes had a utilitarian attitude toward their marriage. Charlotte and JBS had met when she was already married and they had arranged a deliberate scandal to get the law to allow her a divorce. Sister Naomi and her husband were openly polyamorous, and according to some reports, JBS and Charlotte practiced the same, but at a minimum his writings advocate for polyamory for certain types of couples. Spurway was young, motivated, and, more important, mutually obsessed with genetics in a way that could have kept them up talking statistics through the nights during their naturally nocturnal work schedules. She expanded her zoological repertoire to include the noble newt, amphibious, squiggly, and prone to endless genetic variation. She explored every body of water she stumbled across, it seemed, looking for newt specimens to bring back to the lab to examine and breed, impending war be damned.

So, when the lab secretary evacuated, the stubbornly staying-put Haldane needed a new one, and ardent Communist that he had become by then, he insisted she be Communist, too. He thus found Joan Elizabeth Jermyn.

———

Joan Elizabeth "Betty" Jermyn was a wisp of a young woman with a straight nose, piercing brown eyes, and a closet full of long pleated skirts. Her slim figure and quiet demeanor meant she could easily be swallowed by even the calmest crowd, but to those who managed to see her

Joan Elizabeth Jermyn as a
college student.

anyway, her strong, resolute jaw and the sharp feminist bob of her dark
brown hair roared the size of her character. She was born into a world at
war on March 3, 1916, in Norfolk, England, and in the days following her
arrival, gales and blizzards tore through the country, destroying trees,
stalling transport, and smashing storefronts. Elizabeth Jermyn, the
world seemed to announce, was here.

Her father died in a horrific car crash when she was eleven years old,
and she and her mother and siblings were left to fend for themselves. Her
mother moved back to her hometown of Ipswich in Suffolk and took in
renters as a form of income, but because Jermyn's aunt had been behind
the wheel in the deadly crash, her uncle began paying the bills for whatever elite education Elizabeth needed. Her mother had dreamed of being a doctor in her own youth but found her dream at odds with the
strict rules society had for her gender, so she supported Elizabeth in her
fascination with science while her uncle paid for the fancy schools. After
passing the entrance exams with ease, and with strong letters of recommendation calling her "an extremely nice girl" and "above the average,"
Elizabeth Jermyn was welcomed into the Bedford College for Women
at the University of London. There, she would row on the canal, learn

chemistry, physiology, and anatomy, and take voluntary extra classes in the evenings to learn scientific German. There, she would also learn about Communism.

Elizabeth Jermyn and her friends used to stand on street corners downtown, stalwart young figures hawking the Communist paper the *Daily Worker* to rushing passersby. It was their own manner of rebellion against what Jermyn had found to be an infuriating contrast between the extreme wealth of her secondary school classmates and the starvation and unemployment of the economically depressed world outside the school's walls. Jermyn would take her Communist friends and their beliefs back to her mother on breaks, and an older, more disillusioned Elizabeth Jermyn lovingly wrote of her patient mother, "no doubt she tolerated the radical rubbish that we talked."

Jermyn's grades at Bedford earned her a place in the first class on the Honours List, and she was a standout in both organic and inorganic chemistry. But when she graduated, in the midst of the epic global depression, work was hard to find. She applied for a librarian job at London's Science Museum but was told she did not have enough experience. She went for formal secretarial training and moved back in with her mother. Finally, during the start of the evacuations, science, Communism, war, and JBS combined. Because of her hard work with the *Daily Worker*, which often contained writing lectures by Haldane to increase its appeal, she was offered the job as the next secretary at University College London.

In modern times, scientists write their own academic papers and reports. At least, they are supposed to. But in the 1930s, a streak of As in organic chemistry when coupled with the wrong sex was enough to consign a brilliant mind to the fate of anonymously authoring the papers of others. Haldane's upbringing and tutelage under his brilliant father had trained him well in science, but his actual formal education was in the classics, and he had no degree-based qualifications to either work in genetics or teach it. He had picked a scientist secretary so she could write his academic publications and *Daily Worker* articles for him, but at least on paper, she was more qualified than he was. He was not alone; it was a standard practice by male scientists of the time.

Jermyn started her job in 1939 while the world simmered around her and England reeled from the loss of HMS *Thetis*. She slipped into the lab alongside Ursula Philip, Hans Grüneberg, and the audacious Helen Spurway with little fanfare, and she was down the hall from university neighbor Hans Kalmus as he was freshly adjusting with his family to a new life in London. She began to labor at UCL quietly, swallowed by the taller, louder, more publicly renowned sea of humanity around her. Neither she nor any of the other eminently qualified secretaries would be listed on any of the papers they produced. Similarly, in the coming war, her contributions would go unnoticed for decades.

By the end of July 1939, the Admiralty approached the problem of the *Thetis* with the same predictable, clockwork approach that all governments seem to adopt when faced with a failure: They formed a subcommittee. The subcommittee started with exactly two people, a physiologist and a submarine officer, and they began by making a list of the questions that they could not answer about survival underwater.

They named nine separate issues, ranked them by importance, and tidily labeled them letters "a" through "i," but their desperation for insight—any kind of insight—was naked and apparent by the letter "h." That letter begged without specificity for answers about "Any other effects, not included above, arising from the use of self-contained breathing apparatus."

The previous March, the people holding a similar discussion had not thought to mention carbon dioxide, but by July, after the *Thetis*, "carbon dioxide poisoning, its effects, its prevention and its alleviation" became the new top priority, the new letter "a." Vague item "h" was the Admiralty's feeble acknowledgment that each day might bring another *Thetis*-like disaster, which might reveal another new, previously unknown problem. The loss of HMS *Thetis* had been a fatal lesson in humility.

By August the subcommittee had added an engineer, and that engineer asked J. B. S. Haldane for help. "I expect by this time you will have received a letter from the Secretary of the Admiralty asking for your co-operation in investigating the physiological problems which may be attached to the use of the DSEA," the engineer wrote. The letter asked for help figuring out how to keep people breathing and alive underwater. Please.

The engineer described the biggest problems on their list. Carbon dioxide. Decompression sickness. The vague, ominous, wildly understudied tendency of oxygen to cause problems when breathed at elevated pressures. In a separate letter marked "confidential," the Admiralty got more specific. They asked Haldane to "help by drawing up a sketch programme of investigations" and offered to help organize funding for the experiments. They wanted him to repeat what he'd done for the *Thetis*, but for every problem, "a" through "i," like a grocery list for science. Their letters read like the panicked epiphanies of people who have realized that they are in well over their heads.

Haldane's lab was full of geneticists, and most had spent their careers calculating the statistics of genes, not breathing underwater. But war was coming, and as Hans Grüneberg would say many decades later, war "was not the time for fundamental research."

They could do more than pick up guns. The poetic Grüneberg once again described it best: "In war, scientists are interchangeable. It does not matter whether you know anything about the anatomy of the animal on which you are working, or whatever your technology is; technology can always very quickly be acquired. On the other hand, the approach of the scientist, how to tackle a problem—that is something that is common to all scientists."

With the outbreak of war and the current German domination of the seas surrounding their tiny island nation, the points on the Admiralty's scientific shopping list had become not nerdy questions about underwater physiology but real, life-threatening, world-changing problems. And this group of geneticists knew how to tackle a problem. Science is not memorizing the periodic table; science is the method of thought that proved the table was correct in the first place. Without a rigorous, methodical, *scientific* approach, the Admiralty's questions about underwater survival might never get answered.

Haldane began to concoct plans to tackle the list. His lab members were geneticists, but they already knew the difficult parts, which were how to think, how to experiment, how not just to leave bias at the door but to bury it out back behind the building. In Grüneberg's words, he, Haldane, Philip, Jermyn, and the others in the lab would have to "do one's bidding in whatever capacity one was capable of" to help Britain stand strong.

———

In the later days of August 1939, Nazi SS agents planned a clandestine attack on a German radio station in Poland—an attack they would blame on the Polish and publicly trumpet to be evidence of Polish discrimination against German speakers. After WWI, Poland had melded with bits and pieces of the dissolved Austro-Hungarian Empire, and just like they wanted Prague back, German leaders wanted those pieces back, too.

After the Nazis' attack on the radio station, they bellowed about the profane injustice of what was secretly their own work. The lost territories were the goal of this first forcible invasion, and the Nazis said they were "saving" the German speakers there. So they invaded Poland, bombing and burning as they went. Within the span of two days, Great Britain and France proclaimed ultimatums of war if Germany did not retreat; when Germany refused, Great Britain and France declared war. The British draft was escalated, and lines of volunteers wrapped through English hallways in long queues. The Brits were on board to help their allies in Poland in the name of what the London papers called "the building of world order" and "breaking the Nazi yoke of tyranny" for "Britain's fight to save the world." Hans Kalmus made his children sit still and listen to the radio in their London home to hear the declaration of war. Elizabeth Jermyn was horrified that Russia had made a pact with the Nazis to divide up Poland between them.

Hans Grüneberg felt a wave of relief at the news, because he had been afraid that England would come to terms with Germany and with Hitler, which he thought would "be an absolute catastrophe." He and his wife, Elsbeth, had already had one son, Reuben, and she was visibly pregnant with their second. They had permanent permission to stay in England; it was their home now, and they were glad to see it fight back.

America, for the time being, decided to stay out of what they saw as a foreign issue, but, proving that history has a sense of irony, British papers lampooned Hitler's invasion and attempted acquisition of foreign lands while simultaneously declaring that of course the British colonies would be expected to help. And thus, with the Germans and their allies pitted against the Poles and theirs, Europe was at war.

During WWI, Britain had been caught unprepared for the new world of warfare, with their defensive submarines and their relative inability to attack under the cover of water. For this war, however, they were determined not to let the Germans own the seas. With that goal in mind, Haldane wrote a letter, sealed it inside one envelope marked with warnings that it should be opened only "by a thoroughly responsible person," sealed the first envelope inside a second one, and mailed the letter to the War Office. He was suggesting a novel idea to fit into this new world of warfare: free-swimming combat divers.

Diving at the time required a serious physical effort. It demanded a behemoth metal "hard hat" that had to stay connected to the surface via a hefty bundle of cables called an umbilical, with multiple trustworthy friends constantly cranking air pumps up above so the diver did not run out of gas below. The setup and the air pumps needed a support ship; in addition, giant metal boots and hundreds of pounds of lead were required to sink the diver. Diving took planning and orchestration, but perhaps most important, it took so much gear, it was about as sneaky as trying to creep a freight train covered in disco lights through Westminster Abbey. In other words, it was not a stealth operation.

In another few years, Jacques Cousteau and Émile Gagnan, while living in Nazi-occupied France, would contrive a clever little mechanical device that would allow the independent, umbilical-free breathing of pressurized air from a gas tank while underwater. It would not be perfect, it would make lots of bubbles, but unlike the hard-hat diving with the giant surface ship and crew of athletic buddies, it would allow diving that was self-contained. It would strip some of the lights off the proverbial freight train and put back a little more stealth. In 1939, however, that device did not yet exist.

So J. B. S. Haldane thought: *Why not use the same breathing apparatuses that the submariners use for emergency escape, but for deliberate diving?* The only thing stopping the idea was the same list of underwater survival problems impeding the DSEA's use as a submarine-escape tool. With the right testing and the right understanding of the principles of underwater survival, that device could be used intentionally and safely, rather than as a submariner's last screaming gamble. That idea was in the letter he wrote to the War Office.

"Divers using self-contained dresses with their own oxygen supply may play a valuable part." Upon reaching their destination, Haldane wrote, a diver "could either land to carry out work of a special character" or establish a guide rope for underwater navigation "by which several hundreds of other similarly equipped men could assemble below the surface of the water with a view to a surprise attack." "Work of a special character" could also be called a special operation.

Haldane noted, "Such a scheme is, I believe, no more chimerical than the landing of troops by parachute, and is therefore worthy of investigation." That is, the Admiralty had already made combat chimeras, the smashing together of two different worlds, by throwing thoroughly landborne humans out of skyborne airplanes. Why not do the same with the water? The divers could prove invaluable for scouting, he argued: "This would include not only topographical and tactical research but design of a special apparatus, including, if possible, protection against shock from 'depth charges' or shells bursting in the water."

In other words, Haldane wanted to take the Admiralty's list of underwater problems, add "getting blown up"—an item "j," if you will—and study those problems. Then he wanted to send free-swimming, science-backed divers wearing bombproof gear to scout the topography of potential combat sites, support tactical plans, help guide other war fighters, and perform whatever other sneaky amphibious missions that might be required. If the crazy scheme worked, it would make naval history.

It is not certain that anyone in the War Office read Haldane's letter. It is also not clear who else—in what manner or with what timing— might have suggested the same concept and deserved equal or greater credit. But it is clear that in addition to safe submarine escape, a free-swimming diver was the idea that the Admiralty and their newly acquired scientists started working toward, Haldane at the helm, buried under the deepest layers of secrecy and known only to the most "thoroughly responsible" people.

CHAPTER 6

Drunk and Doing Math

Real war had not started yet, but it was already such an inconvenience for genetics research. Shortly after the formal declaration of war on September 3, 1939, University College London got prompt government orders to evacuate the busy downtown campus, and most of the professors followed the instructions in rapid lockstep. The secretary of the university barked that they had twenty-four hours to "dispose of" their research animals, which, aside from the obvious heartbreaking aspect, also meant throwing decades of backbreaking work into trash cans in the form of hastily massacred mice and piles of dead turtles left to rot in the sun on the roof of the zoology department. The secretary was brutal, but he was not totally misguided; in the coming war with its inevitable rations, there would be no spare food to give to lab animals.

Predictably, Haldane tried to argue, but the administration pushed back hard, telling him on the first Thursday after the official declaration that "the college should be absolutely closed after Saturday." To emphasize their resolve, they shut down the electricity, water, and lights in the office buildings and put up barricades. They bolted a medieval-looking hunk of timber across the gates of the main entrance and built a custom "full-height wooden barrier" to cover the archways leading to the south quadrangle. Guards patrolled against wily researchers trying to creep past them, and rumor had it that one desperate scientist accidentally lost his trousers to a guard who was trying to stop him from clambering over the spiked barricades.

But the geneticists squirmed their way in. For some reason, UCL accidentally left the heat on in the mouse-storage rooms, so they worked there instead of in the humans' offices while the weather cooled into the chill of what would become a record-breaking winter. The electric lights were left on in the mouse rooms, too, so after outfitting the windows with dark curtains to hide their furtive midnight science, Helen Spurway, Elizabeth Jermyn, Hans Grüneberg, Ursula Philip, and J. B. S. Haldane could work at night, inspecting the inherited features of the ears and wings of their illegally saved mice, newts, and flies. After all, flies didn't eat much.

By November, even though the bombs had not yet started to drop, belligerent, antiestablishment Haldane was the only member of the faculty who had "succeeded in retaining a foothold." He instructed his visitors to sneak onto campus by claiming they were from a government ministry; "Say the Ministry of Food," he told them. He nailed to his door a poster reading "Freedom is in peril—Defend it with all your might," grumbled about revolution, and continued to pick fights with the administration.

Over the winter Hans Kalmus also found himself left behind on campus, but for different reasons. He had a recurring eye problem that needed regular care, and the eye hospital was nearby. The rest of his lab members evacuated as they were told, but Kalmus, with one eye encased in bandages like a haphazard medical pirate, had to stay.

One day on campus, Helen Spurway and J. B. S. Haldane, the Bonnie and Clyde of clandestine genetics, plopped themselves down at Hans Kalmus's lunch table, asked him why he had stuck around, and invited him to join their motley crew of rebels. Kalmus was puzzling together the phenomenon that would eventually be known as circadian rhythms, but he was doing it with fruit flies, so he would fit right in. And clearly he was not afraid to buck a little authority.

———

By January, 1940, as the stubborn holdouts waded through their undercover genetics projects and picked at organizing the underwater work, the Royal Navy started to sound more desperate. They wrote to Haldane that they were "most anxious to obtain, as soon as possible, information on" their list of undersea issues, and asked "whether [Haldane] could arrange for research on any of these matters to be carried out under [his] direction." But UCL's physiological laboratory was not only closed, it was locked down—so JBS had no human-capable lab. He was squatting in a mouse room.

Sir Robert Davis, the leader of the company Siebe Gorman, which made his eponymous Davis Submarine Escape Apparatus, had received the same pleading Royal Navy letter. He immediately invited Haldane to their London factory to collaborate: "I would provide all the facilities and plant necessary for the tests." Haldane made the arrangements posthaste.

Only one thing was missing. Technically, two. The final two members of the lab—the last two who would join Philip, Jermyn, Spurway, Haldane, Grüneberg, and now Kalmus inside the hefty metal tubes for the experiments being rapidly planned. Two members who would become personal examples of both the necessity and the potential horrors of the work they were about to conduct.

Edwin Martin Case, who went by Martin, was a human manifestation of the word "mischief." He and his two brothers, Bobbie and Ralph, came along to their parents in Birmingham, England, during the first decade of the 1900s, and together they were a three-boy hurricane. They enjoyed, among other things, staging dramatic "accident scenes" for their parents or other unsuspecting passersby to stumble upon, like pretending a bicycle crash had left their three little-boy bodies mangled by the side of the road. In one incident they locked themselves in a bedroom so their experiments knocking one another unconscious with chloroform would go undisturbed by responsible adults. The trio used to tell people that "they did not want common sense, they wanted extraordinary sense"—a philosophy that also led them to do things like take turns in an ice-cold bath when they were curious about hypothermia.

Their father seemed to play along with their mischief—at least the

"Accidents" staged by Martin Case, his brothers, and friends.

safer parts. He papered the walls and ceilings of the boys' bedrooms with arithmetic to help them learn. Then later, as they got older, he replaced the basic math with algebra and more complex problems. They demanded—and got—a childhood riddled with science, math, and exploration. As it turned out, they all survived to become polymaths, so apparently it was the right kind of mischief.

Martin Case as a boy, studying, after an apparent head injury.

By the time Martin Case made his way to Cambridge, he was a noted pianist in addition to an exceptional biochemist. He had grown to become a handsome young man with a strong cleft chin, thick curly hair, and ears that stuck out just far enough to give his face approachable charm. Somehow, during his student years, while still a teenager buried in books, he crossed paths with and began to study under J. B. S. Haldane.

Edwin Martin Case as a young man.

One day, Case got cited by a safety officer for driving his bicycle too recklessly. Alcohol, time, and perhaps human decency have suppressed many details of the story, but according to his family's records, he was profanely drunk, and whatever he was doing was sufficiently hell on wheels that the safety officer gave him a hefty fine as well as a criminal penalty.

But the night before Martin's court date, a large, mustachioed stranger with a prominent forehead and bushy eyebrows walked into the citing officer's favorite pub. Open to the idea of free beers from a generous stranger, the safety officer was happy to chat. Perhaps the safety officer was *too* open, because the stranger turned out to be *very* generous.

The next day, after the safety officer's brief and abortive attempt at giving testimony that came out more like gastric regurgitation than English words, a small team of helpers was required to peel the still drunk man off the floor of the witness box. The court dropped all charges for lack of coherent testimony, and Case, genuinely surprised by the pleasant turn of events, went back to tell Haldane of the unexpected victory. Haldane presented Case with a handwritten bill for the previous night's hefty drinking tab, and stated his surprise that the officer had made it in at all. The bill, while substantial, was far less than the fines and penalties at stake in the trial. With that, the bond between Haldane and Case was set for life.

So when Case lost his job as the war was starting, he wrote to Haldane to ask for a reference to sign up for an officer's commission in the military. He wanted to serve. Haldane pushed back.

"Don't be in too great a hurry [to get a commission]. I think I may be able to get you a job in connexion [*sic*] with research on how to get out of submarines." He included a disclaimer about the research almost as a stray thought: "You might be in slight danger."

He followed that letter up with a similar one directly to the War Office. "I consider that it would be a mistake to grant a commission to Dr. EM Case. He is a chemist whose researches have been of great importance. . . . [His] undoubted abilities should be used in the scientific field. . . . I propose to use him as an assistant." But, Haldane also clarified, it wasn't because he thought poorly of Case: "I have little doubt that he

Case working in a chemistry lab.

would make a good officer. In particular, I think he is likely to display considerable courage if necessary."

Courage mixed with a knowledge of chemistry was rarer than courage alone, and both were needed for what the group had coming, for the list of scientific problems presented to them by the Admiralty.

"My Dear Boss," Case wrote back immediately. He said that letter was "without exaggeration the most cheering communication I've had for months." Slight danger be damned. Mischievous Martin Case was always in for science.

There was only one lab member left to join.

JBS's younger sister, Naomi, had always hovered like an undercredited ghost in the background of his stories, so it is perhaps unsurprising that she deserves partial credit for the last member of the lab, James "Jim" Rendel.

Naomi's adulthood is a story of what happens to genius when it is given money and then ignored. Once grown, she moved through the world with such grace and authority that small children would simply assume she was royalty. She did what she wanted, and she wanted to write. By the turn of 1940, Naomi had married a barrister friend of JBS, established herself as one of the key voices in feminist literature along-side her peer and personal acquaintance Virginia Woolf, and published several books, including her 1935 forward-thinking, sex-positive master-piece, *We Have Been Warned*.

In short, Naomi cared for social mores at about the same level as JBS did, and she would not only fight for her own rights; she would use her status and power to fight on behalf of those around her, too. Naomi, now with the married last name of Mitchison, along with Virginia Woolf took under their literary wings those whom they saw as the next generation of rebellious protégés. One of them was Joan Easdale, who, once married to a geneticist named Jim, would become Joan Rendel. Virginia Woolf would brag of Joan Easdale: "The girl poet is my discovery."

Joan was a totally unsophisticated young woman with smooth brown hair, arched eyebrows, and an elfin mouth reminiscent of classical oil paintings of the aristocracy. She had been coddled and tightly controlled during a childhood plagued by overworrying adults, and as she began to

Joan Rendel, née Easdale.

James "Jim" Rendel as a young man.

taste her first drops of freedom through selling her poems and writings, she seemed to burst into adulthood with the pure joy of one seeing the world for the first time. She wanted to break out into the world, and her clear voice and talent for poetry would let her. In the years before the shortages of the coming war, she used her early earnings from selling pieces and plays to the BBC to buy herself two pairs of vibrant, showy, unavoidably brilliant stockings—one pair magenta and one emerald green—that she wore everywhere with love, pride, and the fluid grace of a ballerina. And while she was living in a shared flat with a female room- mate, she fell poetically in love with the jacket of that roommate's brother, while it was "hanging on the back of the door." That jacket be- longed to Jim Rendel.

Rendel was a geneticist, and by pure coincidence he had already worked with Haldane earlier as a student when he had done research examining how to use genetic theory to improve the hatching yield of chickens. Haldane had even once written him a letter of reference, in which he said that Rendel was "reasonably pertinacious" and wouldn't "necessarily be worried by poultry characters displaying a bit of anar- chism, such as an extra toe." From Haldane, that was high praise.

So when Joan needed a change, when she felt stymied living and working where they were, and when the war seemed to threaten all employment prospects for everyone, including her new husband, Jim, she wrote to her mentor, Naomi's older brother. She asked if JBS knew of a place "where one can go on working and thinking and daily acknowledging the vast void in our understanding, among people who are doing the same." She was originally fishing about genetics jobs in America, but she and Jim wanted a place to be creative among people who understood that drive and who wanted to be creative, too.

Jim Rendel was a tall man, six foot four, with a hawkish nose, round horn-rimmed glasses, and long, lanky limbs that made him look as if he folded rather than sat and loped rather than walked. His sense of humor came through in the wryly ostentatious titles he gave his inventions, like the poultry-related "Eggatron." He spent his waking hours seeking out conversations about consciousness or philosophy or science, or simply someone intelligent and active enough to argue with him. Joan was the same. Haldane understood.

It is not clear whether Joan's plea was the direct reason for Jim's hiring, but within a few months, Haldane was sending letters to find funding to pay Jim Rendel. By the time the Admiralty came calling for diving help in January 1940, Ursula Philip had offered to learn nursing work, Hans Grüneberg had been asked if he could serve as a war physician, and Martin Case had attempted to sign up as an officer. Instead, by April, they would all find themselves serving in a different way from those that they, the Admiralty, or anyone else could have imagined. And it was not long before, just as in the war itself, tragedy would have a turn in the lab.

———

By January 1940, most of Britain prepared for war while Prime Minister Neville Chamberlain tried to negotiate peace in vain. The government issued ration books for butter, bacon, and sugar to address shortages. In February, the Royal Navy began arming all the merchant ships in the North Sea, to the east of England, and Germany immediately announced that it would therefore consider all British merchant ships to be warships. The Royal Navy began laying mines in Norwegian waters in early April, and a single day later, Germany invaded both Denmark and Norway. By April 9, 1940, the Phoney War was over. The tension escalated to violence, and the yearslong simmering anxiety turned to gunfire-driven urgency for results from the scientists, in the case of sunken submarines or in the event of vague, not-yet-planned naval combat that might require espionage. By April 11, 1940, the scientists began to experiment.

On Friday, April 19, 1940, Martin Case, hunched into a sphere, balanced a steel ball bearing precariously on a toylike scoop. He was determined to transfer the ball bearing into the correct hole. With a solid plunk, the metal sphere found its home just before the single minute allotted for the task ran out. Dropping the fiddly scoop, Martin began to pick up the bearings with his fingers instead. That way was much easier.

Case was curled inside the littlest hyperbaric chamber at Siebe Gorman for the fourth day of testing. He passed the "ball game" sideways to JBS, who was also bent over by the white-painted tubular walls. Their

Martin Case (left) and J. B. S. Haldane wearing breathing apparatuses inside the smallest Siebe Gorman chamber, with the door open and visible on the right. This image was taken for intended use in a *Life* magazine article about Haldane, which was never published. The formal clothes are almost certainly for the benefit of the photographer, given the heat involved in most trials.

already small volume was constrained further by the game equipment, pads of paper, stopwatches, and two leather-and-rubber-and-metal breathing apparatuses for use on the way back up. This particular chamber contained no lights, because it could go deep enough that electricity might spark a catastrophic internal fireball. The duo had some small portable electric lamps inside, but most of their vision was extracted from meager bulbs pressed from the outside against the tiny portholes of the darkened tube.

J. B. S. Haldane began his first minute of the ball bearing test of manual dexterity, shaking off a lingering headache caused by the previous day's hyperbaric exposure. He deposited the bearings in the holes using forceps. Case and he began the arithmetic next—mental dexterity, as it were. Haldane went first while Case timed him.

After generating sheets of numbers scrawled in rapid, curling streaks of pencil, they heaved the burly, circular chamber door closed around its pivot of heavy steel hinges, swinging its domed shape into contact with the main body of the steel tube. Now, with the door shut, the carbon dioxide would begin to accumulate inside, just as it had inside the *Thetis*. Air began to intrude through the piping, first with a hiss, then with a roar. The very real physical heat enveloping Case and Haldane began to climb, in an unsubtle metaphor for the rising pressure from the military to achieve results.

The first thing on their checklist, and the subject of today's work, was an explicit request from the Admiralty, from one Surgeon Commander Seymour Grome Rainsford. Rainsford wanted more information about nitrogen narcosis—the recently developed theory that nitrogen might become a powerful, even debilitating narcotic drug under increased pressures. A tendency that, if true, would have been a factor affecting anyone trying to perform complex tasks like espionage or submarine escape deep underwater. Divers had always reported euphoria and a sense of mental stupor at the limits of diving's deepest depths, but the American Albert Behnke had just declared the nitrogen in their breathing air to be the cause. Rainsford wanted to know if Behnke was right. Haldane used the request to get the government to pony up a modest salary for Case: "I should be glad to carry out and supervise experiments on the problem, and should not demand any salary. However I should require an assistant."

So, today, in this small tube, Haldane and Case planned to go to pressure. They would breathe air, which is 78 percent nitrogen. They would see if nitrogen interfered with their ability to manipulate the ball bearings or affected the time and accuracy with which they could complete the written math. In short, they would see if nitrogen got them properly drunk.

In the twenty-first century, nitrogen narcosis is known to be an immutable effect of breathing air under pressure. Divers often affectionately refer to the phenomenon as "Martini's law," meaning that each thirty-three feet below the surface of the ocean can be considered equal in mental deficit to pounding down one martini with gusto. Perhaps given the time and place of Haldane and Case's experiment, a sidecar with a twist would have been a more appropriate drink metaphor. But regardless of the beverage, inside less than four minutes, with the internal chamber air screaming hot from the compression, Haldane and Case reached three hundred feet of seawater, which would allegedly have had the same impact as imbibing nine glasses of straight booze, in less time than it might have taken to drink the actual liquid.

As of this writing, nobody yet understands why nitrogen narcosis occurs. But it certainly does. Martin Case and his old pal JBS were smashed.

Case tried to be modest about it. "Slight feeling akin to what one has always been led to believe is associated with inebriation," the chemist scribbled, coyly selecting words as if he weren't the one who had been cited for piloting his bicycle with such liquid-fueled abandon that he had posed a public health hazard.

Haldane wrote notes, too, or at least tried to. He could not write as coherently. "Reach top. MC says 'We are drunk.' Notices above. Not so [illegible scribble]. JBSH feels abnormal. 'Ringing' in ears. Queer taste in mouth. Looks darker (?) [illegible]."

They were inside a tube, so there was no "above" to be had. The statement was gibberish. Pressure doesn't cause ear ringing or weird mouth tastes, so those were imagined. Even later, neither he nor anybody else in the lab could read some of the words or discern what "darker" was supposed to refer to. Haldane would also say that he had "felt somewhat mystical." In short, the narcosis was winning.

Haldane and Case tried the ball game, followed by the arithmetic problems, but the test results were moot because they both kept forgetting to start the stopwatch. Haldane blamed his distraction on the "sound of bell ringing." There was no bell. When Haldane tried the sums, he began mixing up the sheets of paper or starting a problem in one column but continuing it in the next. When Case noticed how long JBS was taking with the rudimentary addition and asked how much longer he needed, the math genius, who had been first in every math ranking for his entire life and who had literally invented statistical tests, grumbled, "I don't know at all, nor do I care." After incomplete attempts at forcing the numbers, he wrote down, "Feel better," as if he were a credible witness to his own condition.

They had reached three hundred feet at twelve ten p.m. By one ten p.m., they had struggled long enough. They put the mouthpieces from the breathing apparatuses into their mouths and began breathing pure oxygen before the ascent. By removing the nitrogen from the gas they were inhaling and using pure oxygen instead, they could increase the rate at which their bodies dumped the stored nitrogen out of their tissues. They could reach the surface faster, with less time spent decompressing and less risk of decompression sickness. Later, Haldane would have no memory of a small repair he had made to a pipe connection before donning his mouthpiece.

As they reached the surface, Haldane felt a searing pain in his mouth and heard an audible screaming sound emanating from inside one of his teeth. He had had a cavity filled in an upper incisor as a child, but the dentist obviously had not completely filled the hole. The sound was pressurized gas slowly squeaking out of the former cavity. It hurt like hell, but to Haldane, it was all good data. He cheerfully reported: "I believe [the toothache] is a new type of bends." He promptly had the tooth removed to facilitate future tests.

After many more tests, their conclusion was the same as Behnke's: Yes, the narcosis was real. It was caused by nitrogen.

They had achieved their first task on behalf of Britain. Haldane and Case wrote of nitrogen narcosis that "it is quite imperative that no great trust should be placed in human intelligence under these circumstances." It was an understatement.

The first month of testing crept toward May, and the experiments continued smoothly and without calamity. On day one, Haldane and Case had sat upright inside the largest chamber and breathed oxygen and played the ball game at shallower depths, with both subjects developing no more than a mysterious slight cough afterward. Day two, the fit young Case had donned some of Haldane's extra-large clothing and shoes and clambered into a water tub inside the chamber, with water that was 10 degrees Celsius and large blocks of ice chained to the bottom. He let his exhaled CO_2 build up around him to simulate a submariner in a cold, Atlantic-flooded compartment. Government representatives milled around the safe, warm outside of the chamber, muttering observations, until Case crawled out, hypothermic, shaking too hard to complete the manual dexterity test and ready for his rectal temperature measurement. But in the context of the world of science, a little coughing and a little limb shaking were not considered much of a big deal. The scientists were chipping away at the first set of problems, seeing how long they could withstand CO_2 and how long they could withstand cold, and getting a feeling for how long submariners had inside a doomed vessel before they became too incapacitated to help themselves.

Other self-experimenters had put themselves through worse for less. Around the turn of 1800, Sir Humphry Davy invented anesthesia while knocking himself unconscious on repeated testings of his first mixtures. Around 1767, surgeon John Hunter rubbed gonorrheal pus into cuts he had administered on his own penis, hoping to inoculate himself against the venereal disease. Hunter had unwittingly used pus contaminated with both gonorrhea and syphilis, and so accidentally gave himself two diseases rather than one, which was especially bad because syphilis was at that time incurable. In the 1910s epidemiologist Joseph Goldberger was determined to prove that pellagra, a disease now known to be caused by severe vitamin B deficiency, was not caused by infectious pathogens. His method of proof, naturally, was to hold gatherings he called "filth parties," where he and his scientist friends would eat capsules filled with the scabs and scrapings of pellagra victims. None got infected.

Articles from the *Times* of London in the 1920s and 1930s indicate that self-experimentation was considered fairly normal in the era; perhaps not common, but certainly somewhere on the list of the zaniness expected from those who played with test tubes and loved tweed. Scientists were still considered hobbyists, in a way. Science was relatively new as an independent field; its roots were as an activity practiced by middle- or upper-class people at home in their spare time after their real day's work as barons or doctors or lawyers or whatever professions left them enough free time and resources to putter. So the notion of self-experimentation popped up mostly in obituaries describing the lives of such people, usually as stray sentences thrown into otherwise eulogizing articles with the literary equivalent of an affectionate chuckle over their strange hobbies. In 1939, the Royal Institution in London even hosted a Christmas lecture to entertain children with the story of the self-anesthetizing adventures of Humphry Davy.

But the self-experiments were anything but zany to the scientists. With little formal legal structure in place to provide ethical guidance for human experiments, researchers often saw their own bodies as the easiest way—or the only way—to get data, and sometimes the tests were lethal. In 1939, around the same time as the high-spirited children's lecture, the latest in a long list of doctors died during self-trials with anesthesia.

The noble practice also sometimes was—and still sometimes is—used as a loophole by charlatans. In 1939 another "scientist" was put on trial for fraud, and he bellowed in his defense testimony that his "miracle cure" had allegedly fixed his own medical issues. So of course he should have been allowed to market and sell it at high prices without further question! His medical license was revoked, but the same infamous claim of "It worked for me" is still often used today as a dodge by those who wish to avoid regulatory scrutiny. It lets them peddle their "cures" without actual proof, because they're technically not making promises that it will work for anyone else. All good tests need proof from other people.

When the war started, the need for results drove up volunteerism for science. Many who could not sign up for the military for reasons ranging from religion to physical condition signed up to support their countries through experiments instead. Dozens in England volunteered to be infected with painful, itching scabies to test better treatments, because the wars of the past warned that the ailment would soon blaze through the troops abroad. One man named Horace Cameron Wright started plotting out the holes in the Allies' understanding of lung injuries from underwater explosions, and realized that the only way to fill in those holes would be to volunteer to get blasted. In America, pacifist Quakers signed up for voluntary starvation once early intelligence indicated that somewhere in Germany there were people who might benefit from knowledge of the safest way to reintroduce food when starved to the brink of death. And Haldane loudly offered to climb inside and personally test better air raid shelters to protect the people of London once the threat of bombing loomed its ugly head. In addition, of course, to the work he was doing in the pressure chambers. The knowledge of the world moved forward.

By high noon on the eighth day of experiments, April 26, 1940, Haldane had gotten in the chamber four times, Case had gone six times, and one of Haldane's Jewish friends from the Spanish Civil War, who was seeking refuge in the UK, had also climbed inside. They had even conscripted a random entomologist on day five. But now, on day eight, it was Jim Rendel's turn. Jim Rendel would go to depth, and he would make sure the same mind-altering effects happened to him while there.

Martin Case jotted Rendel's physical parameters in his lab notebook: "tall and slim," age twenty-five. Rendel did well on the pretest ball game and finished his arithmetic with zero errors. Haldane curled inside the chamber with them, watching and timing. After precisely twelve more minutes of preparations, they were ready to pressurize. They started slowly at first, "for fear of ear trouble."

Tubes of air run through the human head, connecting the back side of the eardrums to the sinus cavities. It is only because the eardrums are backed by this air that they vibrate properly to let us hear sounds. So, exactly like during the ascent or descent of an airplane, passengers inside a hyperbaric chamber need to let the pressure inside the eustachian tubes of their ears equalize. Otherwise, the fragile skin membrane that is the eardrum starts to scream with pain as it gets shoved outward or inward by gas that is either trying to escape or invade. If people can equalize, they return to comfort, and everything is fine. If not, both the pressure and the pain will build until the gas forces a new passageway by ripping apart the eardrum. It is nearly impossible to reach the threshold required to receive that kind of injury during the modest pressure changes of an airplane, but it is easy to do so in the wild pressure swings of a chamber. In most cases, the eardrum does heal and regrow, but the person is temporarily deaf in the meantime. During his first chamber trip, Jim Rendel had no trouble at all.

He did, however, feel funny. "I feel as if I were going to faint," he gasped to Haldane during the descent. "I feel I can't breathe." The air grew thicker as they pressurized; then, a minute later, Rendel realized the nitrogen was hitting him, too: "I feel crazy." They hit bottom at seventy-six meters, or two hundred fifty feet, of seawater, shallower than before, but Rendel nonetheless reported, "I still feel very unreal."

On the way back up, by the time they reached one hundred feet, Rendel began to feel "perhaps a bit clearer." By sixty feet, he was notably more himself.

Haldane began to itch. Pain bloomed in the tip of his left thumb—a vague, niggling sensation that would not go away with any amount of bending and flexing, a deep pain that slowly began to spread. Another spot appeared in his right little finger, followed by aches inside every one of the many joints of his ten digits, despite the deep, hearty breaths of nitrogen-free oxygen entering his lungs from the leather bag near his chest.

They had spent only thirty minutes at depth, a mere fraction of the time of the Brooklyn Bridge caisson workers hammering away at the wet soil, but they had gone so much deeper that nitrogen had soaked into them like fleshy sponges. Haldane's arms were itching all over. Rendel started itching in the forearms and thighs, and a deep purple rash spread ominously and darkly over one forearm. It was nitrogen trying to escape.

By three p.m. the nitrogen was out, and the pain was gone. The itching would resolve, too, eventually. But the pain and the itching and the rash were reminders of the extreme danger with which they must flirt to get each data point. They were fine—this time.

———

German forces had attacked Denmark on April 9, 1940. Denmark had surrendered on the day of the attack, but Norway struggled against the invasion through April, and it would continue to struggle into June. The Norwegians did their part to fight the Nazis on land while Haldane and Case itched to do theirs.

CHAPTER 7

Everyday Poisonous Gases

Helen Spurway's first test was less eventful than Rendel's. She and Haldane climbed into the chamber on day ten, carrying an adorable guest researcher, a small female canary in a cage. Spurway had a hard time equalizing, leading to a slow, plodding descent that left her clutching her nose to help blow air by force into her ears. Once on the bottom, she thought she might faint but did not, and the fragile, beautiful canary, with its petite beak and its storied family history of alerting humans to dangerous gases, flew about the chamber without concern. In the tiny coal mine that was this metal tube, the gas-sentry canary might have been drunk, but it was alive.

Haldane's joint issues returned, this time with a "violent cracking sound" in his left shoulder. Spurway could not do math silently while drunk on nitrogen, and she read the numbers out loud as she wrote. Her nose, its minor blood vessels traumatized by the continued struggles to equalize her ears, bled slightly at each decompression stop when they traveled toward the surface. Two hours after the experiment, she reported a line of slight aches dancing down the joints of her right arm and wrist, which resolved on their own. The canary was fine. Fruit flies brought to the same depths could not fly while there because of the increased air density, but posttest they also survived to live happy, long-for-fruit-flies lives.

It seemed that people could survive down there. If they knew the rules.

———————

On May 10, 1940, less than two weeks after Helen Spurway's inaugural chamber run, Nazi troops stormed their way across still more borders. The Nazis had originally howled that they simply wanted to reassimilate the territories amputated off Germany after WWI, but this time they invaded Belgium, Luxembourg, and the Netherlands, lands that had in no way been within the country's previous lines. The Nazis were after the whole of Europe. They wanted to reconstruct the Roman Empire and call it their own.

"For 300 years the rulers of England and France have made it their aim to prevent any real consolidation of Europe, and above all to keep Germany weak and helpless," Hitler thundered to his troops, according to an English translation of a speech published in London newspapers. "With this your hour has come. The fight which begins to-day will decide the destiny of the German people for 1,000 years. Now do your duty."

King George VI of Great Britain appointed military-minded Winston Churchill to replace appeasement-focused Neville Chamberlain as the prime minister. Churchill immediately gave the Royal Air Force—the RAF—permission to start bombing.

———————

On an otherwise peaceful Monday in London, on the twenty-seventh day of that same atypically balmy, atypically eventful May, Martin Case and Jim Rendel prepared themselves once more to climb into the chambers. On the southern coast of England, civilian sailing vessels scurried back and forth between England and the beaches at Dunkirk for the second day of evacuating troops from France. At Siebe Gorman, it was time for experiment number forty-seven, on testing day twenty-five. The English military had delayed drafting both Rendel and Case so they could serve in this other way. Haldane had sent a simple letter about Rendel: "Should he be called up for military service, this will delay the conclusion of the experiments and thus cause a certain loss of life among naval personnel." The Admiralty agreed without argument. They indefinitely delayed the call-up of James Meadows Rendel, Ministry of Labour draft registration number 7080.

The previous experiments had served as "controls," meaning they were designed to be sort of "normal" exposures without any additional factors. That way, when a new factor was added to the protocol, its effects could be separated out. Today's new factor happened to be egregious, skull-splitting, profane quantities of carbon dioxide.

Rendel played the ball game with the chamber door open, followed by the arithmetic, as usual. Then the scientists swung the round door closed. Either Case or Rendel reached for today's new tool: a small, pressurized cylinder filled with liquid carbon dioxide. With a twist and a hiss, the cylinder valve loosed its odorless, colorless, poisonous gas into their tiny dark cage. The fervent, high-pitched rate of release turned the metal bottle frosty to the touch. Case began flapping his hands to stir the air, to mix it before he took a sample. The canister had bumped the CO_2 up to a moderate but not yet painful 2.82 percent. The gas release was also enough to pressurize their habitat a bit, and they were now at an accidental pressure equal to a depth of 9.5 inches of water.

Jim Rendel was already hot. The warm weather outside plus their exhaled breath inside plus their body heat needed no help from pressurization to make the temperature levels uncomfortable. Both Rendel and Case began panting, not only from the heat, but also because the carbonaceous beast they had unleashed was starting to wreak havoc on their bloodstreams. Their lungs and hearts were working in overdrive, trying in vain to expel the caustic molecules. They took another sample: 3.32 percent. It was time for the chamber to travel.

But Jim Rendel's left ear would not cooperate this time. He held his nose and he blew, and he blew hard. The moment they finally reached the bottom, Rendel felt the effects of pressure, on top of carbon dioxide and hyperventilation, all building up on him together.

"I don't think I can stand this for very long," he said to Case. Case was panting distinctly while Rendel continued to gulp for breath. Rendel said he felt "indescribably queer" and "very tired." The acidic effects of the carbon dioxide were burning into their brains. Rendel's fingers went numb.

They began their decompression after an interminable fourteen minutes, venting the heat and the headache and the carbon dioxide back out into the warehouse at Siebe Gorman, where the others stood outside and watched and listened. As they passed by ninety feet, with a short stop as a nod to the nitrogen, Rendel's painful left ear began bubbling. He felt and heard the crackling of the pressurized gas escaping slowly past the plug of snot inside his head that had caused him such trouble on the descent. Both of Case's ears began bubbling, too.

At sixty feet, they began sucking down pure O_2 to increase the speed at which the nitrogen left them. Rendel's nose began to trickle a stream of crimson blood. Case's arms and body began to itch, and a few minutes later, Rendel's body followed suit. They exited the chamber forty-four minutes after they had left the bottom, itchy, Rendel's nose still releasing a steady flow of red. But they felt they had accomplished their goal. They had gotten their data point. Rendel could head back home to his wife, Joan, who was now seven months pregnant.

However, the trouble was only starting for Jim Rendel. The data point was "intended to save the lives of sailors," so it was worthwhile and necessary. He had been asked to participate because he was "a young and vigorous man," like many of the sailors for whom he was doing the work. But the test had placed an invisible bomb inside the chest of James Meadows Rendel. The first tick of the bomb's timer was a slight pang in his left armpit, just below his shoulder blade.

For the first few days, Jim Rendel thought the new pang was an undigested meat pie, so, like so many young men convinced of their immortality, he dismissed it. The pain grew. By June 12, more than 330,000 Allied troops had been evacuated from northern France by "the little ships of Dunkirk," Germany had attacked and defeated French forces at the Somme, Mussolini had joined the fray by declaring war on France and Britain, and Jim Rendel finally gave in and went to a doctor. Rendel was healthy and young, but felt alarmingly short of breath, with the left side of his chest in constant pain. The doctor was worried. He asked Rendel to remove his shirt and lie down on his right side, with the aching left half of his rib cage facing upward and exposed.

The doctor made a small incision in Rendel's back for a procedure called a thoracotomy, an exploration of the area inside his chest around the heart and lungs. The doctor inserted a scope to look around. He found that Rendel's left lung was almost completely collapsed.

Something inside Jim Rendel had ruptured during the last test. Something important. Something had torn open, either from the forceful ear clearing, or from too much pressure inside the breathing apparatus, or simply because it was the wrong day to anger an unusual flaw of anatomy already lying in wait inside Rendel's lungs that never would have bothered him had he not traveled to increased pressures. It was impossible to tell. It is still impossible to tell. But trapped expanding gas had torn a hole in the fragile, bubbly pinkish-white shape of Rendel's left lung and created a new passageway out. A small amount of breath from each subsequent inhalation escaped into the rest of his body, where it was not supposed to go.

The first intruding tiny gas bubble had built up slowly but continuously until it consumed the left half of his chest, the left half of the space he needed to breathe. The bubble, which was now more like a cavity, had taken up far too much of the volume inside Rendel's rib cage and pushed his light pink left lung down into a flat shadow of its former self, in addition to shoving his heart nearly four centimeters to the right.

The original gas bubble could have killed Jim immediately if it had traveled to his heart or brain. It was luck that it had not. But as the rogue bubble grew, it could have killed him in other ways, too. Eventually, if its growth had continued, it would have forced his heart too far to the side and stopped it from pumping.

Jim Rendel's doctor was filled with alarm for the delicate medical condition of his respiratory patient stretched out on the exam table, exposed and gasping for breath. So, of course, the doctor handed his patient a lit cigarette.

The flesh of Rendel's back was still open because of the small incision for the thoracotomy. The doctor had noticed that the top corner of the higher lobe of Rendel's left lung had a strange notch in it, a whitish mark with a groove containing what looked like it might have been a small blob of clotted blood. That notch might have been the site where the lung had somehow trapped gas while under pressure and then exploded ever so quietly to release that gas into the chest during decompression. Like an eardrum that had torn rather than equalized, but far more deadly.

The doctor wanted to make sure that the lung was sealed now and that if he stitched Rendel up, no more gas would escape into his chest. The cigarette was a test. Jim Rendel puffed obediently, and when he inhaled, the doctor watched to see if smoke escaped into his rib cage through the insidious white mark. None did. The doctor interpreted the absence of smoke as a sign that the lung had resealed itself. It had. However, as anticlimactic as the chamber experiment had seemed at the time it occurred, the damage to the lung was done.

Two days later, Paris fell to the Nazis.

In between experiments, Haldane continued to give public speeches on behalf of the Communist party, proclaiming to hundreds, then thousands, then at one point more than ten thousand spectators that the British government was not doing enough, not anywhere close to enough, to protect the people of Britain or counter the vile threat of fascism oozing across the European mainland.

It was exactly these types of anti-fascist, pro-Communist lectures that had brought Haldane to the attention of MI5 in the first place. It was his stubborn continuation of these lectures that kept them watching him. He trained his sharp criticisms on what he thought of as the government's lack of action. He had run toward the Spanish Civil War to fight against fascism for Spain, and he now accepted microphones in public squares in England to bellow how he wished to fight against fascism on behalf of the whole world. He understood the dangers of putting control into the hands of one unquestioned leader. Even though war was coming, inarguably at this point, and Haldane was right about the tyranny of Hitler and fascist rule, the British government did not appreciate the criticism.

One MI5 informant finally wrote a letter to the Home Secretary of London to complain. He was being asked by MI5 to listen to Haldane's speeches, to help monitor him. It seemed clear, to this informant at least, that Haldane without a muzzle, allowed to keep questioning why the British government was not doing more to fight the Nazis, was "damaging to the war effort." The informant argued that Haldane should be handled the same way as the other politically suspect residents and jailed. The informant asked directly: "As [J. B. S. Haldane] is a danger, why after this many months has he not been interned?" The answer, which MI5 did not provide back, was the Royal Navy.

Beneath a ragged scar, one of Haldane's old shrapnel wounds was giving him fresh aches on experimental day twenty-eight. Decades before, during the First World War, rough metal had sliced a new entry into his upper back. The wound was throbbing from an experiment a few days earlier. He had sat inside a galvanized iron bath of cold water inside the small chamber. He had plunked his large form into the frigid 1 degree Celsius liquid until it sloshed over the sides of the tub and into the bilge of the chamber; then he pressurized to two hundred fifty feet and checked to see how the cold and immersion affected him in combination. The test was similar to the one conducted by Case, simulating a flooded submarine compartment in the Atlantic. Today, the lingering shrapnel pains and a slight headache notwithstanding, Case and he were going to go back to three hundred feet. They were going to use one of Siebe Gorman's other breathing apparatuses, the Salvus II.

The Salvus device had originally been designed for use in firefighting or to protect soldiers' lungs from poisonous gases in WWI trenches, but there was no reason that it should not work underwater, too, both the researchers and the Admiralty figured. It was similar to the DSEA but larger, so it should have provided divers a longer time underwater. Haldane shook off his old shrapnel pain for the sake of the test.

At 12:17 p.m., Haldane and Case reached three hundred feet. Both of them were panting. Haldane was "sweating intensely." Case was sweating "considerably" but "not more than usual at this stage." Case looked over at Haldane. He worried that the intense but aging scientist looked odd.

"Are you all right?" Case asked through the rubber mouthpiece between his lips.

"No, very far from it," Haldane replied, also maneuvering the words around his breathing apparatus, while at the same time jotting a handwritten note: "Feeling bad. Shall soon feel unless can breathe freely will have to tap 3 time [sic] on wall." With no communications equipment inside the chamber, a small hammer was their easiest means of signaling to the outside for help. Three taps on the metal wall meant that the outsiders should abort the dive. Haldane was struggling to breathe. The man who never quit was thinking about quitting.

By 12:23 p.m., they both felt better, and agreed they had proved the breathing apparatuses could be used at depths of three hundred feet. That was all they had come for today, so there was no point in staying. They tapped on the wall.

Within seven minutes they reached ninety feet. The ascent was going smoothly. Haldane fiddled with the oxygen valve on his Salvus to make an adjustment. Then, when Case looked over at his fellow chamber diver, he noticed something alarming.

Haldane's face twitched. It started as a simple jerk by some unknown but minor facial muscle buried behind the stately mustache and notorious eyebrows. But it escalated quickly. Within seconds, all of Haldane's face and limbs began convulsing, going spastically rigid with the full force of waves of seizure rippling through his body like uncontrolled arcs of electricity dancing from a downed power line, curling him into a stiff, chaotic, relentless contracture of muscle and bone. Frothy saliva began to flow from the mouth of the mighty two-war veteran, around the mouthpiece, drowning the rubber and his chin in a rabid effusion of bubbles. He lost consciousness.

Martin Case jerked the mouthpiece out of Haldane's mouth. He scribbled an urgent description of the situation on notepaper and slammed it against the tiny clear porthole for those outside to read.

The note to the outside was not necessary. Haldane's body's erratic gasps for breath were so loud and rattling that those outside the thick tube could hear the rasping clatter of his inhalations through the metal wall. They were alarmed. Case took off his own breathing apparatus, too, just in case, and he pulled the rest of the equipment's harness off the unconscious, still salivating, still seizing man beside him. Prod and yell as he might, Case was unable to get any kind of response from J. B. S. Haldane.

A few minutes later, the partially conscious and definitely not-OK J. B. S. Haldane appeared to decide he was totally fine. He made some unintelligible attempts at speech and tried to move, get up, stand, enact some unclear kind of motion, struggling against the torpor imposed on his body by the seizures. He failed. He collapsed headfirst into a human puddle against the wood planking. Case tried to pull the professor upright, but limited by the unyielding measurements of their narrow tube, he managed to maneuver Haldane only partway head up.

The outsiders were doing what they could by bringing the chamber up. They pressed a note against their side of the porthole once the chamber reached fifty feet on the ascent: "We are going to give you oxygen." They had been managing the decompression schedule from the outside while Case struggled to manage Haldane. Locked out, with the two divers having just left a depth from which fast decompression would have been an almost certain death sentence, the outsiders were nearly helpless spectators of their toppled leader. They needed to get him to the surface. They needed to avoid DCS in the process. They needed to hurry. Hurrying could have been fatal.

Peace reigned for a total of about four minutes after they piped in the oxygen. Then Haldane lashed out again, bashing his limbs across Case and against the chamber sides and into every object in there with them. He lurched around as much as he could, fighting to talk or yell but managing only to push incoherent gibberish out from between his noncompliant lips. Martin Case fought to restrain his much larger boss.

Once Haldane mastered some portions of his lips and tongue, he began declaring vehemently, with both urgency and desperation, along with more attempts to explode his way straight through the thick steel walls: "I AM DYING." Case managed to shove a flexible hose providing oxygen directly into Haldane's mouth, giving him no choice but to breathe from it for at least a few moments. He did not keep it there long. The outsiders decided to truncate the decompression schedule, whatever the DCS risk. They brought the chamber the rest of the way up, and Case opened the door to get help handling the delirious, powerful J. B. S. Haldane.

As a team, Case and the outside scientists dragged their sizable leader out of the chamber. They left him crumpled on the floor for half an hour, after which they managed to coax him onto a mattress, where they covered him with blankets. His lower back had a new, "very violent pain," he said, separate from the old war wound that had ached that morning. Pain was not surprising, as he had hurled his body against solid steel. The savage seizures of the muscles of his back had warped his spine into foreign shapes. Some piece of him had broken. After Haldane had spent three more hours on the mattress, Case helped him peel himself up, and he got the battered professor home.

Later, Haldane denied that he had been unconscious at any point or that any of his movements had been anything less than voluntary and precise. The rest of the scientists disagreed.

Martin Case had saved J. B. S. Haldane's life. Haldane, while drunk on nitrogen, had meant to make a minor adjustment to his breathing apparatus, but instead he had unwittingly twisted open the knob that controlled the flow of oxygen, flooding the breathing bag with undue quantities of yet another colorless, odorless gas. Case had figured out what happened and ripped the mouthpiece out of Haldane's mouth. Oxygen, too, just like nitrogen or carbon dioxide, can kill you. But how easily was a new discovery.

More than love, or hope, or even potato chips, we are all fueled by oxygen. We can survive three weeks without food, three days without water, but only three minutes without oxygen. Even that diminutive hundred-eighty-second grace period is something we buy for ourselves only by keeping a cache of the stuff circulating in our bloodstreams. Oxygen is the most urgent demand made by our frail, meaty bodies. But just as with sugar, warmth, sunshine, sex, and fried chicken, *too much* of any good thing turns bad. In the case of oxygen, too much can kill you.

Nowadays, the term du jour for the phenomenon is "oxygen toxicity," but the same monster has gone by many names since its discovery in the late 1800s. It has been called oxygen intoxication; the Paul Bert effect when it affects the brain; the Lorrain Smith effect when it affects the lungs; Oxygen Pete by the more playful; and around the time of Haldane and Case's wild chamber adventures, oxygen poisoning. Perhaps more terrifying is that, to this day, nobody has managed to pin down Oxygen Pete and shine a light on why he kills.

In biology's ultimate ironic joke, both the utility and the menace of oxygen lie in its ease of chemical reactivity. The good and the bad are inseparably paired like dark twins, united by this same chemical property.

The two little O atoms that make up the O_2 molecule vibrate together, impermanently locked in unison through the sharing of four electrons. On a molecular level, it's as if each of the two O atoms is one human child, and they've clasped both hands together with another human child. Bonded like this, they can whirl together in circles as a pair, their linked arms and hands creating between their two bodies stability and balance that cannot be achieved by one child spinning alone.

But then, sometimes, in the heat of the spinning, one pair of the hands comes unclasped. Oxygen is exceptionally greedy with its electrons, and it is likely to try to jerk a hand free, even from its twin. Most of the time, the atom children will reentwine their fingers and start sharing electrons again, spinning as a pair, but during that brief moment of hand freedom, they might also reach out to grab another atom child or something else nearby instead. Oxygen does the same thing in its moments of greed. Oxygen is an unreliable playmate. You never know what the free hand, that independent, unbonded electron, will choose to grab during the brief moment when it has the chance. When the child's hand is open and grasping, when the electron is free and unbonded, that electron is called a "free radical."

Our bodies take advantage of this fickle behavior in order to off-load waste materials. We inhale the O_2, and our bodies manipulate it into reacting with chemical detritus we do not need from the sugars and fats we have eaten as fuel. We glean a more raw and usable form of energy out of the reaction, and then we exhale the same O_2 molecules but burdened with additional carbon as CO_2. It's the same general chemical principle as burning gasoline, which consumes oxygen and releases heat and energy but also CO_2 emissions.

Oxygen is a tricky monster, though. It doesn't react only with what we want it to. We don't control it; it controls us. It has low standards for bonding, and it might grab whatever floats nearby. It can also change forms and morph, turning into a menagerie of similar-but-slightly-varied beasts called "reactive oxygen species." These new, similar but slightly morphed oxygen species use the same aggressive, reactive behavior against our cells, our proteins, even our DNA. The undesirable reactions of oxygen and its species, especially against our DNA, are thought to be why we age. We need oxygen to live, to process our food into energy, but as it helps us, it slowly eats away at us, too. Our compromise with oxygen is that it gives us life, but mars our lives with aging and death.

This theory of aging is why "antioxidant" is such a popular buzzword in the pseudohealth community. Certain substances—like tea, red wine, blueberries, and chocolate—contain naturally high levels of chemicals that interfere with oxygen's malevolent plans. These molecules vary in their exact structure, but they get lumped together under the same name of antioxidant because of their tendency to grab the hand of oxygen when it reaches blindly outward. Antioxidants bind with oxygen and neutralize it so that it does not float farther and find other havoc to wreak.

But how well do they guard against the onslaught of oxygen we must breathe in every day to survive? And do they do enough to make a difference when breathing surface levels of the life-giving poison, much less in a hyperbaric chamber, to dodge the mischievous probing grasp of Oxygen Pete? Heck yes, there's about to be some math.

Most oxygen travels through the bloodstream while bound to hemoglobin, which is a special little carrier protein inside red blood cells.

However, some stray, independent oxygen is also always present, dissolved in the liquid part of the blood itself. During normal life, while we breathe air at its modest 21 percent oxygen, not a large quantity of oxygen is dissolved in there. But raise oxygen levels and blood quickly absorbs and dissolves the new free-radical-laden gases without a care for what other tissues might be damaged in the process. A man of J. B. S. Haldane's weight, 108 kilograms, holds about eighty-one hundred milliliters of blood inside his outer container of skin. In other words, if you drained him of his redder juices, you could just about fill a 2.5-gallon fuel can.

At that blood volume, if J. B. S. Haldane started breathing pure oxygen while lounging around at the surface, he would quite quickly have an extra 4,023,000,000,000,000,000,000 oxygen molecules dissolved in his blood. If he wanted to use the power of antioxidants to counteract that, he'd need to consume a bottle and a third of Merlot, or—to keep it British—192 liters of Earl Grey tea, *with every breath.**

J. B. S. Haldane had a limb-clenching, drool-producing seizure because oxygen and its free-radical buddies went rampaging through his lungs, his blood, and, most important, his brain. His body ran out of the resources to neutralize the molecules on its own, and so the unremittent O_2 began grasping its greedy hands onto key bits and pieces of his neurons instead, and onto other crucial molecules doing their normal jobs inside his body. The terrifying part is that to this day, we don't know exactly which parts of the brain oxygen hits. But it sure hits them, and it hit them hard inside J. B. S. Haldane. It hit hard enough that his body could no longer protect his brain or his nerves. They went haywire. He seized. He was at ninety feet of pressure and there's no telling how much oxygen he actually got after his drunken valve fiddling, but if he were breathing a gas of even 75 percent at that depth, to counteract it, he

---

*Dear reader, I don't hate you, so I won't force you through the math required to arrive at these final numbers, and quite frankly, I am not talented enough to make the process interesting. However, if you are in the mood, crack open a bottle of Merlot, unwrap a bar of dark chocolate, and head to the notes section, where you will find the calculations and relevant citations in their full, nerdy glory.

would have needed to ingest with every breath eight bottles of Merlot or 1,564 liters of Earl Grey to protect his brain and stay alive.

Oxygen always wins in the long run. When humans breathe normal air, it wins after an average of seventy-two years or so. After seventy-two hours of breathing pure oxygen, rats and mice start to die, and human patients start to produce crackling, gunky liquid in their lungs as they try to protect themselves from the onslaught. This phenomenon, called pulmonary oxygen toxicity, is why oxygen percentages need to be throttled back for COVID-19 patients after only a few days of the pure stuff. It starts to become a trade-off between giving them enough oxygen to keep them alive, despite their disease-addled lungs, and oxygen-torching the remaining working cells until those cells start dying from the gas faster than they're dying from the disease.

At depths like three hundred feet, the power of increased oxygen makes things that should not burn flammable. Metals like aluminum and titanium become powerful explosives through their willingness to bind with mightier, higher-pressure oxygen. Polyester can start a spark of static electricity that engulfs its wearer in a fireball. And the reserves of the body's natural antioxidants are depleted even faster. Once you're in the danger zones of percentage or pressure, some form of oxygen toxicity becomes inevitable.

The Admiralty had started using pure oxygen during decompression in 1937 to help off-load nitrogen faster. The earliest oxygen seizures in experimental animals occurred just a few months before the testing in 1939. Haldane had his in summer 1940, when there was still remarkably little knowledge about the phenomenon, and there is no indication he was aware of the 1939 example. By the end of July 1940, the lab group had proved, on themselves, through sixty-nine separate experiments conducted over forty days inside the chambers, the theoretical safe levels of carbon dioxide and at what depths submariners could use DSEAs despite the effects of nitrogen, carbon dioxide, and cold mixed together. But the oxygen seizures were new and unexplored. They were something that had only recently emerged from the realm of legend. Nobody yet knew what depths they occurred at or how long people could stay safely at those depths while using pure oxygen before problems would occur.

Haldane was one of the unfortunate firsts to experience the effect, and clearly, as unpatriotic as it might have sounded to the Admiralty, tea was not a solution.

"I think we have now enough data to show what to do during flooding up. The very real danger is, however, that of oxygen poisoning while leaving the submarine," Haldane had written to the Admiralty, unwittingly, two days before his seizure. A few months before, the respiratory expert had also guessed that he thought the Americans had "underestimated the effect of oxygen." He was right.

A week after Haldane's losing battle against Oxygen Pete, Martin Case used the same breathing apparatus at three hundred feet but with helium in the gas instead of nitrogen. It was an idea from the Americans. US Navy physiologists had theorized that helium might be less narcotic, and American rescue crews had used the novel gas blend for the first time during the salvage of the downed submarine USS *Squalus*. Helium had that one reported success under its belt, but the Admiralty wanted their own tests to verify the claim.

Haldane, his back still hurting, was Case's dive tender inside the tube. The breathing device worked, and the use of helium avoided the drunkenness. Royal Navy divers could use these submarine escape-breathing apparatuses safely with helium, and have clear, fully functional brains at a depth of up to three hundred feet, as long as they carefully monitored their oxygen levels.

The lab had checked an item off the Admiralty's list. They had made a step of progress. Any diver, even those bent on espionage, could use the breathing apparatuses at any depth they wanted . . . once someone figured out the mystery of O_2.

CHAPTER 8

The Blitz

Haldane took a month off from being a test subject to recover from his Very Bad Day in the lab. By the end of June, with his lumbar spine still aching, and with Jim Rendel still in the hospital getting his lung reattached to his chest, the battered duo of Haldane and Case climbed into the chamber again. At a depth of three hundred feet of pressure, they gave the valve of a compressed gas cylinder a twist that might have been too vigorous and unleashed enough carbon dioxide to knock themselves unconscious immediately. They later claimed that they had used the hammer to signal they were in trouble before passing out, but nobody outside heard it. Eventually, someone noticed their limp bodies through the porthole. Both scientists had large gaps in their memories afterward. They considered it interesting data.

Helen Spurway tried adding CO_2 at three hundred feet using the same protocol the next day, June 28, because all good science needs input from others. However, despite twisting the cylinder fully open and releasing all the gas she had, she stayed conscious. "Her resistance had been underestimated and not enough CO_2 had been taken into the chamber to achieve such a result," Case scribbled in his lab notebook. The levels used were the same; the young PhD could simply handle her carbon dioxide better. Spurway was disappointed. She had wanted to get knocked out.

With that, Haldane told the Admiralty that their group had roughly figured out the question about tolerance of carbon dioxide—item "a" on the Admiralty's earlier letter with the scientific wish list—along with their questions about nitrogen narcosis and cold tolerance, which had been item "d." Next, the group planned to turn their attention to item "b," oxygen poisoning, to allow the deliberate use of underwater breathing devices for espionage. Divers were no good if they were having seizures. Prior to his seizure, both Haldane and the Admiralty had underestimated the problem, with him even writing at one point, "I think it is an open question whether oxygen poisoning is a real danger when an escape apparatus is used." Clearly it was. Liters of Merlot were not the solution.

"I think that experiments of this kind should occupy us for another week or two, by which time we shall either have been bombed or got instructions to work on oxygen, or both," Haldane wrote. However, other sections of the government were trying to draft the younger male scientists, including the hospitalized Rendel, out of the lab to engage in a more direct version of war. When Haldane pushed back, saying Rendel and all of the crew were needed in the chambers, the draft board suggested that Haldane instead get employees who were not eligible for combat through the Labour Employment Exchange. Haldane managed to fend them off with a bit of his characteristic sass: "Perhaps you will be good enough to tell me at what Exchange I can obtain the services of men who will risk their lives without being paid for it, this being the position of persons participating in this research."

Case was receiving a modest stipend as a salary from the government, but everyone else was doing the tests for free on top of their normal work. For patriotism officially, but given Spurway's open sadness over staying conscious, they were clearly also curious.

———

The rest of England prepared for war. London newspapers advertised war bonds, and on the same day as Spurway's disappointment in her own consciousness, several papers printed small doses of public applause for companies that bought them in massive quantities. Private citizens in London and the surrounding areas dug up their yards and gardens to install government-issued mini havens called Anderson shelters, which were designed to protect people against what bombs might come. The shelters were thin-walled corrugated-metal rectangles with domed roofs over top, and when properly installed, they were supposed to be dug deep into the ground and buttressed by blast-absorbing walls of soil. The Kalmus family received their Anderson shelter for free, since they earned less money than the government's predetermined salary limit, and Hans Kalmus dug the shelter deeper than suggested, nestling it even more firmly than recommended in the secure, dark, fertile soil of their yard. His son Peter would later suggest that the added safety might have been an accident, because his immigrant father was not yet familiar with feet and inches.

Hitler and the Brits had both made promises not to bomb London and Berlin, respectively. Allegedly they would both bomb military targets only. The shelters got installed anyway, because Hitler was not turning out to be reliable about his promises.

Even though a large fraction of the country had died or suffered from the bombs of WWI, and even though WWI veterans now formed a hearty chunk of Britain's middle-aged population, and even though planes had come across to bomb the UK directly during that war, a lot of people nonetheless treated the idea of air raids with a kind of casual dismissal. The government tried to warn them, and Haldane himself tried to warn them through articles, yet many people reported and talked about the possibility of bombings with a kind of flippant humor that suggested they had an unfounded belief in the incompetence of the actually very competent Luftwaffe. The Luftwaffe was not deadly enough, many people seemed to convince themselves, to justify action against a threat they could not yet see. Some were terrified, but another fraction seemed to feel that if they had the right attitude, the right toughness, the right amount of can-do spirit, the bombs would never come, or at least would never hurt *them*. This fraction talked as if shrapnel responded to fear or bravery.

The air combat started over the summer. Woven in among the clouds and the intermittent sunshine of the English skies of July 1940, the exclusively aerial Battle of Britain soared above the heads of civilians. The Luftwaffe tried to chip away at the Royal Air Force in preparation for an eventual land invasion. The RAF fought back, and dogfights overhead became an odd regularity, absorbed into the background normalcy of everyday life. The sky-backed silhouette of a Spitfire aircraft became a mascot to be cheered out loud.

Bombs hit the less populated areas first. German pilots sought to bury their explosives in military targets like airports, bases, and even aerodromes, which were flat areas where planes could take off and land. Most of the civilians, so far, were spared.

"'That Village' Raided Again" ran one mocking English headline in the late summer above an article about German bombs falling feebly among the duck ponds and greens of an unnamed, deliberately written-as-generic southern county. "The curious hiccupping noise of a German aeroplane was heard overhead," the text caricatured the Luftwaffe. The author described the falling ordnance as powerful enough only to knock shutters off their hinges. It was one of many publications to poke verbal fun at the Luftwaffe as its pilots crept their way inland from the coast, hitting mostly military sites and rural areas, because they didn't have permission to aim for maximum carnage. Yet.

Nevertheless, in preparation, police and air raid wardens patrolled the streets of cities at night, looking for even the narrowest slivers of light shining through windows around the edges of blackout curtains, and asking those citizens within to darken things to avoid giving pilots visible targets. Some Londoners drew Christian crosses on the doors of their Anderson shelters, but others drew bull's-eyes. Some adhered meticulously to blackout policies, but others, presumably because they disagreed about the level of danger, whined that the simple requests were government overreach and that the police were drunk on power. In the middle of it, J. B. S. Haldane, Helen Spurway, Ursula Philip, Hans Grüneberg, Martin Case, Jim Rendel, Elizabeth Jermyn, and Hans Kalmus worked quietly into the nights in their still heated mouse labs, in one of the only buildings on campus with working gas and electricity, their blackout curtains drawn firmly against probing eyes.

By mid-August, air raids on cities were clearly possible. But in the minds of many, bombings could happen only to other people, not to them, and the war was on the brink of an easy victory. "There has been a good deal of talk today of the number of German planes brought down over the weekend. . . . The general opinion being 'Germany can't last long at this rate!'" one diarist reported hearing throughout her town in August 1940. The newspapers provided detailed descriptions of air battles and dogfights, but mainly the tallies of pilots killed and planes shot down were all German. British numbers were omitted. The numbers might or might not have been true, but even if true, they were delivered with the tone of overly optimistic propaganda. "89 Raiders Down During the Week-End, Big Enemy Forces Driven Off," one such headline trumpeted. Optimism ruled in the early days of autumn 1940.

Antiaircraft guns had been positioned in the major city parks during the quieter year of 1939, and now, instead of the typical statues of men on horses, these weapons lorded it over the open public spaces. Massive metal artillery beasts patrolled the skies from the western edge of Central London, from the beautiful grassy lawns of Hyde Park. Those guns were close enough to stand careful guard to their east, where Buckingham

Artillery soldiers in Hyde Park practice air raid drills, with civilians watching a mock battle above.

Palace, Westminster Abbey, and the buildings of Parliament and the Admiralty all stood clustered together. UCL was within walking distance of the deadly new monuments, a brief jaunt to the northeast through rows of densely packed townhomes and pubs. Wearing so-called tin hats, eager soldiers drilled to learn how they would use the guns to guard the city against airplanes, often to the delight of civilian spectators.

But despite the changes to the city, the atmosphere remained one of disconnection from the danger. "It's amazing what a capacity most people have for dismissing unpleasant facts from their minds," one Manchester resident wrote of her neighbors, who lived farther north than London. She noted that many of the RAF personnel in her area seemed primarily interested in fooling around with willing and sometimes anonymous women inside the shrubberies close to local pubs. She was not the only one to note that soldiers often tried to "shelter" inside pubs rather than in other buildings when the air raid sirens, which were not yet accompanied by actual bombs, would sound. Some parents brought back children they had sent away from London at the first declaration of war; they had been lulled into a desire for normalcy that conspired with a lack of immediate bombing to convince them it was safe.

Haldane knew better. He had lived through the bombings of civilians and cities in the Spanish Civil War. He campaigned loudly and publicly for better bomb shelters. He grew frustrated with the lack of response from the government and the people, and he grew belligerent. As he grew more belligerent, some listened to his warnings, but others dismissed him as a fearmonger. Some aspect of human nature makes us ignore the experts when their truths are not what we want to believe.

In the early evening of August 15, 1940, German bombers struck within sight of London when they unleashed incendiary bombs over a semilocal aerodrome to the southwest. It went up in flames, sending into the sky thick, billowing clouds of black smoke that did not seem to alarm anyone. It was just another non-London hit to many.

Less than thirteen kilometers north of the blazing incendiary attack, a woman recorded only as Mrs. H. struggled and failed to get her husband to take shelter during the raid, because he wanted to eat dinner first. It was the same for her deaf neighbor, who was upstairs in the middle of a cup of tea and wanted to finish it. When the sirens first yelled their warning, a few people bolted for cover, but most, instead of taking shelter, went to their doorways to watch. They clustered at the gates of houses to chat, their eyes on the sky. At least one man on a common kept working away peacefully in his garden allotment. The bombers were less than a minute and a half away by air. In the world of aerial combat, the bombs were next door.

One soldier on the ground, standing in the streets among the watching crowds, pointed out the lack of concern. "Look at us all. Standing out here after the sirens have gone. Still, just shows you what we think of raids over here, doesn't it? Look at the old warden—if his own family don't go into shelter, how can he expect anyone else's to?"

A twenty-year-old male student heard the sirens and took them as a cue to leave home to try to get some milk, in case he needed to stay home for a long time during the raid. "It was not till I had left that I thought about the possibility of bombs dropping near <u>me</u>," he wrote after his errand, carefully underlining the word "me" in his diary. "But of course, it's impossible to believe that anything will really happen in your own neighbourhood. It always seems bound to happen somewhere else."

The people in Croydon, near the aerodrome hit by the bombs from a reported fifty German bombers, were surprised, too. Many of them had kept doing what they were doing while watching the planes overhead with curiosity and fascination. One woman reportedly remarked to her sister as they kept gardening, "Look, they're dropping something." It was not until they were hit by the ground-shattering blasts from the resultant explosions that the sisters ran for shelter beneath a staircase.

Despite the preceding solid year of warnings about air raids, the sisters later reported that "apparently everyone [in Croydon] was rather stunned by the suddenness of the whole thing." Bombs, until you felt them yourself, were a surreal abstraction. Bombs were a thought, mere words in black and white in the papers. Bombs were too different from normal life to process until they came. The Londoners were not alone; people have shown the same pattern of response to danger throughout history. Something about human nature makes us feel that if we ignore danger hard enough, then it will come only for other people.

By the next day, the mayhem at the aerodrome was forgotten. "I heard more than one prediction of victory for us by Xmas," wrote the same Manchester woman who had been concerned about the frisky soldiers in the bushes. She said of the following weekend that it was free from all talk of war, and "so quiet that I couldn't find anything to say worthy of record."

Like a slow and fatal spiral, however, the bombers circled London on the map, inching ever closer. On August 23, stray bombs destroyed two cinemas and some other buildings in a suburb a bit over twenty kilometers to the northwest of the city. No one was killed. So still, nobody paid much attention. The next day, German bombers with what they claimed were slippery fingers and mistaken aim accidentally dropped their munitions even closer. They said it was an accident. They still say it was an accident. They had not meant to violate their promise. But this time, nine civilians died. This time, it was too close to dismiss.

Britain and Winston Churchill replied with an immediate scream of force and fury, launching pilots straight for Berlin to shock both Hitler and the unwary citizens below. They sought to unleash hell on the people of Germany in retaliation. No German humans died; only one of the elephants in the zoo. Germany in turn, though, escalated back. Diplomacy was dead. At 12:15 a.m. on August 25, 1940, the first bomb fell in London proper. It plowed into the pavement and masonry of Fore Street, less than five kilometers east of Buckingham Palace, three kilometers east of the UCL campus. Then, in response to what might have been sparked by a single mistake by a single pilot in an era when proper aim was known to be difficult, both countries continued to escalate, to begin what remains one of the greatest mutual bombardments of civilians in recorded history.

The Admiralty asked the scientists to wait to start their experiments with oxygen toxicity until "the personnel are fully insured against life and limb, i.e., injury to health or a fatal result." But after honest disclosure of his seizure and Rendel's accident, Haldane was struggling to find an insurance company willing to take on the risk at a price that he or even the Admiralty could afford. So, the Medical Department of the Navy, now in a greater hurry, agreed to insure the scientists themselves, saying "payments of compensation on an ex-gratia basis will therefore be paid in the event of these experiments resulting in the death or permanent disablement of any of those taking part." Haldane was not as worried about payment in the event of death, but he wanted to make sure that his people were covered if they got so injured they could no longer work. The government also removed Jim Rendel's name from the draft pool for as long as he kept working on diving questions.

Even the government will bypass bureaucracy when the need is urgent enough. *Get started, please hurry,* they pleaded when faced with the new bombs. If the scientists could figure out how, given that their lab and the Siebe Gorman factory were where the bombs were falling.

Young Peter and George Kalmus began sleeping every night in their family's Anderson shelter, joined during the air raids by their parents, Hans and Nussy, and sometimes also by frightened neighbors who lacked sufficient shelter of their own.

The raids kept increasing in frequency and violence, and the diaries of Londoners reflect the rising tempo of anxiety among the people. By August 27, an evening still early in the marathon of suffering, a Mr. M. Paine wrote, "Last night we had the longest raid of the war," before he described the German planes, bomb explosions, and "thousands of searchlights" amid a strange absence of defensive machine guns or RAF fighters.

The next night, a woman named Mariel Bennett described the noise from her home, which was a short walk north of UCL: "Did not sleep very well, the planes were the nearest they have been to us and seemed to be swooping overhead." Someone else, an unnamed forty-year-old veteran of WWI, put the same event more graphically. "Blimey they didn't half make a row—sounded as if the whole band were falling. . . . I've had bombing in France, last war, but nothing like that. I thought it was coming right down on the house." Even for a seasoned veteran, the shellings were getting intense.

By the weekend, M. Paine said some of the bombs were "close enough to shake the windows," and Mariel Bennett wrote that she was "getting used to [her] cupboard" where she sheltered during raids, before she also wondered "whether the Jews as a race are cowards" and whether that cowardice could account for their relentless persecution by "bullies" like Hitler. Londoners dragged mattresses down to the lower levels of their houses to sleep where a building collapse presented less risk of death, and they learned to identify the types of bombs being dropped by their sounds.

The papers quietly announced that the only place the dead would be named was on sheets of paper posted at public buildings, so the Germans could not count their victories. The true death tolls therefore stayed largely unseen to those not yet touched by the blasts. Those who wanted to ignore them could.

The papers refused to publish the locations of buildings and people that had been hit because that information could have made its way back to Germany and informed more attacks. So information was left to spread through the town by a quicker but less reliable method: gossip. Some curious people could not find the reported damage where they heard bombs had allegedly fallen, creating an atmosphere in which official information itself was believed to be unreliable.

One unnamed London resident described how, as more munitions fell while August cooled to September, attitudes and mentalities evolved. "From the start people seem to have fallen into two diametrically opposed categories in regard to warnings and raids," she wrote. "The first is definitely nervous, timid, and inclined to be panicky. The second is casual, more or less fearless, and inclined to be foolhardy. Age, sex, and class seem to have little or no importance as determining factors; though of course men, by training and education, do not display their nervousness openly as women may do." As the bombing moved north by September 3, hovering near the UCL campus and the formerly peaceful London Zoo where Haldane had taken Grüneberg to cheer him up, she wrote that everyone had simply become entrenched in their former attitudes rather than changing.

"The casuals have become more casual than ever; the nervous, far from becoming contemptuous with familiarity, are on the whole getting more and more nervous as a result of cumulative strain and loss of sleep." The "casuals," as she labeled them, begrudgingly began to ac-

September 7, 1940. Smoke from bombings of the East End neighborhood on the first official day of the Blitz, behind London's famous Tower Bridge.

knowledge that the bombs were real. However, they found ways to bend logic to dig into their belief that their actions had no effect on their risk of injury. The casuals were the people, as she reported, who said of the bombs, "if it's coming to you it's coming—fate kind of thing."

On Saturday, September 7, 1940, with the arrival of three hundred escorted, guarded German bombers over London, the period officially known as the Blitz began. "The worst area to suffer was the East End," described one formal report of that first harrowing weekend, "and by the following Monday and Tuesday many East Enders, mostly women and children, were leaving their homes." People from the densely packed, traditionally impoverished area of London fled for their lives, by lorry, by train, by whatever means they could. Up to twenty thousand of them fled from the East End northwest to Oxford alone, leaving behind what possessions they had, bringing themselves, their children, whatever clothing they could carry. They took over a large cinema on the outskirts of town, where they created an improvised refugee camp. London, the former host of refugees from mainland Europe, was now creating them.

That Sunday night, a couple walking their dog in South London was surprised by an air raid as if it were an unexpected rainstorm. They took shelter inside the home of generous strangers who had spent two years fortifying their house against bombings. The walking couple said they had done almost nothing to prepare themselves or their home, as they had never believed that there might really be a raid. Their hosts could not figure out how the couple had arrived at that naive conclusion. At least four more air raids happened in the same neighborhood that same day. Local schools were canceled because of unexploded ordnance embedded in the ground near the buildings.

By Monday, the north of London was on fire. Rescue workers labored to pull an eight-year-old boy out of a massive bomb crater while his injured mother, freshly retrieved from the same rubble, refused to leave the scene to get medical care for herself.

On Tuesday, an unexploded bomb close to a factory meant a laborer could not go to work that day. Therefore, he would not get paid, even though he was already behind on the rent. The same day, a medical student named D. E. Marmion remarked that because of the bombings in his area, it was "obvious that evacuation of the hospital would have to take place." He carried patients on stretchers until he lost count of them, until his hands and arms ached, until his skin poured sweat with the effort of saving lives in a way that he had never expected. Almost exactly halfway between the UCL campus and Siebe Gorman's test lab, a fire lit by incendiary bombs continued to burn into its third day unquenched.

Peter and George Kalmus were among the many children who ran outside after air raids, looking for interesting pieces of shrapnel to keep. Most of the children who lived through the Blitz, Peter Kalmus included, later reported being unafraid. With no direct experience of mortality yet, the blissful joy of youth looked at the whole thing as "a great adventure." Diarist and mother Mrs. L. C. Tavener wrote that her young son, Kenneth, "clapped his hands" when he heard the firing of antiaircraft guns from the makeshift bed she had built him inside the cupboard where she forced him to shelter.

School interruptions became frequent, usually because of unexploded ordnance or damage to buildings. Peter and George Kalmus's elementary school took a direct hit one night. Since it was at night, there were no injuries, but the building was destroyed. For them, school shifted, adapted, became private clusters of students working to learn in the homes of select parents instead of in one consolidated building.

Shrapnel hunting escalated to a sport among the children, and during the frequent school shutdowns, they would run through the streets of the city hunting for pieces to compare with their friends'.

The streets that slice through the campus of UCL are long and wide, with broad, pedestrian-friendly sidewalks paved by stones hewn into massive rectangles. These imposing spaces force the campus to conform to the shape of the city, and the impressive domed and columned academic buildings and labs are tamed by these roads into tidy organizational lines. The oft-cloistered world of academia is forced to meld with the city as the city dictates.

The western windows of many labs of UCL look across the broad stones toward narrow townhomes capped with ancient brick chimneys, and the density of these innumerable chimneys reveals the density of the people inside the buildings themselves. With one chimney for each room that has a fireplace, the rooms are stacked one above another in the hallmark construction pattern of people trying to pack themselves together.

For the lab members to walk from the mouse rooms of UCL to the site of the test chambers at the lab of Siebe Gorman required starting with a trip south along those densely populated lines, along the stones and through the vibrant thrum of living London. During the Blitz, the homes along this walk began to disappear. Individual homes in the lines dissolved under the force of the bombs, like teeth rotting at night out of a neglected smile. London teenager Eric Arthur Hills described how normal the changes became: "It was a house that was there once, and it wasn't anymore, and there was a great pile of rubble." The scientists would walk to the chamber lab past these newly formed piles of bricks pooled between other still intact homes.

In the decades following the war, constructions of glass and steel filled in the missing teeth of the city, marking through their stark architectural contrast the locations where homes had crumbled, where people had died. The trees along the walk today are all young. The city has healed, but with subtle scars.

From the neighborhood of UCL, today's quickest walking path to the former site of Siebe Gorman includes a bridge across the River Thames that was not finished until 1945. So, during the Blitz, for lack of a more efficient way to cross the water, the scientists were forced along a longer route south, past the city's missing teeth. Past the imposing and famous National Gallery building with its impressive colonnades. Past Trafalgar Square with its fountains and broad public space named for a victory of the British over Napoleon. Also, past one particular statue.

Edith Cavell was a British nurse who had worked in Belgium during the early months of WWI, and she was known for her insistence on treating all wounded personnel regardless of their nation or allegiance. In addition to treating medical wounds, she also helped several hundred Allied troops escape German-occupied lands. She got caught. On October 12, 1915, Edith Cavell was executed by a German firing squad. Now she stands fixed in marble off the northeast corner of Trafalgar Square. Her gaze points toward the river and she stares off into the distance, wearing an austere nurse's apron chiseled into white stone on a pedestal.

After walking through freshly ravaged neighborhoods, Haldane, Case, Kalmus, Grüneberg, Rendel, Spurway, Jermyn, and Philip had to pass by Cavell's unwavering stone gaze. Faced with the reminder of her, they might have wondered what a seizure or a sore back was in comparison.

As the road slopes downward and farther south toward the Thames, the final turn before crossing the water loops left around the edge of imposing carved-stone buildings bedecked with statues. These majestic structures host the offices of state, the War Office, the Admiralty House, and the buried, reinforced bunker where Winston Churchill and his staff planned while the air raids swirled above their heads. This swift left hook around the seat of government brings the soaring pinnacles of Westminster Abbey suddenly into view, with the famous Clock Tower that holds Big Ben completing the Abbey's distinctive silhouette.

By mid-September, Buckingham Palace had suffered bomb damage at least twice. Each walk to the Siebe Gorman chambers was an inad-

An American WWII poster by artist Dixon Scott with a photograph of London taken from the County Hall building, looking north. The image shows Westminster Bridge, with Westminster Abbey and the Big Ben Clock Tower on the left. The Siebe Gorman laboratory was directly behind the building from which this photo was taken.

vertent tour to check many of the nation's monuments for newly chipped stone and blown-out windows.

Siebe Gorman's test lab, at 187 Westminster Bridge Road, lay almost directly across the Thames from Churchill's bunker and Westminster Abbey. It hid behind stately County Hall, which still stands today like a stalwart, obtrusive stone layer cake greeting pedestrians after their last few steps across arched Westminster Bridge. The long brick Siebe Gorman lab building was nestled behind this guardian County Hall in the neighborhood called Lambeth, known for being "a poorer area" at the time but nonetheless considered a "lovely" home by many of its families.

Lilian Emmeline Welch was a resident of Lambeth. She was a student at Westminster Bridge Road School, which was a brief walk or even a jubilant skip down the road from the Siebe Gorman chamber lab. She spent the whole of her thirteenth birthday, September 15, 1940, singing

A portion of a WWII poster by artist Walter E. Spradbery showing St. Thomas's Hospital and the portion of Westminster Bridge that would be damaged by bombing.

songs loudly with her family in the basement of a Lambeth shop so that they and the others sheltering "wouldn't hear the bombing" of the aerial battles raging overhead. Later, that date would be dubbed Battle of Britain Day.

The same day, directly across the street to the southwest from the Siebe Gorman lab, medical student Arthur Walker emerged from an operating theater inside St. Thomas's Hospital. He and fifty or so hospital employees had stayed there, working resolutely, refusing to evacuate through several weeks of bombing and two direct hits to the hospital already. They carried on as a skeleton crew and slept on cots lining a bare concrete corridor in the basement. The operating theater remained intact, so they kept using it to patch up blast traumas.

On September 15, 1940, walking out of that operating theater was the

Bomb damage to University College London from the air raid of September 1940.

last thing Arthur Walker could remember. A third bomb hit the hospital, lit the dispensary on fire, turned the whole corner of Westminster Bridge into "just a wreck," buried Arthur Walker under debris, and cracked the base of his skull. He woke up a week later in a hospital far outside the city, where he eventually recovered.

Lilian Welch's home, her school, Arthur Walker's hospital, and the Siebe Gorman lab all shared one thing: They were clustered near the major transportation hub of Waterloo Station. The Germans were aiming for that hub. They wanted to cripple the infrastructure, take out the transportation. So, late one evening in mid-September, it was a surprise when UCL—not Siebe Gorman—was first to be shattered.

The strength of the blasts stripped the refined exterior off the dignified dome overlooking UCL's main quad. Behind the dome, the bombs crumbled to rubble the upper stories of the physics laboratories. Fires from incendiary munitions rampaged through the library, scorching walls and finding eager food in the bindings and pages of innumerable books. The internal structural lines of the buildings tipped, cracked, and tumbled, making precarious mazes out of once loved rectangular hallways. The exact level of damage to the mouse rooms themselves is not clear, but what is clear is that, finally, the gas lines to them were destroyed. Also, the electricity went out.

Hans Grüneberg was the first remaining scientist to say *Enough*. He and his wife, Elsbeth, were raising their bubbly, active four-year-old son, Reuben, and the previous spring they had welcomed screaming, pink-cheeked Daniel into the world. Hans and Elsbeth Grüneberg had become naturalized citizens before the declaration of war, which meant they were exempted from the internment practices the government was using to round up other German refugees. Hans had his own research funding, and Elsbeth ran her own dental surgery practice. In total, they had the independence to evacuate without waiting for permission from anyone else.

Hans Grüneberg later said: "When the college was badly damaged in September 1940 it became impossible to work [there] anymore." It's not clear whether he and Elsbeth decided to leave because of fear, loss of lab space, or both, but when they did decide, it was quick. On the morning after the major hit, within half an hour of his asking, Hans and his mice were offered a new research home at a hospital in a more rural area, far northwest of London. He and Elsbeth jumped at the chance.

Hans Grüneberg left the chamber-diving group after the trials with carbon dioxide had provided the first step toward allowing submarine survival and underwater reconnaissance but before the bombing schedule allowed the beginning of their work with oxygen. He and his family evacuated to what they thought was relative safety, but by 1942, Grüneberg felt too disconnected from the fight that his new homeland was conducting to save his own people. So the refugee medical doctor joined the Royal Army Medical Corps to, again in his words, "do one's bidding in whatever capacity one was capable of."

Despite the widespread destruction of their campus, the rest of the lab plowed forward. Haldane and Case planned to do one more capstone experiment with carbon dioxide before moving on to oxygen. The vice admiral (submarines) (VAS) of the Royal Navy invited them to use for their final test a hundred-cubic-foot tank at his facility at the submarine naval base along the southern coast of the country in Portsmouth.

Helping to fend off German invasion from the coastlines, submarines had taken a leadership role in the conflict, and VAS Max Horton had his finger on the pulse. At any given time, the savvy WWI veteran knew where every one of his boats was in the world and what they were doing. In September 1940, every Royal Navy submarine available was at sea, and Horton was clamoring to get more built. Most boats were tasked with patrolling the waters between England and Germany. Their hub was Portsmouth, a beachside city full of low brick buildings and infinite low brick walls wrapped around the sides of twisting roads, which closer to the shore were interspersed with relaxed minor dunes of rocky sand and patches of long, waving seagrass. Horton needed more submarines. He also had his own ideas for what they should look like. He invited Haldane and Case to the windy seaside town of Portsmouth to help him.

VAS Horton had his own very practical top secret agenda. At his request, Haldane and Case crammed themselves into a small metal tank suspended in the water near Portsmouth naval base. They breathed inside it in the cold atmosphere, with the tank fully immersed in water, to confirm that what they had measured inside the Siebe Gorman lab also held up inside the salt water of the ocean waves.

On their second attempt inside this small tank in the water, on September 22, they brought an oxygen supply and CO_2 absorbent with them. They were trying to see if they could intentionally control the atmosphere inside their restricted volume. It was, essentially, a model of a tiny submarine. However, midexperiment, the weights holding the tank underwater came loose, resulting in "continued and violent motion" from any tiny perturbation outside. For four hours and twenty-four minutes, the curious duo "were thrown about and bruised" by the rolling and rocking of the tank in response to the waves and nearby movements of ships. They stuck it out, clenching tightly onto every handhold "to prevent themselves from being thrown about." They incorporated the physical exertion into their calculations of the carbon dioxide they produced.

They emerged panting, as usual. With searing headaches, as usual. They referred to that state as "perfectly fit." In another hour, they guessed based on experience, they would have vomited. However, they proved that, even in a tiny, enclosed space, they could use bottles of oxygen and CO_2-absorbing material to control the breathing gas enough to survive. VAS Horton must have been thrilled. The conclusion fit into his goals perfectly. Of course, they would need to repeat the experiment to be sure, and they did. Several times.

Portsmouth was a useful test site, and it was outside the main brunt of the Blitz, but it was not immune. The whole of the parade ground on the military base was "honeycombed with air raid shelters" where the soldiers and sailors had to dive and take shelter should the sirens sound. Locals installed Anderson shelters, too, and young teen Doris Eileen Connolly described spending "every evening" after her father came home from work in their shelter, as a matter of routine. They slept on wooden benches. She was more afraid of the spiders than of the bombs.

Bombs fell inside the submarine base, around it, along the seashore, and within eyesight. Not daily like in London but certainly weekly. On September 22, three of the small houses opposite VAS Horton's office were reduced to rubble a few hours after the end of Haldane and Case's experiment. At some point, in the middle of one of the experiments, crane operators raised the test tank out of the water in response to a raid but ran for cover before bringing it down on land, leaving the two scientists trapped inside yet also suspended in midair among the falling ordnance.

Haldane and Case, however, were sticklers as ever in their monomania for results, and so they considered the details of being bombed too irrelevant for entry in their lab notebooks. Experiments that had to be aborted were simply removed from their records, because incomplete data points were not useful. However, their dodging of the bombs can be read indirectly through the calendar of events. They conducted tests on September 21 and 22 and locked themselves inside the tank for three straight days of fun from October 2 through 4. Meanwhile, bombs peppered the base and its neighborhoods in between, punctuating the days of the calendar the way they punctuated the rows of buildings, with confirmed major bombings in Portsmouth on September 17, 22, 26, and 29, as well as on October 7 and 9. It was never specifically written down as such, but it is likely that the experiments had to wait for the gaps when munitions stopped raining from the sky.

Wrapping up the CO_2 bit of the science, Case wrote down that in a space 3.2 times as large as that little tank, "two men could stay in it for 13 hours without serious discomfort, and for 16 hours or more if they could lie down and rest."

Haldane and Case did not include in their report why the Admiralty wanted to know how long two men could stay in a volume of three hundred twenty cubic feet, and specifically that volume. It seems to have been a number dictated by Max Horton. The precise purpose of the science was described even in the lab's letters and notebooks as "of a secret character." However, VAS Horton had been suggesting miniaturized submarines since shortly after his time in WWI. Horton's idea for top secret so-called midget subs was gestating. The UCL lab had proved midget subs were survivable, and provided the science needed to survive them.

Meanwhile, undercover police officers continued to attend Haldane's lectures. MI5 officers intercepted and copied documents sent to the lab, filing in their bulky folders the photostats of letters graced by the compact vertical signature of scientist-secretary Elizabeth Jermyn. The thickness of Haldane's surveillance file continued to grow.

———

Work in London was becoming impractical. The government raised more explosives-filled balloons on long tethers to try to dissuade airplanes from flying over the most densely populated areas, including around Lambeth, but it is unclear whether the balloons had much deterrent effect. German gunners sometimes used the balloons as target practice, loosing uncounted bullets over the city in addition to explosives. The lab's records show few experiments happening during the daily bombings, despite the scientists' letters to the Admiralty stating that they were trying.

Londoners who traveled to unscathed cities and towns began to describe a feeling of disconnection from their fellow countrymen who had not been through the same thing. They said that it felt as if those who had never been in an air raid had a certain emotional "distance" from the war. Diarist Emmeline Cohen, on one trip outside the city, wrote that she had "long and uninterrupted" sleep for the first time since the beginning of the war but also that she found "a gulf seemed to separate [her] from these folk." Outside the direct-impact zone, people had sleep, peace, and fewer food shortages. Some of the diaries of Londoners who tried to explain the Blitz to others read eerily like the words of modern hospital workers trying to explain the COVID-19 pandemic to those who lived a different version of the same reality in the comfort and safety of their own intact homes.

By the height of the Blitz and the end of the carbon dioxide trials, about 40 percent of those remaining in London slept in their damp yet

Londoners sleeping in an Underground station. *(NARA 195768)*

private Anderson shelters at night. Nearly a quarter of the city slept in the train stations of the tube, the London Underground. Many workplaces gave up having night shifts altogether, as the work was too often disrupted by raids.

The city was adapting to war. Women abandoned traditionally feminine clothing in favor of practical trousers and flat-heeled shoes for sitting and sleeping in public. Some of the tube platforms and staircases were so packed with improvising humanity that one observer wrote that so many people condensed underground reminded her of the imagery in Dante's *Inferno*.

Almost everyone reported a form of unity through suffering, a kind of "Blitz spirit," but diaries also contained numerous complaints overheard in the streets about Jews and foreigners taking up space in shelters and on train platforms. At least one Belgian woman was denied access to a tube station until she proved her husband was in the British military. The Jewish refugee scientists labored in and for a city peppered by resentment against them.

Most of the tube residents described the lack of sound as the primary appeal of the Underground stations rather than the lack of danger. Citizens lined up for hours to get a spot in the silence of the tube, not away from the risk, but away from the "cruel warble" of the sirens.

"It's wonderful down here. You can't hear anything and we can sleep."

"I feel lovely and safe. You can't hear the guns or anything."

"I feel completely safe there. No noise. None at all. That's what gives you confidence, I think."

"You can't hear that plane going round and round when you're down there."

"Bit hot and dirty, but you feel secure, and you can't hear bombs or planes or anything."

Budding author Celia Fremlin, who would go on later in her literary career to win awards for her novels, described the psychology most poetically: "It may be the attitude of the ostrich which makes people dislike hearing what is going on; but even the ostrich wants some sand in which to bury his head. He is not satisfied with merely shutting his eyes." The bombs were coming. The people could not stop them. So people did what they could to take cover. Beyond that, they wanted to be able to shut out the rest mentally. Eventually, the government issued free earplugs.

Bombs penetrated the road and water pipes of the Balham tube station in southwest London. Sixty-six people died trapped underground. Hopefully, they died in peace without hearing the bombs coming.

One of the most nefarious inventions of the war was the SD2 or "butterfly bomb," an elegant-sounding name for a type of cluster bomb, which was an aerial machine released in large numbers. The SD2 armed its fuse through the spinning of expanded metal wings while it dropped from the plane to the earth. Once activated and on land, it posed a menace to those around it, blowing apart anyone foolish enough or unlucky enough to be nearby. The butterfly bombs, and the other miscellaneous munitions embedding themselves in the streets and buildings of London, were a threat to the ability of the determined, unified citizens to carry on as much as possible during the day while sheltering at night. As a result, much of what we know today about the safe disposal of explosives can be traced back to the Blitz. British soldiers signed up to try to defuse and dispose of the new threats to reclaim the city, its buildings, its schools, its streets.

In the city of London in general, during the daytime, people frequently took mental refuge in cinemas, seeking a break from their own reality in comedies. Fred Astaire and Ginger Rogers tapped across the screen during air raids. However, when people tried to engage in their normal, casual activities like having lunch with their sister, sometimes they would find their regular waitress at a favorite café was out of work because she had "had both of her legs blown off" by a ground-level bomb. And so, to do what they could, weaving among these fervent daytime efforts toward normal life, busloads of soldiers drove through the streets of London with "bomb disposal" painted on the buses' sides, accompanying ambulances and fire engines to the worst blast sites from the night before. The crews in these units would also often chalk their own nickname for themselves on the sides of their vehicles: "Suicide Squad." They, too, accepted the danger and did what needed to be done.

Eventually, the bombs became too much for the rest of the UCL lab. The research group had tried to do their experiments with rats instead of humans because they didn't want to be trapped inside a pressurized and vulnerable chamber in case they needed to take shelter, but Siebe Gorman flew into a state of disarray when one of its other factory sites took a direct hit. Human experiments simply were not an option at this moment in London. Hell, walking to the lab was barely possible at that moment in London. Haldane wrote to his navy contact, "Air raids have hampered our work very considerably."

The scientists needed a new home. Luckily, through contacts with collaborators, they learned they could wedge into some bench space far north of the city at the Rothamsted Experimental Station, the agricultural research establishment in Harpenden, Hertfordshire. Rent in the area was at a premium because of the astronomical number of evacuees. Haldane rented a home that seemed arguably large enough for the entire lab, if they squeezed in, and he offered to foot the bill.

On July 17, 1940, Jim and Joan Rendel had welcomed bubbling, happy baby Jane to their family a few short weeks after Jim's accident, while he was struggling to recover and during the scream of an air raid siren. He was not allowed to lift her for months. Hans and Nussy Kalmus also came with Peter and George. Ursula Philip decided to plead with a friend to share an apartment with her, so she would join the group in exile at Rothamsted but retain some privacy, at least at first. Martin Case would also find his own lodgings. Hans and Elsbeth Grüneberg had already escaped. However, Elizabeth Jermyn would join the others in the shared home for lack of other options. Helen Spurway and the married J. B. S. Haldane were, at that point, deep in the throes of their genetics-sparked love affair, so Spurway would come, too. The group of ten, including the children, would move in together. The Kalmuses offered to bring their piano for entertainment, if it would fit.

Haldane's offer to pay the rent and get them out of London was financially generous, but the group would soon learn a dangerous lesson about living with their quirky, entitled, notoriously stubborn boss. They settled into their new home in Rothamsted as the fall of 1940 plodded along to become the winter of 1941, and they would give their new home a nickname that wasn't exactly affectionate: "the Piggery."

CHAPTER 9

The Piggery

Joan and Jim Rendel's bedroom was in the attic. Technically, they had the entire floor of the Piggery to themselves, along with eight-month-old baby Jane, but that was only because the second, smaller attic room was crammed full of the landlord's personal belongings. Their modest space came with a bland carpet the landlord had generously described as "fawn," a watercolor painting of a dog, and a gilt-framed print of squirrels. Those images hung in their new living quarters alongside a damaged cuckoo clock and stained window curtains.

Because the Rendels' bed and bedding had been provided for them as part of the house rental, they'd been able to limit how much they'd had to carry during their evacuation, but the landlord had acknowledged in advance that the painted iron headboard was chipped and that the bedding and mattress were marred with rust marks. Joan was even less flattering in her description, calling the whole building "a hideous house, full of pigstye [sic] brown." The tank of their upstairs toilet leaked, and its water oozed through the ceiling of the room below.

Since his right lung had ruptured now, too, Jim Rendel still could not pick up his growing daughter, who had thus far lived only in a world of rationed food and abundant air raid sirens. Even a "very ferocious sniff" was enough to recollapse either lung, so normal parental actions like carrying baby Jane up the stairs were out of the question. Instead, Rendel did what he could to be a loving father. He got busy at work, continuing his genetics research at the lab's new guest location at the Rothamsted Experimental Station. He was still too sick for the military draft, but with Haldane's help in guilting research sponsors into funding a modest salary for him, Rendel was able to keep their little trio fed. As a result, however, much of the day-to-day work of adapting to the new communal lifestyle fell to Joan.

At first the adult "girl poet" and new mother had treated the situation with optimism and looked forward to living in a house full of other intellects. She wrote to her dear friend, Haldane's sister, Naomi Mitchison, that she was happy in the rental home because "the atmosphere [was] thick with creative activity." Joan was excited for the mental reprieve from living in the war zone of downtown, where "London conversation and London letters are of nothing but the bombing." Near Siebe Gorman and the UCL campus, her London life had consisted of "trying to get a little sleep when one can and in arranging everything as best one can to give more of a chance of preserving life from moment to moment." There, because the most basic tasks of survival had to be shoehorned between randomly timed air raids, she said, "the elaborate bother of filling one's stomach" presented a constant obstacle, which she had to sneak into her family's days while dodging screaming bombs.

Their rental home, an hour's drive northwest of campus, had dozens of cracked windows that hinted of bombs. The two-story house was nonetheless lower risk because of its distance from the city, and the lab site at the nearby biology station Rothamsted would allow the scientists to continue puzzling out the truths of nature in rooms that had working heat and electricity. The flies' food, after all, needed gas to get cooked. As did the humans'. Those basic features made evacuation a dream of normalcy.

The reality did not quite match the dream. Haldane was as difficult to deal with as ever, but now they had to live with him full-time. Despite paying Ursula Philip and Hans Grüneberg the same salary for equal work back at UCL, he nonetheless held the common but sexist expectation of the time that once women got married, they had no choice but to devote themselves fully to homemaking. He considered this duty such an innate obligation of the sex that he assumed his own life chores would be covered by the married women in the house, Nussy Kalmus and Joan Rendel, even though they weren't married to him. And, of course, the same rate of scientific progress in genetics was still expected from the scientists, despite the voluntary war work now on their calendars, too.

Therefore, instead of Joan Rendel's dream of an intellectual oasis replete with relaxed, nightly, salonlike academic chats near a roaring fire, she and Nussy Kalmus suddenly found themselves the forced housekeepers, cooks, and general caretakers of eight other people, some of whom were often cantankerous. The women did not have much choice. Haldane was paying the rent here. It was either live with him or return to the bombs.

The women started out alternating weeks with the cooking and the chores, but as the days grated on, the living situation became more tense. Baby Jane's needs became difficult to balance with the massive workload, and Joan began to crumble.

The kind, born privileged but boundlessly resourceful Nussy Kalmus, with her slender build and her short, wavy hair, somehow built a pseudo-comfortable home for the scientific team within the cramped, battered rental house. It was not as effortless as it might have looked on the outside, and her husband's memoir includes a trail of bread crumbs indicating her profound exhaustion from all the work. But she did it. On one occasion, she magically multiplied an insufficient number of precious, heavily rationed lemons into a lemonade treat with enough for everyone. Every day she wrangled an oversized tray into the lounge to serve the group tea while they sat amid the worn carpet, cracked ceiling, and blackout curtains.

Haldane always sat in a studded, cozy-looking leather armchair tucked into the corner of the communal lounge near the fireplace, and that chair was for him alone, even if he was not home. He rather belligerently declined to help pay for an increase in the gardener's salary so the gardener could keep paying his own rent, which had risen because of the price spike in response to the desperation of the refugees. Instead, Haldane declared to the landlord that the garden did not contain vegetables or any other practical plants, so the gardener should make himself useful and contribute to the war effort rather than tending to pointless flowers. On another occasion, Haldane demanded that a guest visiting the house mow the lawn. When the guest refused, Haldane did it himself, but he performed a loud, petulant display of indignant huffiness while doing the entire task.

However, there were delightful moments, too, among the madness. Haldane and Joan Rendel bonded when they had to burn some meat infested with maggots, and she was worried the bugs would suffer. He comforted her with the scientific insight that maggots lacked the neurological development necessary to feel pain and fear, and she was touched that he took the time to give her that explanation. In the bay window of the crowded, gloomy sitting room, Helen Spurway kept an aquarium of crested newts named in tribute to Wagner characters, and young George Kalmus caused a stir when he used a few judicious drops of red ink that convinced the biologists that unicellular algae must have been blooming in the tank.

George, who slept with his brother, Peter, in a laundry cupboard, also rattled the adults when he counted up the beds and the bodies and determined that there were not enough of the former. Spurway and Haldane were sharing, but Haldane had sworn all of the adults to secrecy over that fact, as his divorce from Charlotte was not yet final and he was anxious about being busted by the very real morality police. In the rental paperwork, he had slyly agreed only "not to bring extra beds."

Scientist-secretary Elizabeth Jermyn joined the local Air Raid Precautions (ARP) group, which that year changed its name to the Civil Defence Service, and she got out of the house in a productive way by volunteering to watch for incendiary fires lit by bombings at Rothamsted. Ursula Philip settled into her own apartment nearby but continued her genetics work at the new remote lab. To help with "the meat rationing problem," Haldane's sister, Naomi Mitchison, shipped the group occasional pheasants and woodcocks hunted on her estate. Jim Rendel and Helen Spurway plucked and prepared the birds for all to enjoy. On the surface, everyone seemed OK.

J. B. S. Haldane and Martin Case traveled to Siebe Gorman in London and there pressurized to 130 feet, to 168 feet, and once more to 168 feet time and time again. They flushed their breathing apparatuses with oxygen. They watched each other in the chamber, waiting for air raid sirens or facial twitches and trying to figure out the elusive rules for when Oxygen Pete struck and why.

On March 28, 1941, Joan's mentor and idol, Virginia Woolf, filled her pockets with rocks and waded to the bottom of the River Ouse. Her final suicide note—she had written two before, but aborted her plans both times—contained characteristically poignant phrases about her depression, despair, inability to cope, and hopeless desire to be a burden no longer to those around her.

Woolf's words echoed, eerily, Joan Rendel's letters about her own feelings at the time. "You see I can't even write this properly. I can't read," Woolf wrote in her final explanation to her husband. Joan Rendel, three days later, with likely no knowledge of the contents of Woolf's note, described the same emotions, and it was not the first time she had expressed them:

> I don't know quite what is happening to me. I feel as if my thinking powers have been bled so that they have become weakly and can only concentrate on anxieties. I am in one of the most intelligent environments I have ever been in, but my intellectual, or cultural life, has died within me ... now I cannot talk. I cannot concentrate properly on what is being said around me. . . . I feel sobre [*sic*], stupid, closed in, dimmed, with nothing more to say.

She wrote to Naomi, "one takes oneself away with the baby upstairs for several hours each day to the bedroom alone, leaving the voices downstairs." Joan was terrified of failing the people in her life, and she shut down more and more as she found the workload too much to process.

Joan and Nussy suffered apart. Nussy carried the burden of the communal home downstairs. Joan stood upstairs for hours, feeling alone, grasping her infant, listening to the disembodied voices waft up through the stained fawn-colored carpet.

After a few months, when the Blitz became slightly less incessant, the scientists could increase their rate of trips into London to Siebe Gorman. Haldane had to get a special dispensation from the government for extra fuel for his car to make the journeys, which they took in his jaunty 1937 Jowett saloon roadster with its inadvertently appropriate license plate of EGO 848. He drove like a banshee and incurred frequent damage to the car's body paneling, and he always seemed to have a version of events that made the dents someone else's fault. Later, internal damage to the cylinders of his engine would give mechanical testimony to the number of trips they had made during a time where there was no fresh motor oil to be had.

The calendar of test days reveals no apparent pattern; nobody knew when the Luftwaffe would strike, so the group tore into town when they could. At least one experiment found bombs raining from the sky mid-science, and the chamber was evacuated in haste. But as with the bombings they'd endured in Portsmouth, the bombs drove the schedule, but they were not data. And so the bombs did not deserve entry into the formal lab notebooks.

On experimental day fifty, for experiment number eighty-three, J. B. S. Haldane climbed into the chamber and positioned himself near the back. It was Friday, and the weekend loomed. Martin Case climbed in next and closed the door. They hit bottom quickly at two hundred feet, also known as sixty-one meters, and Haldane popped in the mouthpiece to breathe pure oxygen. The experiments were now focused exclusively on the effects of oxygen. Haldane, Case, and their group had made short work of the rest of the Admiralty's list.

When swimming underwater in real combat scenarios, military divers prefer to stay below the surface to avoid detection. Harbor sentries who skim patrol lights across the waves and sharp-eyed spotters on surface craft pose a danger to anyone attempting stealth. However, nobody knew how deep the combat divers could safely submerge while breathing oxygen, should they need to dodge their enemies. So there, in the chambers, in the illusion of safety, free from enemy eyes, within easy reach of medical help, the chamber divers would try it first.

Haldane's pulse began to drop immediately after he breathed from the mouthpiece. After 4.5 minutes he pulled the mouthpiece out and reported to Case that he did "not feel all right." The oxygen tasted weird to him, super salty, and Haldane suspected that his lips were twitching. Case replied that he could not see any twitches.

Case tried the mouthpiece himself, flushing the bag first with fresh gas to make sure that pure oxygen was in there, that there were no contaminants. He felt nothing odd. He could not taste anything like salt. He handed the rebreather back to Haldane.

"I definitely don't feel OK," Haldane gasped after another four minutes of breathing. Case could hear bubbles popping deep inside Haldane's chest with every breath, a gurgling, bubbling sound that punctuated his words and suggested fluid in his lungs.

"I feel as if my lungs are full of water," Haldane confirmed.

Haldane lay down flat, facedown, apparently in the hope that gravity would help sluice the fluid out of his breathing passageways. Case watched with concern and let him reposition himself, but also tapped on the chamber wall four times with the hammer to signal for quick decompression.

The first seizure hit the large professor hard when he was lying prone on the wooden planks. The adamant power of oxygen bent Haldane's long body backward like a bow, and the clenching of the strong muscles beside his spine contracted him into a macabre backward arch. His back and arms and legs and all his other muscles seized taut. Then his limbs began to jerk, a rhythmic spasming that alternated between arms and legs. He tried in vain to grab a wooden stool to still his own erratic motion.

His "eyes staring and protruding, mouth half open and rigid, salivating profusely," his hands tangled in the stool, the famous scientist could not seem to hear the nearby shouted words of his longtime friend Martin Case.

So Case made a dangerous choice. Haldane's seizures continued in two- to three-second bursts of rigid limbs. Case signaled again to those outside to get the chamber pressure to the surface as quickly as possible, decompression schedule be damned this time. Bends be damned. His own joints be damned. Haldane needed help. Case could not go through this again.

The large man was lurching about. Case could not control him. Once Haldane found control of his tongue, he began bellowing once again that he was about to die, he was going to die, he was *actively dying* and that he had too many scientific papers to publish to be able to die. It was a classic professor concern and would have been comical in another scenario. Case managed to gulp in a few breaths of oxygen for himself near the surface to try to reduce his risk of DCS. It was not enough, but it was as much as he could do. Eventually the door swung inward. They made it out.

Afterward, Haldane did not remember a thing. Case escaped the bends, somehow, although according to all reason and theory he should have been twisted into a pretzel shape by pain. The team added another oxygen data point to their ledger. At two hundred feet, more than five minutes' breathing oxygen might be a problem, they decided. But again, the key to science is repetition.

———

Less than two weeks later, on the same day that Joan Rendel wrote about her brain fog, Joan Elizabeth "Betty" Jermyn climbed into the hyperbaric chamber for nearly the same test. They compressed her to 170 feet. The twenty-four-year-old scientist-secretary with the dark bobbed hair and the determined jawline took the mouthpiece of a pure oxygen rebreather in her mouth and began to take deep, full gulps. She knew what had happened to her boss. After five minutes her dive tender decided to end the test. She reported feeling "giddy & faint, fidgety," but she was OK. She returned to the surface without calamity. It was her second ride in the chamber.

Ursula Philip went through the test, too, for her first dive. She also made it five minutes without incident. Ursula felt fine. Clearly, oxygen was a bit random in how and when it took its victims. That was part of the problem, part of what the researchers could not solve.

Ursula Philip did not know why she was testing oxygen in the chambers, at least not officially. The military missions were secret, and the immigrants in the group were allowed to know even less than the others. She knew the test could hurt her. But she did it anyway, because she knew it was for the war. She knew it was for her new country. As of March 1941, Ursula Philip and Betty Jermyn had officially joined the in-chamber team.

The group began to collect new oxygen data on a daily basis to try to solve the mystery of how oxygen toxicity hit and when. After his second seizure, Haldane's back would hurt for the rest of his life. But still they repeated the test. Again. And again. And again. They tested once or twice a week at first. Soon, they increased to three or even four times a day.

They made small changes for each experiment—to depth or to time—to probe the boundaries of safety. Sometimes they added in physical exercise. They rotated the victims, so nobody had to go through the ordeal twice in one day. Haldane and Case went one day. Maybe Jermyn the next. Sometimes Philip. Sometimes Kalmus. Over and over. Never knowing who might be next to seize. Never knowing when one of their bodies might become permanently broken by the oxygen. It was the only way to determine the bounds of safety for the combat divers who would do it in the real world.

––––––

One MI5 wiretap caught one of Haldane's friends saying over the phone that Haldane was working himself to death because he was "too bloody afraid of us losing the war." He was. But that did not mean that Haldane could not make the situation work for himself a bit, too.

Research work had kept at least Philip and Grüneberg out of the refugee internment camps that the Allies had set up for suspected German spies, even though the Allies had no reason to suspect them except that their native language was German. So Haldane decided the tests could also work to help his buddies from the International Brigade from the Spanish Civil War. Their lives were in danger from the Nazis for a different reason—not because of religion, but because of politics.

The Communists fighting in the Spanish Civil War had been fighting on behalf of the established Spanish government against an insurrection led by would-be fascist dictator Francisco Franco. When the old government fell to Franco's forces and he established himself as the new, iron-fisted despot of the country, he also set up mechanized, organized killing systems for his fallen enemies. Naturally, everyone who had fought against him, including every Communist, was on his wish list for rolling heads.

Hitler shared Franco's hatred of differing political opinions, including Communism. And so, by the end of 1939, Communists in many of the countries of mainland Europe found themselves hunted, their citizenship rights revoked. The Russian propaganda machines were still convincing many outsiders that Communism was working, that in fact it was a paradisiacal solution to the world's ills, and so the ideology was considered far less of a fringe political stance than it is in the twenty-first century. The lists of Communists wanted dead by Hitler and Franco, therefore, were long.

One of those people was Hans Kahle, German by birth, rambunctious by choice. Just over forty years old and roughly six feet tall with deep brown eyes, dark brown hair, and a slight patch of thinning baldness where his hair should have swirled, Kahle listed himself on his immigration paperwork as a journalist, but he had also been a fighter. Kahle's most lasting claim to fame was squiring Ernest Hemingway around the battlefields of Spain, a task that got him memorialized in Hemingway's *For Whom the Bell Tolls* as the thinly disguised "fictional" character named General Hans. After the Spanish Civil War ended in 1939, Kahle came to England as a political refugee, but he was also, according to MI5, a high-ranking member of the OGPU, the notorious Soviet secret police agency. From the moment Kahle touched British soil, MI5 agent Milicent J. E. Bagot made careful notes about his activities, just as she was already making careful notes about Haldane's. She monitored them both, sent officers to listen if they spoke in public, read their mail, and tapped their phones. Over the course of the war, MI5 placed more phone taps for Communism than they did for Nazism. They seemed to think every Communist was a spy.

Milicent Bagot suspected that Kahle had not merely fought while he was in Spain. According to the records she had compiled on behalf of MI5, Kahle had also willfully gunned down unarmed civilians "wholesale." According to MI5, Kahle was a war criminal. Their file on him uses the word "murder." He was classified as being of serious concern for espionage.

But he was also willing to get into the chamber. So, after Kahle was picked up by the London Metropolitan Police late on the night of May 16, 1940, after he was arrested and his apartment ransacked for Communist contraband, after he was given certificate number 708001 for processing, after he was formally interned—after all that, Kahle's friend J. B. S. Haldane exploded with fury. J. B. S. Haldane exploded like only J. B. S. Haldane knew how.

Haldane began writing letters to everyone, leveraging his work on the Admiralty's underwater experiments to demand his friend's release. He leaned on the claim that Kahle's age—older than the rest of the test group—made him a unique and valuable data point. He was presumably counting on the fact that nobody understood the science enough to see that this claim was nonsense.

"The work would be more valuable could I employ middle-aged subjects," Haldane lied to one Commander R. T. H. Fletcher, an MP—member of Parliament—in the House of Commons. "Of course, if you can get me a fat Tory M.P. aged about 50, that might be equally good! But frankly the work is unpleasant and rather dangerous."

He also played up Kahle's weight and said that a British citizen would have to get in the chamber instead if Kahle was not released. "If Mr. Kahle is not used, some British subject will be exposed to danger. . . . I may add that I do not know of a fat middle-aged Briton temperamentally qualified to take his place."

Haldane hammered on the extreme danger of the testing. "Of all the subjects he [Kahle] is the most likely to be killed or injured." That part was true, as both age and high body fat percentage are risk factors for severe DCS. "In the national interest it seems better that he should run this risk rather than a British subject." He also emphasized to the Admiralty that Kahle's continued internment could jeopardize the timeline of the urgent work: "with Kahle interned our position as regards personnel is not too good."

Finally, after a hailstorm of letters that lasted months, the government relented. Milicent Bagot pointed out, with an air of resignation, that letting Kahle participate in the experiments might work out for MI5, too.

"If the experiments are successful it will be of value to this country, while if something goes wrong there is a good chance that you will not have to worry any longer about Kahle," she wrote when she assented to his release, after saying she would be "happier if he were interned." In other words, he might die and remove himself from their to-monitor list. One less potential spy.

Without further objection, Hans Kahle was yanked from his internment camp in Canada back across the Atlantic—where Hemingway had written him to tell him of his impending literary cameo—placed on a ship to England, and ordered "to proceed direct to London" with strict instructions that "on his return to the United Kingdom he will report to Professor Haldane, University College, London." The assessment ranking that indicated his estimated danger to British society was also downgraded a level. Haldane had to promise that "this man is rigidly precluded from access to confidential and secret information while undergoing these experiments." Haldane promised.

"He [Kahle] has no access to any documents, nor does he know the detailed reasons for the experiments, except that the DSEA is filled with pure oxygen."

Haldane pulled the same trick to secure the freedom of Juan Negrín, the former prime minister of Spain who had been ousted by Franco. However, according to the lab notebooks, Negrín never participated in any experiments, no matter what Haldane might or might not have reported to the government.

Before he had been arrested and interned, Hans Kahle had participated in exactly one experiment. After he was released, he participated in three more, for a total of four. After that, he disappeared from the lab records. MI5 continued to follow him, but he was never again interned, and he never again participated in Admiralty science. He got a get-out-of-jail-free card after a total of about eight hours of work as one of Haldane's "rabbits." He needed to get a dental bridge replaced after one chamber decompression cracked something inside his mouth, and the British government paid for it. Otherwise he had no physical damage. In comparison, Jim Rendel needed yet another lung surgery a few days after Kahle's release.

The closing lines of Hans Kahle's MI5 file read, "The whole tenor of Kahle's career suggests that, in addition to many other political activities, he was working for the Soviet Security Police. There is, of course, no question of Kahle being a high grade spy himself." Four experiments and some dental work were not a bad trade for a suspected high-grade spy to make to get out of prison.

But Kahle's lack of participation was not entirely his fault. Rather, in the spring of 1941, just as the research team was regrouping, as Kahle was reengaging in the trials, the Luftwaffe finally found Siebe Gorman's laboratory.

———

By spring 1941, the Admiralty was writing reports with alarming phrases like "precise information on this question [oxygen poisoning] is still lacking, but Professor Haldane is undertaking research, some of it of a hazardous nature, on the subject." And "our investigation of these principles has surveyed existing information available and has disclosed serious gaps in present knowledge."

The Admiralty also acknowledged, "The elicitation of this information has been possible only by the subjects of the researchers voluntarily submitting to the utmost personal discomfort and, in many cases, to the risk of personal injury," followed by the names of the UCL lab members and descriptions of their injuries to date. Rendel's lung. Haldane's back. All the times Case had been knocked unconscious.

These reports might have been dramatized to a degree to get the financial and material resources from the government that their Royal

Navy authors were asking for, but they made it sound as if the UCL lab members were the navy's only hope. The small number of government-employed researchers who might have been able to do the same work had been blitzed into huddling in a rented cottage, and they had no physical lab space at all. So, dramatic as the cries for help were, they seem accurate.

The government had three urgent goals remaining for the battered UCL lab members. First, Vice Admiral (Submarines) Max Horton had already documented that he was driving to use the UCL results for midget submarines, which at that time were limited in status to being his pet project rather than a mainstream Admiralty goal. He wanted more data about safe gas limits to make living inside the minisubs possible.

Second was the concept of amphibious divers, which was a more tightly held secret. More data with the rebreathers was necessary to making amphibious diving safe, to make sure divers were not risking seizures or other kinds of effects from oxygen. It was still not clear how these divers might be used.

On top of these two lofty missions, those worried about disabled submarines also wanted more results—results that showed how to escape submarines that had been turned into waterlogged death traps.

These three military applications were technically separate, yet the demands were made at the same time. And the navy knew they had stacked the workload high. One high-ranking officer interested in submarine escape admitted it himself in a plea to the petroleum board to get Haldane more fuel for his car to drive between the Piggery and Siebe Gorman: "Since January 1940 Professor Haldane has carried out under its auspices"—with "it" being the Physiological Sub-Committee on Saving Life from Sunken Submarines—"researches of the utmost importance; these researches are entirely separate from the work which he has carried out for the Vice-Admiral (Submarines)."

In the middle of the navy's clamoring about gaps in their scientific knowledge, and in the middle of their fervent use in government reports of words like "anxious" and "fear" and "unsafe" related to diving, the luck of the downtown Siebe Gorman lab ran out. Finally, the little brick building with the hyperbaric chambers was hit.

The London Necropolis railway station, less than two hundred meters from Siebe Gorman, was bombed on April 16, 1941.

Around the lab, schools, homes, a hospital, and a private train station had already been impacted, some much earlier and others mere days before. Across the river, on May 10, 1941, the faces of Big Ben's internationally famous Clock Tower had been cracked by a blast. Siebe Gorman was hit on May 17. It was an incendiary bomb, and the roof, the wooden flooring, and all the other flammable materials burned quickly down to nothing. The windows were shattered; the files of corporate documents were burned to ash. However, the two-story brick husk of the building stayed standing. It was sturdy, charred, but intact. So, too, thankfully, were the chambers.

With the chambers undamaged, Siebe Gorman and the UCL lab could rebuild. But the setback was huge. Meanwhile, the Italians shared the Allies' same goals, and they were able to keep moving forward on behalf of the Axis. The Italians were way ahead.

CHAPTER 10

The Littlest Pig

Gino Birindelli crawled like a metamorphosing fish out of the blue Mediterranean waters onto the tip of the pier. Dawn was breaking on a new day in October 1941, and he was cold and tired. As he lay on the wooden planks above the lapping waves, he clutched the oxygen rebreather strapped to his chest and struggled to regain control over his rebelling body and brain. His mind and thoughts were foggy. His legs were cramping. He could not swim any farther.

After he caught his breath, he stripped off his rebreather and full-body rubber suit. He stood and walked the length of the pier, across the beach, and into the streets, where he attempted to blend in, even though he was sopping wet, with the rest of the pedestrians. British soldiers and sailors were among them.

The soggy Italian military diver passed through the crowd mostly unnoticed—or at least unremarked upon—as he made his way toward a small Spanish merchant ship nearby. If he could get to Spain, he would be safe. He thought that if the innate British sense of privacy let him pass unnoticed—or at least unremarked upon—then he could squelch his way onto that ship, then to neutral Spain, and then to freedom, even though he had just emerged unexplained from Gibraltar Harbor like a rubber-clad creature from the deep.

The British territory of Gibraltar is a small peninsula jutting downward from the southern edge of Spain, and its most famous feature is its large, looming rock with a steep cliff face. Gibraltar forms a natural nautical choke point across the mouth of the Mediterranean because of how closely it flirts with kissing the northern coast of Africa, which makes it a strategically key place to keep a lot of military ships. Gino Birindelli had washed ashore there after his aborted special operations attack on one of them.

Birindelli kept walking. He made it onto the Spanish merchant ship, which meant he had almost made it home. Almost. A sentry noticed him, and when he saw Birindelli board a ship that was Spanish, the sentry took action.

Later, from inside a British prisoner of war holding cell, Birindelli wrote to his mother a letter that he was certain she would find to be gibberish. He hoped that she saw it was the right kind of gibberish and passed it along to the Italian military. The gibberish, if deciphered, explained what he had done. It told the Italians not to stop trying.

Birindelli's tiny underwater *Maiale* vessel—essentially, a torpedo that he rode and steered with a partner—had failed him just short of his target. He'd had to dump the massive bomb he had been carrying into a benign part of the harbor instead of beneath a fat, juicy British ship and swim to the pier. However, he had set the timer of the charge before dropping it. The bomb had gone off. This attempt, which he made along with the crews of two other *Maiale*, was their last effort, they had decided beforehand. It was their third attempt with the *Maiale*, and the first two had failed. If they failed again, there would be no more such plans, and "the war of the *Maiale*" would be over. But he had not quite failed. He had simply missed.

His mission proved that the plan could work. His Italian-model rebreather had left him dizzy and gasping for breath, but the rebreathers could work. The *Maiale* had lost depth control and threatened to plummet down into the sea without his permission, but it could work, too. Because he had set his bomb off, he had proved that the idea of an attack with *Maiale* vessels had potential. Using them, the Mediterranean could belong to Italy. As long as Birindelli's mother figured out his gibberish.

Birindelli's clever mother showed the letter to her son's compatriots in the Italian Navy. They figured out the riddle he had buried inside his made-up story about a nonexistent professor. They figured out his real meaning: that his mission had not quite worked, but he had proved it was possible.

The Italians decided to keep trying. Their next plan aimed slightly farther east, in a quiet maneuver that would rattle the world.

———

The night of December 18, 1941, a jet-black new moon hovered over the port city of Alexandria, Egypt. With a dark moon in the sky, the world was lit by stars.

Alexandria lies perched on the northern rim of Egypt, on the western edge of the green delta where the Nile dumps its life-giving water into the Mediterranean. For the British sailors stationed there, the city was an oasis, a respite from the war. They had to go to sea at times to dodge and toss bombs, and to look for and hide from submarines, but when they got back to shore and walked into Alexandria, it felt temporarily like the world was at peace. They were foreigners there but not enemies. The city was not being bombed. In the Egyptian markets, all the supplies they could not get at home because of wartime shortages flowed as freely as the juice squeezed from a plump, readily available lemon. There was also a hell of a nightlife, according to some.

Perhaps that was why the Allies didn't see the Italian divers coming. Even though the Allies had intel they might, and even though they knew the Italians had tried three times before, including Birindelli's efforts in Gibraltar. The Allies knew there were divers riding on the backs of steerable torpedoes like terrifying maritime horsemen. Those divers gripped with their legs to stay astride their belligerent, bomb-carrying steeds, their *Maiale*.

Perhaps the blissful reprieve from combat was why, on the black night following December 18, 1941, the Allied sailors decided, against all prudence, to open the boom that was the harbor's main defensive barrier.

The word "*maiale*" is Italian for "pigs," and it was a nickname slapped on the little person-ridden torpedoes because they were difficult to steer. Their operators reported that you could either ride them on the surface or plummet them nose first into the dirt, but there were few options in between. The formal name for the seven-meter-long craft, which was a narrow cylinder a mere fifty-three centimeters in diameter, was *siluro a lenta corsa*. That lilting Italian poetry translates to "slow-running torpedo," and gets truncated to the efficient moniker "SLC."

The riders of these SLC pigs were the highly specialized, intensively trained divers of the Decima Flottiglia Motoscafo Anti Sommergibili, otherwise abbreviated as the X MAS. No relation to the holiday. The divers of the X MAS were pioneers of Italian special operations.

On the evening of December 18, near but outside the Alexandria harbor boom, Lieutenant Luigi Durand de la Penne, known to his friends as Gigi, floated in the forward pilot's position with one leg curled around each side of his aquatic pig. Chief Petty Officer First Class Emilio Bianchi rested in the copilot position directly behind him. They wore thick rubber suits and carried oxygen rebreathers on their chests. Bianchi's position as copilot lowered him more in the water, so he was immersed to his neck, because of the reared-back angle of the torpedo. Their mechanical swine had a three-hundred-kilogram warhead attached to its bow. They floated, perfectly still, cautiously waiting outside the boom of Alexandria's harbor. They squeezed processed food out of tubes and into their mouths and called it something like breakfast. They had known their opportunity might take hours to come, and they were prepared. Two other pairs of X MAS divers, riding two other *Maiale*, floated nearby.

A large motorboat skimmed back and forth on patrol, periodically dropping small explosive charges into the water to deter enemy divers. With each detonation, de la Penne and Bianchi felt the reverberation of the underwater explosion rattle their lungs. De la Penne would later describe the explosions as "rather worrisome." But they remained undeterred.

Twenty-four minutes after midnight, a theater's worth of lights sparked to life across the boom, illuminating it and the other defenses like a stage. A small group of destroyers needed to pass through the barrier. The easiest way to permit them through was to open the boom.

De la Penne made a decision. Given the stubborn nature of their pig during depth changes, the easiest travel was either on the surface or into the bottom, and so he picked the surface. De la Penne and Bianchi, their oxygen rebreather mouthpieces in their mouths, their heads huddled low above the water, decided to try to slip through the open boom with the destroyers. The other two teams would come, too.

When the torpedo riders traveled alone, the wash of waves against their heads caused a slight iridescent glow to erupt from the microscopic creatures of the ocean. This bioluminescence is common and nearly global, and explodes bursts of blue-green sparkles throughout the water anywhere these aquatic creatures are disturbed by motion. The potentially telltale lights caused by the heads of de la Penne and Bianchi were swallowed by the more massive visual displays provoked by the wakes of the destroyers that they followed through the boom.

By 0200 de la Penne and Bianchi were close to their target, HMS *Valiant*. De la Penne tried to submerge. The *Maiale* bucked and heaved instead, plummeting to the muddy bottom at seventeen meters, equal to fifty-six feet. They were beneath the *Valiant,* but their *Maiale* was stuck in the soft muck. De la Penne turned around to look for Bianchi. Bianchi was no longer seated behind him. De la Penne tried to drag the *Maiale* through the mud to a position closer underneath the battleship, but between the soft nature of the seafloor and the burning sensation burgeoning in his lungs, he made little progress. The oxygen had caught up to him. He could feel himself starting to become overwhelmed by it. He knew he didn't have long underwater, working hard as he was, without a serious oxygen problem. He returned to the surface to find Bianchi clinging to a mooring buoy, having somehow floated to the surface after passing out. Bianchi, because of how the backseat placed him physically lower in the water, had had to breathe oxygen longer than his partner. Thankfully, he had drifted up rather than down. Luigi Durand de la Penne joined his copilot on the mooring buoy.

De la Penne and Bianchi were scooped from their mooring buoy by the crew of the *Valiant* and taken prisoner. They kept silent and refused to explain to their new captors what they had done, what they had placed, what they knew was ticking down one second at a time. They refused to talk while hoping that their explosive payload had crashed into the bottom close enough beneath the hull of the *Valiant* to achieve their goal, and that its timed fuse was still counting down the seconds until it would explode. The *Valiant* crew tossed their prisoners down deep into the bottom rooms of the ship, hoping terror through proximity would motivate them to spill their secrets. It did not.

Midshipman Adrian George Holloway was asleep inside the *Valiant* when the bomb went off. He scrambled in his pajamas up the quarter-deck ladder, and when the ship began to shift, he felt the disaster ripple through the entire body of the hull. He nearly got heaved clear off. The officers on deck, he noticed after he regained his clasp and finished climbing, were in all kinds of attire—pajamas, clothing, really whatever they had been wearing when they were woken up with zero notice and yelled at to muster posthaste. The crew was told to gather at the waist of the boat, the middle section of its deck, because nobody knew if their stern would be blown off next. They did not know if there were more charges. The *Valiant* had a hole in its side as large as three semi-trucks.

Suddenly, a second blast came, set by a second team of Italian divers. The bombs had been timed to go off together, and they almost did. But the explosion was not from the *Valiant*, which was already settling to the shallow harbor floor. Rather, it was from the ship's immediate neighbor, HMS *Queen Elizabeth*. A whoosh erupted out of that battleship's funnel, flinging detritus into the sky before it rained back down. The force of the blast burst rivets from the hull, shooting them across internal rooms and embedding them in opposite walls. The Norwegian oiler *Sagona* went next, courtesy of the third *Maiale* pair. The *Sagona* was moored next to the destroyer *Jervis*, and both were soon on fire.

Across the harbor, sailors on other ships saw the masts of the battleships shake from the force of the explosions. If they watched later, they were able to see them "listing quite badly" in the locations where they settled with a lurch onto the shallow, muddy seafloor.

With six divers, three bombs, and the ability to breathe underwater, the Italians had destroyed the core of the Allies' Mediterranean fleet. The three teams of Luigi Durand de la Penne and Emilio Bianchi, Vincenzo Martellotta and Mario Marino, and Antonio Marceglia and Spartaco Schergat were all captured, but they had regained control of the Mediterranean for Italy for the next several months. They had done it with zero casualties for their side. Midshipman Holloway called it "the most perfect attack." Leonard Alfred Dunn, a seaman aboard HMS *Euryalus*, one of the ships for which the protective harbor boom had been opened, said, "Bang went the Mediterranean Fleet, more or less." The attack was the most perfect demonstration of the developing concept that was special operations.

———

"Please report what is being done to emulate the exploits of the Italians in Alexandria Harbour and similar methods of this kind," Prime Minister Winston Churchill wrote to his Chiefs of Staff Committee a month later. "Is there any reason why we should be incapable of the same kind of scientific aggressive action that the Italians have shown?" His infamous temper emanates through the words.

When this letter reached Vice Admiral (Submarines) Max Horton, the man obsessed with midget subs, "the need for an immediate underwater striking-force became apparent," recalled William O. Shelford, one of the other submarine officers involved in the innovation and planning. "British human torpedoes were accordingly conceived as a stop-gap, awaiting the completion of the midget submarine."

On a cold, sunny afternoon in February 1942, VAS Horton was finally able to present his pet project to a long conference room table populated by staff officers. In the words of William Shelford, who sat facing Horton from the far end of the table, "The Admiralty, [Horton] explained, had decided that such a weapon could not be ignored, and must be exploited by the Royal Navy." The goal was for them to execute an operation similar to the one in Alexandria against German battleships by that summer. It was an ambitious deadline. The midget submarines would take longer to build, but because the human torpedoes would use already studied Siebe Gorman rebreathers for underwater breathing, they could be modified and fielded more quickly. The Royal Navy had Gino Birindelli's abandoned *Maiale* from Gibraltar. They started copying it immediately. As a longer-term measure, Horton's midget submarines were going mainstream. The UCL group's research about survival underwater, in the wake of the success of the *Maiale* vessels and at the urging of Churchill himself, was officially considered an emergency.

They also knew that, given that they had been throwing explosives in the water to deter the *Maiale*, the Axis could be expected to do the same to them. To make such divers possible, the Allies would also need to know how to protect themselves against underwater bombs.

CHAPTER 11

Lung Maps

Horace Cameron Wright donned a life preserver belt and jumped overboard. An unrecorded number of Royal Navy sailors followed suit, each also putting on a fluffy, khaki-colored life belt before they took their plunge into the English Channel. British ships were sinking all over the Atlantic after being torpedoed by wolf packs of U-boats, and they were dotting the map of the seas with lost lives. But this particular ship was fine. It was not sinking. These volunteers had jumped for science.

The small group clustered near where they had dropped, floating in ninety-six feet of water off the southern edge of Portsmouth, England, with only their heads poking above the salty chop. Their skin prickled in the sharp, cold ocean as, despite the warmth of the summer air, the frigid water forced them into a pattern of shallow breaths. Small pressure gauges hovered underwater near them at a depth of three feet.

The volunteers rotated so they were all looking in the same direction. A little motorboat puttered fifty yards in the distance, slowly lowering the point of today's experiment down into the water. The volunteers watched. And floated.

It was a bomb. Only five pounds of TNT, slightly larger than a softball. But in water everything is magnified, and a tiny bomb can pack a giant wallop. This charge was smaller than the ones the Allies had been throwing into the Mediterranean to deter the riders of the *Maiale*. Once in place, the bomb dangled at a depth of fifteen feet.

The British were using similar small charges to warn off potential saboteurs from harbors at home, and the Germans were using the same to guard ships and harbors, too. For any and all future underwater plans, in between any diver and any beach, there would probably be an obstacle course of explosives.

Yet despite the ubiquity of these little bombs, neither side understood in a scientific way how much damage they could do, nor what the safe standoff distances should be. What size charge was needed to deter enemy divers. What size charge could kill them, if they were enemy divers. On top of that, thousands of sailors in the water had been blasted accidentally after abandoning sinking ships during combat, and nobody knew how far they had to swim to be safe.

Horace Cameron "Cam" Wright, civilian Admiralty employee and freshly minted PhD, wanted to find out.

Cam Wright, now waiting to be blown up off the coast of Portsmouth, had a prominent cleft chin, straight white teeth, and classically strong facial features that, in a less war-torn world, could have made him a movie star. He was known for being shy, quiet, and private, but also straightforward, direct, and blunt when needed. He loved hunting on horseback in Ireland, racing his Alfa Romeo, and science that did not wait for any bureaucracy, even that of his bosses in the Royal Navy. To get things done, he had a tendency to hoist heavy objects, gear, or even farm animals over his broad, wiry shoulders and carry them with ease to wherever he needed them to go. He had specifically requested to work with the military as a scientist, and then, when granted leadership over these blast experiments, he had insisted on being the one in the water.

Wright knew that the subjective effects of the blast were key. If a diver survived the initial explosion but was in too much pain to swim or keep his head up out of the water, then he would still drown. As with the UCL lab's trials of oxygen toxicity, it was impossible to measure those subjective effects with pressure gauges or even animals, so Wright, too, got the data from the most reliable source he could think of: himself.

The experimental countdown ticked away as Cam Wright and the other volunteers floated a meager fifty yards from the freshly placed five-pound bomb. The countdown reached two, then one. An experimenter flicked a switch or pressed a button. Fire blazed to life within the TNT—a fire that traveled through the explosive material and burned it into superheated gas faster than the speed of sound could provide any audible warning to the volunteers floating by their necks.

TNT stands for trinitrotoluene. The "nitro" part is a chemical compound containing an extra, extremely unstable oxygen atom. It is an oxygen burdened with a bonus free radical, a child spinning with one hand outstretched and grasping at all times. It's just dying to donate that extra electron elsewhere, to grab onto fuel, any fuel. Each molecule of TNT has three such nitros, which is where the "tri" comes from. TNT, in other words, is a fire eager to happen. The pressure created by the now blazing-hot ball began to ripple outward into the water.

The TNT released a shock wave. Each molecule of water kissed more intimately against its neighbors, forced together by the might of high explosives. The wave rippled outward, traveling toward Wright.

In science, engineering, and most of life, people usually assume water is incompressible. That it cannot be smashed down, made smaller, increased in density like most of the other materials in our world. But it can. Shock waves cannot exist in or travel through materials that cannot be compressed. They can travel through air, human flesh, even steel because they are strong enough to compact all of these in a ripple as they pass. And they can travel through water, too. With enough pressure, enough force, a shock wave is mighty enough to compress the very ocean.

Wright and the others saw the explosion before they felt or heard it. It started as a slight fizzle on the surface of the water when the shock wave broke through, followed by a much larger plume as the gas released by the bomb ruptured free.

They saw it first, but not with enough time to regret every choice that had brought them to that moment before the shock wave hit them square in the gut, in the lungs, in the spine—all within the same half of a second that they saw the water fizzle, all too quick for conscious thought. It felt like a punch or a stab, depending on the person, and depending on where the gas bubbles lay that particular day inside their own intestinal tract. The shock wave traveled with ease through the skin and meatier flesh of these minimally clad bathers, but came to a screeching halt at gas bubbles, at lungs. There, where it slammed to a stop, it wreaked havoc and pain.

Some felt as if water were rushing uncontrollably into their rectums, although later they would learn it had not. Some of their bodies provided a powerful command to urinate; they did. Some of the volunteers felt as if they had been kicked in the balls.* They were kept afloat by their life belts until they could be plucked from the water and placed back on board the ship, where Wright quizzed them about what had happened.

It was not the most powerful blast Cam Wright and the other volunteers had experienced—that experiment was at least their fifth self-inflicted explosion—but it was sufficient for them to decide they did not want to go any closer to a five-pound ball of exploding TNT than the fifty yards of that last test. Had they been enemy swimmers, they decided, they would have been deterred from their mission.

The experiments had started out small. One and a quarter pound of TNT at a range of seven hundred yards. It produced no deterrent effect. Five days later, the researchers upped the charge size to five pounds. They worked their way closer to the bombs. Two hundred yards. One hundred yards. Fifty yards.

Some of the volunteers sat astride wooden logs to mimic the human torpedoes under construction. They breathed through submarine escape rebreathers, DSEAs, one of the same models being tested by the UCL diving group.

They did most of the experiments, including riding the logs, inside a modest test pond named Horsea Lake, which was off the same harbor as the submarine naval station in Portsmouth, on the southern coast of England. It was a brackish, spring-fed, chalk-bottomed rectangle of water with a clean-cut shape obviously not carved by nature. Horsea Lake was in the center of a small island of the same name, and Wright and his

---

*The sensation of a devastating kick to the balls is a commonly reported experience following exposure to an underwater blast. However, to soothe the concerns of readers who have such equipment: Beyond the unpleasant feeling, damage to the testicles is highly unusual at any exposure levels below ranges that are lethal anyway.

compatriots cobbled together a makeshift lab in a trailer on the shore. They might have been somewhat "playing" at war by straddling logs inside a safe, enclosed lake, but the bombs—and therefore the dangers— were real.

The British human torpedoes had been dubbed Chariots, the midget subs X-craft. The officers of the Royal Navy had promised Winston Churchill they would have "six operational Chariots with trained crews" ready to conduct shallow-water attacks by the end of September of that year, 1942, and X-craft with crews who were also qualified for rebreather diving by "the early Spring of 1943." In Wright's view, the word "ready" included knowing what would happen if these divers got hit by explosives and how far they had to swim or dive to escape those threats.

After one experiment with divers riding fake Chariot logs, Wright recorded, "The DSEA afforded protection to the chest." In other words, using the rebreathers provided an extra layer of insulation between them and the shock wave and lessened its *thunk* against their lungs.

In the urgency of war, the names of the volunteers were not written

A diver wearing a suit and breathing apparatus and sitting in the pilot position on board a Chariot. The explosive charge is missing from the front of the Chariot. *(IWM 22119)*

down for each day's trial. But the group's surviving records make it clear that Wright insisted on going first for every new experiment, because he wanted to make sure he was not jeopardizing the lives of others. In addition, he would often get in the water with blast-inexperienced divers to stay near them during their first explosion, as assurance. Each time they wanted to try something new with an explosive, Horace Cameron Wright felt it first.

———

In between blast experiments, the personnel training to become the new amphibious submariners had their first casualty. Lieutenant Patrick Charles Annesley Brownrigg was "undergoing instruction in diving" in Horsea Lake as part of his training for riding the human torpedoes when his diving attendant reported that the lifeline tethering Brownrigg to the shore had come free. Brownrigg could have, in theory, surfaced immediately under his own power and survived. However, even though sound signals were sent through the waters of the lake telling him to come up, for whatever reason, he never did. His body was pulled out of the water later that day.

His mother, Mrs. Aileen Maud Brownrigg, received a simple telegram with no further information about his death: "From Admiralty. Deeply regret to inform you of the death of your son Lieutenant PCA Brownrigg RNVR."

He was the first casualty of the top secret group of amphibious submariners whose deaths, if they died, would remain a mystery to their families. He would not be the last. Wright, Haldane, and all the others were working separately, but they were united by their common goal to minimize that final body count. To minimize the final number of families left wondering.

———

On May 13, 1942, UCL experimental day sixty-four, experiment number 125, the clock on Churchill's Chariot deadline in September ticked as loudly as ever. On the same day as Wright's blast test in the Channel, shortly after lunch, Elizabeth Jermyn locked herself inside a newly relocated hyperbaric chamber with Martin Case. The chambers had been moved to Siebe Gorman's new factory in the southwest of London, where there were fewer bombings and a freshly constructed, intact building. The lab members had finally restarted testing earlier that month after a yearlong gap to rebuild. Jermyn and Case each carried a breathing apparatus, a nose clip, and a set of goggles with them.

Military divers would need to operate at night. The chamber divers were learning that breathing pure oxygen under pressure could affect their eyesight, but they did not know if it would affect a diver's visual acclimation to the dark. The scientists were reporting seeing flashes of hallucinated color during some experiments. Hans Kalmus usually saw golden yellow, Helen Spurway hallucinated in purple, and Martin Case's oxygen-addled eyes flashed him spots of red. They called these visual disturbances "dazzle."

Jermyn and Case breathed from their rebreathers and alternated wearing the goggles while sitting in the pressure chamber with the lights off. Jermyn periodically wrote down both of their pulse rates, since a drop in pulse was the only predictor they had worked out to check if a seizure was coming. It was a poor one.

Case began to see dazzle by the end of the test, but most of the dark adaptation of his eyesight remained fine. Jermyn tested his field of vision by waving her hand around the perimeter of his face to see if he could detect its motion. No seizures this time.

Case had lost some peripheral vision, which could be concerning for combat divers. As with all other tests, they would have to repeat it—both on themselves and on the rest of the group. But most important, it seemed as if military divers should stay adapted to seeing in the dark at night, even when breathing pure oxygen.

Assuming they avoided seizures. And survived the explosions. All of the scientists needed to accelerate their work to meet Churchill's ticking clock.

―――

Horace Cameron Wright's metaphorical path to Horsea Lake was indirect, and the lack of concrete information has twisted his story into a legend more reminiscent of a mythical creature. There are some contemporary newspaper clippings quoting him describing being blown up as "great fun," but they contain few personal details. The current generation of scientists at his former government lab still have his original lab notebooks in storage, but nobody working there now was able to remember his first name. One of his closest friends, in the address he gave at Wright's memorial service, included the line "Loved by many, he was known by few."

However, two common threads shine through the little documentation that is available. One, he was a genius. Two, he lived a life burdened by unfortunate circumstance.

Horace Cameron Wright was born smack in the center of London in the borough of Islington in 1902. His father was a lithographer and an artist, but he scratched together the money to sustain the family by replicating prints for other, much more famous artists. One of those was the prolific painter, mosaicist, lithographer, and designer Frank Brangwyn. At least one source states that Brangwyn adopted and raised young, orphaned Cam Wright after the tragic, premature death of both of his biological parents. However, Cam Wright's mother died in 1932, when Wright was thirty. His father died in 1948, when Wright was forty-five. He was hardly an orphan. Brangwyn was a close friend who sent Wright regular letters, but it is not clear how the adoption rumor got started, or if Wright started it himself. It has contributed to his legend.

No records are known to exist describing how Wright got into science. Nobody else in his family had any such interest. Before marriage to his father, his mother, Laura, had been described as a spinster. Before landing his career making prints, Cam's father, Edward, tried to live as the artist son of a housing agent. At the time of the 1921 census, nineteen-year-old Cam was living at home with his parents, and by 1935, he had earned a degree in physics and was working as a physicist researching radiation treatments for breast cancer at a hot spring clinic and spa. Then, inexplicably, he organized a scientific conference and jumped a ship to France.

From 1936 to 1937, Horace Cameron Wright researched carcinoma of the breast with Professor Irène Joliot-Curie, the elder daughter of Marie and Pierre Curie, at the prestigious Sorbonne in Paris, France. Born of relative struggle but now in the Curie lab at the Sorbonne, Wright had somehow reached the pinnacle of elite scientific prestige. One description of his work there as "extramural" implies he was enrolled as a student, likely in the medical school, but it seems no other records remain.

By 1938, he was abruptly back in England, where he was awkwardly, embarrassedly scrambling to find someone who would hand him a medical degree without any additional work. He wrote to other universities, including the University of Lausanne in Switzerland, asking if he could get a medical degree by passing their qualifying exams without having to attend classes. The University declined, and in his translation of their letter from its original French for someone else, Wright omitted the part where it was suggested he might have been better suited for a PhD than for a medical degree. It seems as if he quit—or was forced to leave—the Sorbonne without graduating. It also seems that he spent the rest of his life scrambling to recover from the shame.

War broke out a year later, and at age thirty-seven, Wright was on the older side of the preferred range for soldiers. However, he wanted to contribute, and so he started writing letters offering his services as a physicist.

"I have explained to him about your diffidence regarding the degree," one of Wright's friends wrote to him after advocating on Wright's behalf for employment with the Admiralty. He was telling Wright not to be ashamed over the still fresh wound of the Sorbonne. "This appointment should enable you to recover any status which you think that you have lost in the past two years," another friend wrote in 1941 after Wright was accepted for government service. The Admiralty was Wright's chance to heal from whatever had happened.

By 1942, the navy's physicians were stumped. Sailors kept dragging their brethren out of the ocean dead, but with no external marks on them, after the explosions of nearby depth charges. World War I had been the first major global conflict with high explosives, but despite the fact that horrifying numbers of people had died from blasts, the subject had received relatively little formal academic study in English-speaking circles.

"It is a surprising fact that it was not until the recent hostilities had commenced that the true seriousness of the injuries found amongst blast casualties were appreciated. The medical history of World War I contains a limited amount of clinical material on the subject," one unnamed World War II physician complained about the lack of information.

Then, as the ships began to go down and bombs began to go off in the water, "many of the survivors reported that immediately after the explosion occurred they lost all power in their legs and occasionally the anus, and as a result of this temporary paralysis men had been seen by their colleagues to gasp, cease all efforts to remain afloat, and slowly sink." It was clearly a problem that needed study.

The medical director general convened an urgent meeting in London. The problem was one that would require a knowledge of physics, mathematical waveforms, and the anatomy of the torso, the most easily blast-injured part of the body. Wright—a physicist, a researcher of the complex mathematical waveforms of radiation, and a specialist in the anatomy of breast cancer—stepped up.

Wright was not known to have ever expressed a religious affiliation. "His striving to make discoveries that would benefit mankind was religion enough," said one friend and former colleague. His prayers took the shape of waveforms sent out to save others from blasts. Helping was his god. To help, though, he needed a more advanced degree.

In January 1942, Cam Wright furtively wrote to a now defunct correspondence school called Intercollegiate University. For a small fee, Acting Registrar Sidney E. P. Needham was thrilled to mail Wright a piece of paper declaring "Degree of Doctor in the Faculty of Philosophy." It was supposedly based on Needham's academic evaluation of Wright's preexisting research in the field of functional capacity, which is the measurement of a person's ability to perform physical work. Needham, however, was a bishop, not a scientist, and it does not seem that he had experience in any scientific field, including physiology. On February 2, 1942, Wright received his mail-order PhD, added those three dignified letters to his résumé so that he would be allowed to lead projects, and then hurriedly buried the story of how he had gotten them.

Intercollegiate University closed a few years after Wright's postally issued degree, and no university records are known to survive. However, unlike for most other universities, only one person working in any non-religious field can be found who publicly discussed their affiliation. It is Leon DeSeblo, Doctor of Divinity, graduate of Intercollegiate University and self-proclaimed "world-famous scientist." DeSeblo spent roughly a decade wandering the United States on a traveling lecture tour to promote his book, his "amazing discovery" miracle cures, and general ways to "prevent and destroy all your troubles!" as he put it in his newspaper ads. He claimed to be able to prevent aging itself. "Dr." DeSeblo had at least one major run-in with the American FDA over the contents of his questionable bottles and pills.

Both graduates came from the same institution, which seems to have been nothing more than a degree mill. Leon DeSeblo is proof that letters after a name are not the same as credibility, and that letters carry less weight than actions. Horace Cameron Wright is proof that academic track records and vaunted pedigrees, or lack thereof, are poor predictors of brilliance, courage, and heroism.

Cam Wright stopped his murky forward plodding at the end of the long underwater arc of line at the point that indicated he was directly beneath the marker buoy floating above him on the surface. He stood there, below Horsea Lake, the top of his rubber-hooded head six feet and one half inch above the soles of his heavy metal diving boots. Small whorls of sediment rose from the hard chalk bottom as Wright turned around to face the coming shot.

Today he was wearing the same protective dry suit that was planned to be used by the Charioteers and amphibious submariners so he could be fully underwater and feel exactly what they would. A thick, black, durable, waterproof rubber skin encased both his legs beneath the laces of the heavy metal boots that helped him stay oriented upright. The same stiff rubber nested in passive folds over his stomach, where the suit telescoped to let him get in and out more quickly. Siebe Gorman had built this Sladen "shallow water dress" in the same factory where the

UCL scientists were working. A circular visor, which let Wright look forward while allowing his face to stay dry, was attached to the black rubber hood with a thick, watertight gasket. The breathing bag for an oxygen rebreather rested on his chest. A corrugated rubber hose ran up to his lips. Only Wright's hands remained unprotected, poking out into the cold blue water of Horsea Lake through seals around his wrists.

Five pounds today. TNT again. Wright stood at the bottom of the roughly twenty feet of water. The charge was fifteen feet beneath the surface. The bomb hovered, suspended at his eye level, fifty feet away, out of eyesight for him inside the brackish depths of the military training pond.

He waited in the same waters where Lieutenant Brownrigg had drowned a few weeks before. He used the strength of his chest, the diaphragm beneath his lungs, the small muscles beneath his ribs, to suck the oxygen in through the corrugated hose, then push it back out on repeat.

The shock wave hit Wright, rocketing across the lake and ricocheting in deadly zigzags off the hard bottom, slamming against his chest, his gut, his spine, and now, since he was fully underwater, his head and ears, too. The other scientists and he would calculate the pressure later. It was an estimated 280 pounds per square inch (psi), equal to roughly twenty-seven hundred adult human men standing on the front of his body. It hit him for the length of a mere few thousandths of a second. Not long enough to move him. But long enough to damage him.

"Spinal concussion occurred," the test notes recorded. Reports from other days clarify what that means: His legs went numb; he lost control of his body. The effect was brief, but it marked his tolerance limit for distance to the charge. His ears, however, had been protected for this trial. After he crawled out of the water, he removed the "ear defenders" he had been testing from over his ears. They were slabs of rubber, each one the thickness of a hearty steak. In his opinion, they had helped. He could hear better than after the last trial, when he had been blasted without them. He took a hearing test afterward, and the audiogram showed minimal hearing loss from the explosion.

Those pads should work for divers in action, he decided. The testing had been rushed, but Wright and his team had gotten what the amphibious submariners would need to know for their missions. He had also worked out a new type of vest made of kapok, a fluffy and naturally occurring type of material that seemed to help reduce the impact against his chest and abdomen.

After nearly every one of his blast trials, Wright placed the cold, metallic business end of a stethoscope against his chest. He breathed in deeply, then back out, shifting the polished circle of the medical instrument to a new region of his skin with each breath. He was listening for crackling. When the narrower, daintier passages of the lungs fill with liquid, that liquid seals those passages shut when they shrink during a normal exhalation. When they are forced to open again during the next inhale, the plugs of liquid are gradually pulled apart by the expansion until they separate with a pop, like a bubble. The thousands of tiny pops in the tiny passages of the lungs are heard through the walls of the chest as crackling. It sounds like a little internal bag of popcorn bursting to its edible format inside the body. Except it's destructive. The popping damages the walls of the breathing passages. The sound of fluid in the lungs is the sound of illness.

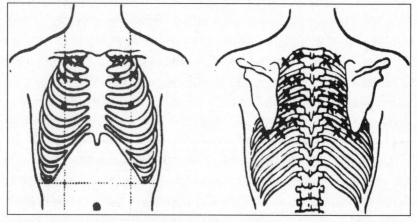

One of Horace Cameron Wright's maps of the damage to his own lungs. The X marks denote locations where he heard blood.

Or, in the case of Horace Cameron Wright, it was the sound of blood. After the experiments he did on himself, he drew maps of where in his lungs he heard crackling. Maps of where there was blood.

———

The Americans had the same questions about underwater blasts. While Wright was wrapping up his experimental series, American divers stood on the bottom of a similarly shallow body of water off Stump Neck, Maryland. They, too, gradually inched their way closer to the charges. However, they stopped at 187 yards at their closest, over triple Wright's closest distance.

The American researchers had less data with which to make recommendations because they had done fewer trials than Wright. They provided a loose estimate that they thought the surface of the water "shredded," or made a plume, at a pressure level of about 500 psi. Therefore, they reasoned from the warmth and comfort of chairs in an office, that was probably the pressure level where the inside of the human lungs begins to "shred" and spray with blood. In other words, they assumed it was the pressure level that would cause damage and injury.

Five hundred psi is roughly the threshold where everyone begins to die from an underwater explosion. It is nearly double the level where Wright experienced "spinal concussion" and lost control of his legs. To estimate where people in water might be safe, the Americans simply divided that first number by ten, and arrived at 50 psi. They left no indication of why division by ten was the magic number. They do not seem to have had any data to support it. Now we know that 50 psi is about the level where serious injuries begin to occur.

However, the guidelines that the risk of probable injury starts at 500 psi and that there is safety at 50 psi remain, as of this writing, in the *United States Navy Diving Manual*, more than eighty years later. Not one piece of data supports them, because they are wrong. The American injury-likely guideline for underwater blasts was—and is—a near-certain death sentence. It is not difficult to imagine what Cam Wright would have said.

The Nazis wanted answers, too, for blasts in air and water. Toward the end of the war, they executed a thorough purge of their own scientific records, and after that purge, their official documents and experiments seem ethically squeaky clean. Their remaining blast records make it seem as if they tested exclusively on inanimate glass windowpanes, although they acknowledged blasting a small number of dogs. Yet somehow, magically, as if produced from nowhere, they also had recommendations for human risk, with no traceability back to original data. Their human-risk guidelines do not seem made up.

It is hard to imagine that, given the earnestness with which the Nazis scrubbed their experimental records, their human subjects were volunteers.

CHAPTER 12

Nerves

Scientist-secretary Elizabeth Jermyn was puttering about in circles. She walked in slow motion around the bottom of the largest water tank at the newly relocated Siebe Gorman factory southwest of London. It was a hair past noon on a Monday, and the gigantic, sunshine-filled pool was almost four meters tall and roughly ten meters in diameter. When they had relocated after the bombing, the big bosses at Siebe Gorman had brought the intact old chambers, but they had added some new tricks, too. Bigger tanks. Chambers that could be filled with water. Jermyn's current giant circular pool. She was a veritable marine creature walking about her own underwater zoo. She was preparing for experiment number 170.

J. B. S. Haldane, Martin Case, and famed physiologist Sir Leonard Hill, all of whom stood on a dry platform above, peered down at the Jermyn fish in the water. The tank had a sturdy curved rail around its open upper edge, where the light from the factory windows could stream in and where the observers could lean as they spectated.

When Jermyn decided she was ready, she climbed the rungs of the ladder of piping bolted against the side and back into the air. After spitting out her mouthpiece, she reported that the DSEA seemed to be working well enough for her to use it for a real test. She suspected a slight leak in one valve, but that was normal for this configuration. She walk-sloshed across the second-story platform to the top rim of a smaller, darker tank: a vertical pressure chamber. Haldane, Case, and Hill attached a rope to the back of her harness, which led to a system of pulleys so they could haul her back out in case of emergency.

The Jermyn fish climbed down into the vertical cylinder. She slid through the narrow round hatch at the top of the chamber, first through a compartment of air and then farther into a tall column of cold, 65 degree Fahrenheit water, where she continued to breathe like she belonged there. Only a meter or so wide, the hatch itself wasn't a squeeze for her slight frame, but neither was it spacious. Haldane followed and seated himself on a small platform in the air compartment above her head, letting his feet and legs dangle in the water, his intruding lower limbs protected from the cold by rubber fishing waders. If he stretched his arms out, his fingertips would have been forced to bend by the dimensions of the cylindrical container, the upper edge of which was rounded "like one of Mr. Churchill's famous hats."

Someone bolted shut the hefty, circular pressure-proof door at the top. Case was taking notes outside. Haldane had easy access to Jermyn's rope tether. Her DSEA rebreather was delivering her oxygen as needed. They were ready to start.

Gas began to hiss into the air compartment, pressurizing the space around Haldane and therefore also the water around Jermyn. They started slowly. They paused at twenty feet. Jermyn signaled that her ears were doing fine.

They reached their goal depth at fifty feet, and Jermyn drifted about happily in her confined space. After six minutes she passed up to Haldane a writing slate with a note saying she was cold but otherwise all right. For another four minutes, Jermyn breathed into her DSEA, feeling the heft of the water push back against her lungs each time she exhaled. She produced no bubbles or sounds that could be seen or heard from the surface.

At 12:44 p.m., Haldane bellowed the order to decompress. Now. He bellowed the order to decompress *NOW*.

The chamber had a small round window in the side, and through it, Case could see the twitching, spastic body of Elizabeth Jermyn. She was having a seizure.

A seizure *underwater*. Her legs kicked; her hands flailed. She made erratic, ineffective motions like she was trying to climb up Haldane's legs. Worse yet, she made clutching, desperate gestures that could knock out her own mouthpiece.

Before it began, she had started writing something on the waterproof slate. It wasn't clear what. She let out a "squeal" or a "squeak." Then the oxygen took control. And unlike the team's past experiences with seizures, rather than simply let this one pass on its own, they had to get her out of the water before she could knock out her mouthpiece. If she did that, she would almost certainly drown.

Her limbs continued to thrash. They had to get her out *now*.

———

Joan Elizabeth Jermyn was young, active, curious, and kind. She left a written record that she had volunteered for the trials of her own free will, but a member of her family later also pointed out with a dark chuckle that "if [Haldane] wanted you to volunteer, you volunteered." Jermyn and Spurway's being actively engaged in war work exempted them from being drafted into the Labour Exchange and forced into clerical service, so the tests at least had that perk.

Outside of her time in the pressure tank, Elizabeth Jermyn also led a more definitively optional midnight life as a bomb watcher. The main research building at their group's evacuation site of Rothamsted Experimental Station, a hundred miles north of Horsea and north of London, was a long, rectangular two-story slab of vibrant red brick, and it was perforated down the length of each floor by a row of identical vertical windows. Sculpted spindles marched around the perimeter of the building's upper edge, no doubt originally put there to provide safety for scientists and guests on rooftop strolls in peacetime. During the smoke and flashes of a bombing raid, the same spindles would have looked like military crenellations installed to provide cover for defending forces.

While the group did their diving experiments in the chambers at Siebe Gorman in London, their official lab away from home was a shared space in Rothamsted, and Rothamsted was the site of war-relevant scientific research. Botanists trying to increase crop yields to supply more food to the troops. Chemists trying to develop better fertilizer to do the same. The UCL group had evacuated there for the working electricity and gas, but it was not exempt from the war. They regularly drove the hour south for the chambers, but the lab at Rothamsted, too, was a high-value target for Nazi bombers. The amount of war work that could have been destroyed by one judiciously placed incendiary bomb, if left to burn amok, was huge.

Elizabeth Jermyn, the athletic, stalwart, and ingenious former queen of the field hockey and lacrosse pitches in her school days, wanted to help with the war effort however she could. She had already been volunteering as a "guinea pig" for the scientific trials and patrolling the grounds of Rothamsted as a warden for the Air Raid Precautions group. Members of the ARP sounded the air raid sirens. They stood outside in the dark, unprotected by shelter, and watched munitions fall from the skies so they could mark the locations of unexploded bombs that would need disposal. They fought the fires lit by incendiary devices. They also looked for injured people and rendered first aid.

Jermyn's sister Molly learned how to drive an ambulance. Her brother, Bernard, was in the army. To do her bit, the indomitable "Betty" got in the water tank and walked the roofs of the buildings during raids, her keen, dark eyes sweeping the grounds while she wore a flat-brimmed, government-issued steel military helmet to protect her brain from shrapnel. Her helmet would have been emblazoned with a white "W" to indicate her status as a warden. A leader. Jermyn was eager to help, and she was not scared of anything. Plus, for a few hours a night, it got her out of the communal house where she had to live with her difficult boss and her less difficult but still cramped lab mates and their families.

One night, while watching for blazing deadly munitions streaking across the sky, Elizabeth Jermyn crossed paths with fellow volunteer Nariman "Nari" Bamji.

He was a botanist by training, and he had been nominated the chief fire officer for Rothamsted. He worked on compost formulations during the day, but he watched the skies and organized ARP volunteers at night, and when incendiaries threatened their lives and their science, he operated the hoses and pumps and threw sand buckets himself. He was of

Parsi heritage, born in India in the state then called Baroda but now called Gujarat, and he had deep, large, curious brown eyes paired with striking high cheekbones. He had come to England and Rothamsted on a botany scholarship, and he had Communist leanings because that political party agreed with his passionate belief in independent rule for India. His life in India had been marked by efforts toward independence from British rule, and according to his family, at one point he was imprisoned alongside Mahatma Gandhi. His torso bore a vicious scar from a knife wound he had received from putting himself between a group of protesters and their religiously motivated attackers. Despite his feelings about British rule of his homeland, though, when the war came, he took the side of the Brits over the Nazis without hesitation.

Nari Bamji was denied the ability to join the Allied military because his birth certificate had been lost long ago, somewhere back in his homeland. However, because of his work in war-relevant compost research, he was allowed to escape internment. He had a dark, thick forest of gorgeous, wavy black hair that he kept pushed back off his forehead, and he was just tall enough to need to look down into Elizabeth Jermyn's eyes.

They fell in love quickly. Neither left behind much record of their whirlwind romance, amid the war with its rationing of paper, camera film, and all other recording media. Bamji had been staying in a rented room, but by the spring of 1942, he was spending substantial time in the Piggery with the rest of the UCL lab members, and somewhere around that time, the two got engaged.

Nari Bamji was fiercely independent and motivated by a strong sense of right and wrong. J. B. S. Haldane was, too, but in ways that sometimes morphed to support his unique conception of what he was owed by those around him. Now they were together often in the Piggery. Haldane sometimes made demands of Bamji, like commanding him to mow the lawn; demands Bamji refused. The two loathed each other.

Elizabeth Jermyn might have volunteered for the tests freely, but her untimely drowning in a water tank would have torn their research family to shreds in more ways than one—and long before their work was close to finished. Long before the Allies were done needing them.

The chamber reached surface pressure quickly after Haldane's command to decompress. Martin Case, Leonard Hill, and he opened the hatch. They hoisted the limp body of Elizabeth Jermyn up by her rope, through the narrow opening that she had climbed into of her own volition only twenty-two minutes before. She was pale. Not moving. Specks of froth coated her inanimate mouth. When Haldane peered closely at the delicate features and strong bone structure of her elegant face, he swore he could see small twitches of muscle dancing along the rim of her lips.

Leonard Hill felt her pulse. She was alive. Hill deemed her pulse "good" with what seemed like a shoulder shrug. Her heartbeat felt faint, feeble even, Hill said, but he had thought her pulse feeble to start with. He therefore declared her to be fine. Everything was normal. Normal-ish. Haldane thought the knighted physician "pretty incompetent."

Elizabeth Jermyn recovered slowly. Later, she remembered nothing. After about two hours had passed, Martin Case donned a piece of breathing equipment to try the exact same test. At the same depth. In the same water. On the same day. Experiment number 171. Because good science demands repetition.

The human spinal column is buttressed to the left and right by the erector spinae, a collaborative group of muscles and tendons that work together to unfurl our upper bodies backward and upright. That includes pulling up our bulky heads, which are gargantuan from a zoological perspective, basically lumpy bowling balls of skull and meat. In other words, those muscles are strong. Elizabeth Jermyn's erector spinae, during the throes of her oxygen convulsion, contracted so hard that those two beefy columns of muscle crushed one of her very bones. Her fifth thoracic vertebral bone, in roughly the center of her spine near the level of her armpits, was fractured.

Over the next few days, she tried to power through the pain. She tried to ignore it. Sir Leonard Hill assumed the pain meant she had a collapsed lung, like Jim Rendel's, which was still healing and holding him back from lifting any weight. On a visit to Jermyn's mother's home, Nari Bamji badgered her into a proper medical exam, like a good partner should, and X-rays confirmed the diagnosis that her back was literally broken. The doctor said she could still do secretarial work and help with the tests. She would also continue to aid with the math and statistics, but she was not to get in the water anymore, not as anything more than an observer or attendant. Definitely not as a test subject. The risk posed by a second seizure was too great.

Jermyn's bone fractured in place, and none of the bone fragments shifted or displaced, so she recovered fully, if you don't count the disabling back pain that would plague her for the rest of her life. (She didn't.) However, it takes only one half inch of compression to explode an entire bone in the spine. The human spine is so flexible in bending that a person's face can fully smack into their chest without damage to the bones or spinal cord. It will cause a hell of a case of whiplash and muscle strain, but no damage to the bones themselves. In pure compression, though, one half inch will fracture an entire vertebral bone into ragged shards. Those ragged shards often sever the spinal cord in the process, disrupting the traveling electrical impulses that dance down that long, bundled string of nerves to allow our brains to control our bodies as puppeteers do with dancing dolls.

This specific fragility of the spine is why automotive rollover accidents have a high rate of paralysis. We did not evolve to land on the tops of our heads. We did not evolve to survive torn spinal cords. If Elizabeth Jermyn had had another seizure and that cracked bone had given way completely, she would probably never have walked again.

Haldane and Siebe Gorman had insured the group against injury with the assistance of the Admiralty, and while Nari Bamji helped Jermyn recover, Haldane asked the insurance company for money for the medical bills. The insurance company attempted to squeeze out of it by claiming a woman should never have been included in such trials in the first place. Haldane blew a gasket in defense of women in general and Jermyn in particular, pointing out that other women were participating in the tests without injury and that men had gotten injured, too. He said that in the previous battery of tests, which had taken place in air, Jermyn had done "better than the majority of naval ratings tested," and he pointed out that also-woman Helen Spurway often dove harder, deeper, and longer than anyone else, so much so that Spurway had "held several of the world's diving records, and still holds some."

Haldane hinted that if the company denied Jermyn's claim, it would face a loud lawsuit stacked with scientific experts from his universe. The insurance company reluctantly doled out a meager two hundred pounds for her medical bills. It was about one year's salary for her, but her salary was poor.

Elizabeth Jermyn, unwittingly, through her seizure, was responsible for one of the most harrowing discoveries in the history of diving. She uncovered a key phenomenon for the first time. A phenomenon that, had it gone undiscovered, could have killed uncounted Allied divers and destroyed innumerable Allied missions. Even more stunning, the reason why that phenomenon occurs remains a mystery.

Jermyn had had her seizure while breathing pure oxygen at a depth of fifty feet after only ten minutes. Less than a month earlier, she had dived for twice as long at a depth of ninety feet without so much as a sparkly visual dazzle or a tiny lip twitch. There was only one difference between the tests. The first time, she had been sitting inside a dry chamber, and the second, she had been immersed in water.

Two months before Jermyn's seizure, Charioteer Brownrigg had died in shallow water while breathing pure oxygen without a clear cause of death. Three weeks before Jermyn's seizure, Chief Petty Officer and engine cleaner Redmond Vaughan Rowkins of the Royal Navy had volunteered for a chamber test during which he crawled out of the water and collapsed after 9.75 minutes of oxygen at seventy feet, looking "rather blue" in appearance, teeth clenched and frothing at the mouth.

Almost everyone who knew about Rowkins's bad dive had assumed he'd had CO_2 poisoning. At least at the time it happened. The Admiralty analyzed his breathing gas and disagreed even when Haldane calculated from the samples that it did not contain problematic levels of carbon dioxide. As for Charioteer Brownrigg, they assumed he had simply drowned. Both incidents had logical alternative explanations. But Elizabeth Jermyn was a seasoned chamber veteran who had proved she could tolerate oxygen under pressure. She did not drown; she did not lose her mouthpiece; she did not suck down any water. Before her seizure, she had already done dives in air that were deeper, longer, bigger, badder, worse.

In the world of science, Elizabeth Jermyn was a controlled data point, and controlled data points are the best way toward the truth. Meaning that the only variable that had changed between her multiple dives was that she went underwater. The water, therefore, had to be the problem.

It was. For some reason, the symptoms of oxygen toxicity occur harder, faster, and worse when someone is underwater than when in air. Inside hyperbaric chambers in the dry, people breathe oxygen at pressures over double the modern limits for doing so underwater, and they do it with a perfect safety record. Something about immersing ourselves in water changes us.

Hundreds, if not thousands, of researchers have tried and failed to figure out the reason why over decades of painstaking work. It is a principle that has dictated the mission planning for every military diver since Elizabeth Jermyn's Very Bad Day in 1942. Especially for those divers who breathe pure oxygen.

As of the dawn of 1942, the United States Navy gave a limit of two hours for oxygen at fifty feet, and the Royal Navy copied it. Jermyn had lasted ten minutes.

So few divers had used oxygen for free-swimming diving that, luckily, those air-based guidelines had gone largely untested in water. As recently as six weeks before Jermyn's seizure, Haldane had written: "I have also heard at last from Lieutenant Commander Behnke in America. It is fairly clear from his letter that half an hour [of oxygen] at ninety feet would be safe for most people at any rate." Jermyn's seizure proved they were wrong. All of them.

But for the well-placed German bomb that had forced Siebe Gorman to relocate, the military divers of World War II would have been sent out into the cold, deep ocean with oxygen guidelines built around tolerance levels in air—guidelines that would have killed them.

Yet the name of the one scientist who discovered the truth—the one person who volunteered and experienced the issue in a controlled fashion that pinpointed the problem for the first time—has previously gone unmentioned in connection to the widely studied phenomenon, even in the dustiest, most archaic references. Joan Elizabeth Jermyn. Betty.

Elizabeth Jermyn upended the world of Allied diving. The Royal Navy had been paying attention to the UCL group before, but now they sat bolt upright in terror.

———

Some of the officers training the aspiring Charioteers had written a letter asking for help after Rowkins's saliva-heavy accident in the tank. They wanted to dive riding the Chariots while breathing oxygen to a depth of fifty feet to stay safe from scouts and deterrent bombs. They had previously thought that fifty feet would be fine for divers, based on the official guideline of two hours, but after the accidents of Brownrigg and Rowkins they had begun to question their assumptions.

"In conjunction with [Lucan] Pratt I have made out a memorandum (attached) of what we urgently require an answer to," wrote Commander Sladen of the Royal Navy, the inventor of the underwater Sladen suit being blasted by Cameron Wright and intended for use by the Charioteers. Officers Sladen and Pratt said they would "be most grateful if Haldane and his team could work on it for us . . . the sooner we know the safe time limit the better." They wanted the UCL lab to tell them if they could start training the Charioteers more aggressively underwater.

Evaluating the safety of their equipment at fifty feet had been Jermyn's task. Clearly, it had not gone well.

The pressure for results was coming not only from the Charioteers but also from those who would dive out of the X-craft, the miniature submarines. In addition, Rear Admiral Edward de Faye Renouf invited J. B. S. Haldane to his offices to discuss Haldane's suggestion about using the DSEA breathing apparatuses for offensive diving. Renouf held the intriguingly unspecific title "director of special weapons," and by the end of the war, he would earn a reputation for his creativity in inventing novel military tactics. He asked Haldane questions about "at what depth" and "for how long" the DSEAs could be used by divers. The day of that meeting, Haldane, Spurway, and Case all climbed into a chamber for more experiments. They were trying to get the answers.

The military demand for their science was growing right at the moment that their oxygen science was upended by Jermyn's revelation. Their old data points, in air, were no longer useful and would have to be repeated in water. Also, with increased demand comes increased attention, and along with increased attention follow those who are motivated by fame or glory—people who might make choices that benefit themselves rather than the mission. Little did the UCL group know, but in their efforts to get more data, they were about to train the very person who would one day incite their eviction from their own field.

On June 18, 1942, Surgeon Lieutenant Kenneth "Ken" Donald strode into the Siebe Gorman factory with twelve Royal Navy divers in tow. Donald was tall and trim with short brown hair. He looked sharp in his Royal Navy uniform, in the handsome yet generic way that could have made him a darling for throwaway stock photos and forgettable billboards. Deep down, he knew he had no real experience with dive medicine. He admitted it himself. He had never been inside a chamber. He had no recorded diving experience. However, he was determined; he was cocky; and he was not about to let a minor obstacle like lack of knowledge stop him.

He later claimed that, on behalf of the Royal Navy, he "took over" the entire research wing of Siebe Gorman in the summer of 1942. When he arrived, according to him, he simply declared the test lab had the new name of the Admiralty Experimental Diving Unit, AEDU. He made an official stamp for it and everything. He also declared he was its sole leader. From there, he single-handedly led the way to underwater victory and glory, at least according to his own retelling.

"There I was, a Surgeon Lieutenant, aged 30, with no research and little hyperbaric experience, with a not inconsiderable research unit on my hands. I was, by some strange and splendid accident, in no way responsible to any senior medical or scientific person or committee," he claimed of that summer when he was a low-ranking junior officer. In reality, he did have a commanding officer named William O. Shelford, who was both well-informed and experienced in submarines and under-water life, but Shelford did not seem to factor into Donald's memory of his own self-image. Shelford knew what was happening with the Chari-ots and midget submarines, and he was the one in charge of the mili-tary side.

By the time Donald walked through the door at Siebe Gorman, the UCL group had conducted on themselves 180 separate, meticulously documented experiments in the chambers. In the world of human test data, that number is massive. They had spent eighty-two total days run-ning tests while locked inside the metal tubes, in addition to some un-counted number more for setup and equipment checks. Donald made one brief, loose reference to the fact that at the start of "his" experiments, there had already been some level of expertise swirling about the fac-tory, but he did it without attribution: "expert technical help was imme-diately available."

The UCL group knew they needed more bodies to get enough data points. They needed the Royal Navy to get those bodies. Donald margin-alized them in return.

> A further piece of extreme good fortune was the discovery
> that JBS Haldane (fils) was working in one of the dry cham-
> bers for two days a week completing a contract to investigate
> the physiological factors relevant to the submarine HMS *The-*
> *tis* disaster in 1939. During the next year, while Haldane was
> completing his programme, he gave most valuable advice.

But the UCL lab notebooks show the truth. Martin Case ran the chamber dives for Ken Donald and the Royal Navy divers, at least until they learned how to do it on their own. J. B. S. Haldane analyzed their gas samples. The first several rounds of experimental notes for the new AEDU's trials are recorded in Case's handwriting. Helen Spurway, J. B. S. Haldane, and Hans Kalmus at a minimum, if not all the others, too, acted as their dive tenders. During Donald's first trial on himself, when he finally climbed into a tank a week after the rest of the Royal Navy divers had gone, it was Helen Spurway who looked into his eyes to monitor him for twitches.

The UCL group had been testing inside the chambers for at least thirty of the forty-seven calendar days between when Siebe Gorman reopened its doors and when Kenneth Donald showed up, an average of between four and five days per week. They had run more than ninety experiments in that time span alone, in both the wet and dry chambers, not counting their previous work at the old location or the time it took to set up experiments or analyze data. They were in no way—as Donald dismissively described them—puttering about a maximum of two days a week and wrapping up an outdated, unclassified project about the *Thetis*.

J. B. S. Haldane, Martin Case, Jim Rendel, Elizabeth Jermyn, Hans Kalmus, Ursula Philip, and Helen Spurway taught Kenneth Donald and the Royal Navy divers how the chamber setup worked, and then in late June 1942, they helped the first twelve of the military personnel dive at fifty feet in the wet pot to build up the safety data set for the Charioteers. The UCL group seemed to welcome the additional bodies. However, because of Donald's methods, those bodies came with ethical complications.

"These divers were operational personnel and not experimental divers," Donald wrote. "They considered, commendably, although quite incorrectly, that oxygen at toxic pressure was another trial to be undergone before selection was completed. Symptoms were therefore minimal." The word "selection" refers to the military process of selecting personnel for specific operating specialties. These divers were the hopefuls to be the first straddling the Chariots and swimming out from inside the midget submarines, so they pushed themselves to the limit, past the limit, beyond any warning signs, without reporting any warning signs, well into the danger zone, because Donald had not bothered to inform them what the experiments were actually for. His comment that "Symptoms were therefore minimal" is a wry joke that he thought the divers were not reporting symptoms because they did not want to be kicked out of the training program. The divers thought if they confessed signs or symptoms of oxygen toxicity, then they would not be allowed to be divers

anymore. In other words, the divers could not have provided true consent for the experiments.

They all climbed into the tank. Five of the first twelve Royal Navy
divers seized at fifty feet within less than thirty minutes. Those dozen
tests confirmed what Jermyn had stumbled upon. The Royal Navy sent
in more divers, one after another, until the wooden planks of the flooring
above the test chamber became a water-drizzled assembly line of human data, with a crowd waiting to be fed one at a time into the square-
cut hole above the vertical chamber entrance. "On occasions, the next
diver stepped over the last casualty [recovering on the platform] to take
his turn," Donald recorded. Out of a hundred total human in-water tests
by the Royal Navy divers at the Charioteers' requested depth of fifty feet,
twenty-six dives ended in seizures after less than thirty minutes in the
water. The earlier guess of two hours as a safe limit had been disjointed
from reality.

The Admiralty needed a new plan. One that would avoid seizures, but
one that would allow divers to bolt toward the protection of the surface
quickly should explosives start drifting down around them. Their current breathing-gas choices to dive with were either oxygen or air. Oxygen
was suicide for a full operation at fifty feet, but so was air, given the goal
of a quick ascent. Because air, after all, is 79 percent nitrogen, and even
when divers were faced with explosives, the lessons of the Brooklyn
Bridge could not be forgotten. The Charioteers were trapped between a
seizure and a bomb.

———

The UCL group had dealt with decompression sickness before, but in the
chambers, it had always been mild. A sore shoulder here, an itchy forearm there. Usually those who got it were the tenders—the people who
hunched inside the chambers and stared into test subjects' faces, watching for twitches or seizures or loss of consciousness. They didn't worry
about it. Their symptoms almost always went away by the next day after
a good night's sleep and a heat pack or two.

Decompression sickness, however, has another, more nefarious manifestation. The minor nigglings of pain, the throbbing yet self-resolving

joint troubles, the itchy, rashy skin—they all get categorized under the more benign DCS label "Type I." But there is also a Type II, and it is far, far worse.

If divers spring to the surface too quickly after deep, long, or even cold dives, bubbles invade even more tissue types, and they accumulate, blocking blood flow, causing cells to scream in pain as they die off by the legion from this weird, invisible, internal kind of strangulation. In Type II DCS, the bubbles lodge in body parts that don't always grow back. In Type II DCS, the rampaging bubbles block the blood flow to choke out nerves. The spinal cord. The brain. Fractured vertebrae like Jermyn's are not the only way to disrupt the delicate white thoroughfares of nerves connecting the brain to the limbs, and itching and death are not the only possible outcomes.

Type II DCS often starts as patches of numbness on the skin as the first, tiniest bubbles begin to lodge. Small, delicate, possibly undetected portions of the body's outer shell go numb. A few inches of forearm on the left. Maybe the area outside the kneecap on the right. Perhaps a stray oval across the back. Sometimes it stops there. Maybe the deadened, choked nerves regrow slowly, if they are the peripheral type that can, and over a time scale of months, everything goes back to normal. Usually, however, it does not. Usually the DCS spreads.

One case report decribed a diver known as Miss C, who was experienced in the aquatic world and had made literally thousands of dives. On a warm, salty afternoon off the coast of Indonesia, she surfaced after her second dive of the day. When she looked up from bobbing on the surface to hand a member of the boat crew her hefty diving weights, she felt a slight twinge in her neck. She made it on board, but over the next twenty minutes, her legs went numb. She lost control of her right arm. Her symptoms continued to escalate. She tried to walk. She was clumsy, the internal distress of her nerves made evident through her uncontrolled and awkward attempts to move with her usual gait. There was blood and bubbling in her spinal cord, in the portion under the forward arc of her neck that controls breathing and the motion of every body part below it. She could not control her limbs.

In another case, a twenty-nine-year-old diving instructor was working his second job as a baggage handler on a night shift in New Zealand. He had been diving that day, and called it "lovely." Then while he was hauling luggage, the region of his brain behind his left eye exploded with pain, his vision blurred, and his speech slurred as his tongue rebelled against voluntary control. His left calf bloomed with internal fire, and he collapsed to the ground, shaking and hyperventilating. By the time he reached a hospital, he was conscious again, but patches and swaths of his arms and legs had gone numb to touches and pinpricks. His doctors never reached a consensus on whether he had Type II DCS or another, unexplained neurological disease.

A fifty-year-old man diving off the island of Guernsey, in the middle of the English Channel, felt chest pain and had double vision, vertigo, nausea, vomiting, clumsiness, and memory loss minutes after finishing his dive and reaching the surface. He slipped into a coma. Upon recovering some semblance of consciousness, he could not control his limbs or the motion of his eyes. He struggled to speak. He had amnesia. According to his case report, "he was cold, pale and in peripheral circulatory failure."

Terrifying reports of suspected Type II decompression sickness have been published for divers breathing air from as shallow as thirty feet, only nine meters, less than the theoretical (and likely) limit of possibility, even less than the goal depths of the Charioteers and X-craft crews, and shallow enough to glimpse the midnight pallor of the moon radiating down through clear ocean waters. The traumatic death of nervous tissues means that the corresponding body parts will feel a crippling pain that cannot be numbed by any drug. Sometimes, after treatment by recompression in a hyperbaric chamber to let the bubbles reabsorb more slowly, *sometimes* these wounds can be healed. Other times, the damage and the pain are permanent. Miss C survived, but only after long hours and multiple straight days of treatment in a chamber. The New Zealand instructor got better, too. The Guernsey man lived for another few years but without the use of his legs.

Type II decompression sickness was not a risk worth taking for military divers operating in hostile foreign waters far from informed physicians, far from chambers to recompress them. Such divers might have been forced to choose between temporary paralysis from an explosion or permanent paralysis from rocketing to the surface. The only conclusion for the military to draw was that neither air nor oxygen was a realistic choice.

Perhaps both would work. It was not a new idea to mix air and oxygen together and thus finagle the better bits of both, but it had not been thoroughly tested before. Someone would need to make sure that the two gases combined as simply as hoped, that the oxygen would bring with it no more than its calculated, proportionate fraction of oxygen risk, and that the subtracted chunk of nitrogen would take away no less DCS risk than it should. Someone would need to do the math and hand proof to the Admiralty before they could give such mixtures and their life-saving instructions for use to future combat divers.

On July 7, 1942, Martin Case squirmed into the closely fitted black rubber Sladen suit of the Charioteers and X-craft divers, although the suit was now rebranded the Shallow Water Diving Dress or SWD. Others nicknamed the suit the "clammy death." Each of Case's shoes was fitted with a lead plate weighing five pounds; a corrugated tube led from his mouth to his oxygen breathing apparatus; and a shatter-resistant viewing plate provided him a panoramic view of his blank-walled chamber. At a steady, preset rate, a shallow trickle of oxygen entered his breathing loop from a high-pressure bottle that had been liberated from the cabin of a downed Luftwaffe fighter plane and repurposed. The German gas bottles had proved superior to any being made by the Brits, and collecting these aluminum trophies was a silver lining to living among the bombings.

For experiment number 205 Case spent thirty minutes breathing at ninety feet. Later gas analysis showed he was breathing 71.7 percent oxygen, a custom mixture of pure oxygen and air. He never felt any symptoms; he never had a seizure. He did not get decompression sickness during the intentionally quick ascent. The absence of both was a good surprise; the theory of mixing the gases seemed to match the experimental results. The plan just might work.

Edwin Martin Case, however, would never finish the test series, because he was going off to war. After wrapping up his final few experiments, he traded his diving suit for a Royal Navy Volunteer Reserve uniform and became acting temporary lieutenant commander of a special branch. He would still be doing science, but with epaulets on. His brothers, Robert Alfred Martin Case and Ralph Martin Case, were already engaged in the same force, so the three wild brothers were back together in spirit, making good trouble with their "extraordinary sense." Helen Spurway took Case's place as the coleader of the chamber investigations. Haldane worried about her slight build and about who could control him "during a fit."

Elizabeth Jermyn climbed into the chambers several more times after her seizure but before her doctor forbade her to do so. She spent a few hours at sixty-six feet in the dry, and then hours more at fifty feet in the dry, as further data points for comparison against her wretched ten-minute dive in the wet. On the advice of her doctor, she and Nari Bamji finally took a holiday for her back to recover, but only after she refilled Haldane's desk-drawer stash of chocolates and peppermints.

The brothers Case in their uniforms: Ralph and Martin, with Pat Morris in center.

CHAPTER 13

The Inevitable Solution

The botched raid on Dieppe happened while Jermyn was on holiday. As she tried to heal, her broken back throbbing, its chronic ache intruding on her sleep, some 6,086 troops mustered in the waxing August dawn and tried to land on unfamiliar, Nazi-occupied French beaches code-named "Orange," "Yellow," "Red," "White," "Green," and "Blue."

By nightfall on August 19, 1942, a total of 3,367 Allies had been killed, wounded, or captured. Mostly Canadian. Those killed in the brief raid fell at an astronomical rate of one every thirty-five seconds, roughly equal to the pace of death at the Battle of Gettysburg. Except at Gettysburg the rate was for both sides' dead combined.

Captain Patrick Anthony Porteous and his Group II of F Troop, British Number 4 Commando, had relied on luck for the German-held lighthouse to stay lit and guide their journey into the correct spot onshore. They used the blissfully still luminated lighthouse to stay on the route previously cleared by minesweepers, whose tight paths gave little grace to any who deviated. Luck had also been with their counterparts in Group I when the rain washed clear the dirt covering the mines in their planned route up the gulley—the route they needed to climb to reach their target battery.

But luck abandoned the rest. It did not protect them from the nests of machine guns in the cliffs. It did not tell them where onshore there might be mines or other types of hidden traps or gunners. Luck did not tell them in advance that the beaches would be stones rather than sand and that those smooth, unremarkable gray rocks would become the primary nemeses for their tanks, above any threat presented by more animate enemies. Luck had not warned them about sharpened steel obstacles underwater.

Luck had given them unprepared commanders. Commanders who would marginalize the difficulty of running across a bare, exposed beach under a barrage of crossfire from determined enemies. Commanders who would stay safe so far offshore that they would not even be able to see the battle, much less redirect it or make any meager decision that might stem the rising flood of dead. Commanders who were so ignorant that they kept sending boat after boat filled with human sacrifices rather than concocting an espionage-based plan to recover the information they wanted.

It was while Elizabeth Jermyn waited for the cells of her bones to re-knit themselves across their cracks that a British physician named Cyril Joseph Thornberry bobbed in the waters off the beaches of Dieppe, his ship freshly sunk. His life jacket held him upright in the water, and in those moments, he tried to suppress within his own mind the horrifying knowledge that his vertical position "was a bad thing if there were underwater explosions." He had seen the case reports of the underwater blast victims, but he did not know exactly how much risk he was truly in.

Soon after the raid, Doc Thornberry was scooped out of the water by another vessel, and he started hastily injecting morphine into the arms of the "vast numbers of Canadian casualties" who had been "virtually thrown on board" the same ship. There were so many of them. Too many for anything more than "glorified first aid" during the interminable wait to reach British shores. Yet at the same time there were too few, because others lay immobilized on that calming but hostile shore.

George Cook woke up thirsty. He blinked his eyes open to meet the cheerful sun shining down upon his face. When he tried to get up and look around, however, he realized he couldn't move. But he could hear the drone coming from the thick black cloud of flies intoxicated by the blanket of blood that soaked his paralyzed body.

Seeing that he was awake, three Wehrmacht soldiers made their way across the quieted Dieppe beach toward him. One of the Nazis pressed the slim, tapered edge of a bayonet against Cook's throat while a second asked him in clear English: "Are you wounded?" followed by "Can you move?"

"Yes," Cook responded to the first question, and then, "No." Bending over his helpless body, one of the Germans rifled through Cook's pockets and pulled out his cigarettes, offering the injured man one of his own smokes before asking what they could bring him. He was, after all, now their prisoner. Cook requested water. The English speaker told him to stay still, and they walked away, assuring him they would be back.

Two Frenchwomen from the town had ventured onto the battleground once the sounds of war subsided, and they plucked their way among the barbed wire and the remains of fallen men, looking for survivors. Presumably directed toward Cook by the departing Germans, the elder of the two rushed over to him and lovingly cradled his bloodied head in her lap while the younger prepared to feed him water from a cup. But Cook's jaw was almost entirely missing. Without bone or skin to hold the fluid in, the water sloshed down the front of his neck. His shoulder had been shattered, too. The pattern of his injuries suggested a machine gun had traced its brutal arc across his upper body.

Unable to make the cup work, the young woman ran off and returned instead with a glass of wine and a spoon. She carefully transferred the wine one spoonful at a time directly into the open hole of Cook's exposed throat. His thirst sated, he once again slipped into blissful unconsciousness.

Cook would wake up more permanently in another five weeks to find the remains of his jaw wired shut and most of his upper body in a cast. Cook, of the British Number 4 Commando, survived the failed amphibious raid on the beaches of Dieppe. Over the ensuing thirteen months, German doctors in Paris would cobble together new pieces of bone to reassemble his face, which would later be polished through another year's worth of untallied operations and skin grafts by British doctors back home. He would spend the rest of the war as a prisoner, as would Laurens Pals and a small handful of others who had somehow dodged the bullets.

"And what did they achieve?" Captain Porteous later wondered about the raid. "Absolutely nothing."

The Allies needed new plans based on solid, robust information. The UCL group's grueling efforts had been helpful in gaining knowledge about the human body and survival underwater, and it was time to apply that knowledge. Preparation must take over where luck might fail.

On the same day as the raid on Dieppe, Royal Navy diver Donnelly, no known first name, had a convulsion in the wet pot at Siebe Gorman, at a depth of fifty feet.

Across the European continent and beyond the Mediterranean Sea, Lieutenant Commander Nigel Clogstoun-Willmott also buzzed into action following the news of Dieppe. He knew scouting was the way forward for beach landings. He had for a long time. In the wake of Dieppe, Clogstoun-Willmott finally got permission to start collecting, equipping, and training a handpicked "party of officers and ratings" to develop better techniques for beach scouting.

Clogstoun-Willmott was tall and trim, with large, strong hands and an overstated facial bone structure with a firm, thin mouth. He had been working on the beach-scouting idea for about a year without official permissions with his partner in machinations, Lieutenant Roger Courtney. Before the war, Roger Courtney and his wife, Dorris,* had spent their honeymoon paddling a collapsible type of canoe, a folding boat or "folbot," named *Buttercup* down the River Danube and through several mainland countries to end up in the Black Sea. Courtney still had the

---

*Dorris E. Courtney served seven years as an agent with the famed Special Operations Executive (SOE).

Buttercup, and such collapsible canoes were the centerpiece of his and Clogstoun-Willmott's first plan.

In Egypt, in the spring of 1941, after practice on friendly shores, Clogstoun-Willmott and Courtney loaded themselves and their collapsible canoe onto the submarine HMS *Triumph*. With them was Lance Corporal James Sherwood, who had been picked from the volunteers because he had no wife to widow and no children to orphan. The submarine carried the trio to a target for a future Allied assault: the island of Rhodes off the southern coast of Turkey. On a "pitch dark night" with no moon in the sky, Courtney and Clogstoun-Willmott balanced their tiny civilian canoe on the horizontal foreplane of the surfaced submarine. Clinging with one hand each onto a taut cable between the hull and the foreplane, the duo managed to balance their way through the darkness and into their seated positions in the folbot. As they pushed off into the water, they scraped the keel of the canoe along the submarine's smooth wing. They had roughly a mile to paddle to shore.

Once close enough, they took turns plunging over the sides of the small boat and swimming the rest of the way through the warm Aegean Sea toward dry land, then clambering quietly up the beaches. They made sketches of the terrain, noted the times and levels of the tides, and scouted navigational marks that could be used to direct a landing. During the daylight hours, they hunkered back on the *Triumph*, compiling what they had learned. Sherwood helped maintain an infrared beacon to guide them back for each return trip. That beacon was critical to finding their way back to the submarine, which was barely visible above the rim of the large, dark skyline. Each day until the scouting was finished, Sherwood patched the folbot for the next night's excursion.

The invasion of the island of Rhodes was ultimately abandoned as a plan. However, because of that clandestine beach mission in 1941, "the true value of such a reconnaissance was proved, and the seeds of an idea were sown." When the Dieppe raid happened in August 1942, there were people like Clogstoun-Willmott, Courtney, and Sherwood ready and waiting to do more such work, ready and waiting with ideas for beach scouting.

By mid-September 1942, less than a month later, Clogstoun-Willmott had mustered beach scouts at the British command post in Portsmouth.

They called themselves "Party Koodoo-Inhuman." "They were hastily and partially trained, equipped with very improvised special stores and craft and were finally sent out to Gibraltar" for their first mission. Once they were at Gibraltar, amid a scattered flurry of improvised equipment, self-contradictory orders, and noncommittal officers, it was decided that the teams were not ready to go onto enemy shores. They used their mapping techniques to survey the North African beaches intended for the next landing, but only through the limited lens of an offshore submarine's periscope.

On November 8, 1942, Allied troops landed on the northern shores of Morocco, across the narrow strait from Gibraltar. The first troops were surprised to find the terrain was sandstone, not sand. That had not been visible through the periscope. The assault lasted eight brutal days.

However, that time, before landing, despite their limitations, Party Koodoo-Inhuman had mapped out the landing locations enough to deploy infrared marker beacons. The beacons led the landing craft to the correct spots, no lighthouses required. No luck required. It was a step forward, and leadership noticed. The officers at the top of the planning pyramid wanted to invade the island of Sicily soon and use it as a launching point to invade the nearby Italian mainland. They wanted to fell Mussolini, to cut off one of Hitler's legs by taking the Italian boot, and both steps in that process would require beach landings.

"Beach reconnaissance should become an integral part of future amphibious operations," the Admiralty declared in December, opening up to Clogstoun-Willmott's idea. "The Chiefs of Staffs ruled that there were to be ten [reconnaissance] units formed, equipped and trained to be able to undertake this work." They were going to expand Party Koodoo-Inhuman, eventually shortened to Party Inhuman, with Clogstoun-Willmott at the helm. The new units would be named COPP, or Combined Operations Pilotage Parties. But—if anyone asked—the acronym officially stood for Commanding Officer Police Patrol. This alternative name was a pretense that the operators in those units were simply law enforcement patrolling the shores, and that was their cover story if caught. Clogstoun-Willmott was in charge of the party organizations and supplies, and would be the leader of COPP 1.

The COPP teams recruited professors to teach them to collect beach samples and how to analyze the soil and sand. They were going to practice to death the skills of navigation, scouting, marking, and analysis. However, they still needed better ways to get onshore, especially if they would be swimming through the frigid gray chop of the wintry Atlantic instead of in the balmy Mediterranean. The water on the northern shores of France was cold enough that the same methods of getting ashore would not transfer directly. So Clogstoun-Willmott and his staff called Siebe Gorman to ask for some new aquatic toys.

In the same month that the COPP teams were scrambling together, in December 1942, the UCL lab members completed experiments 294 through 317 on themselves. They tested the new idea of oxygen-rich blends in the rebreathers, and they probed the limits of where the novel mixtures would give them DCS.

The Charioteers, in their parallel and equally hurried work, were approaching combat readiness on Churchill's demanding schedule. They were ready to try out their toys for the first time outside of practice, and they had the perfect target in mind. The German battleship *Tirpitz* was a brand-shiny-new gem in the diadem of Hitler's fleet, and it was the sister ship to the infamous *Bismarck*. *Tirpitz* prowled the North Atlantic with its four turrets and eight massive guns. Rumors and observations about this beast of war had been a subject of fixation in Allied reports since its completion a year earlier. The Allies wanted it sunk.

On October 26, 1942, at 12:14 in the afternoon, a stolen fishing trawler labeled as the MFV *Arthur* puttered out of the harbor of Lunna Voe on the tiny Scottish island of Shetland, headed straight for Norway and the *Tirpitz*. On board were ten men. Four were native Norwegians, four were British Charioteers, and two were dressers, attendants specially trained to help the Charioteers don their bulky waterproof diving suits and rebreathers. The mission was dubbed Operation Title.

The cargo area below the deck of the fake *Arthur* had a trapdoor to hide the Brits if needed; that secret door could be covered by bags of peat when passing through German inspection points. The Norwegians would pose as fishermen hauling said peat, and they wore civilian fishing clothes but carried special waterproof maps printed on silk. Tied beneath the trawler, however, below the depth at which they could be spotted from the surface of the murky ocean, were two Chariots on long ropes. The Chariots that would be used to attack the *Tirpitz*.

The ruse worked. Once they reached Norway and its harbors, with their forged papers and layers of obfuscating peat, the phony *Arthur* and its crew slipped past a German guard who climbed inside the hold and saw nothing awry during his inspection.

After that, however, luck once again abandoned the Allies. A swift turn of the weather sent the crew bucking with the huge waves of the unforgiving Norwegian waters. In between two particularly notable crests, the crew on deck felt a menacing snap shudder through the hull. Once they found shelter from the fearsome wind and surface chop, Able Seaman Robert Evans donned his diving suit and oxygen rebreather and swam beneath the *Arthur* to check on the Chariots.

They were gone. Both Chariots had snapped free. "The towing wires were still attached to the boat and at the end of the wires were the shackles." At the end of the shackles, there were some lead weights. Then there was nothing. The Chariots, the entire point of the mission, were somewhere at the bottom of a fjord.

Meanwhile, as the group tried to formulate a new plan, the fishing boat MFV *Arthur* gave up on its own battered life. For the third time that mission, the overtasked little engine on the mislabeled boat shut down completely, and no amount of elbow grease from the Norwegian sailors or the English divers could resuscitate it. The Chariots were gone. The *Arthur* was dead.

The ten members of Operation Title scuttled the fake *Arthur* poorly in a shallow watery grave and, splitting into two parties of five, set off to hike through wintry, hostile-held Norway to neutral Sweden for rescue. But the Germans found and raised their boat and pieced together that it had been a failed attack after locating the confused owner of the actual *Arthur*, which was still afloat elsewhere. Nine of the ten crew made it across the Swedish border, but they starved on the way. They lost toes and other pieces of flesh to frostbite from the thigh-deep Nordic snow and frigid river crossings. Charioteer Robert Evans was shot by a border policeman and then nursed back to health so that he could be executed by the Gestapo. The German harbor guard who had waved the fake *Arthur* through the harbor entrance died right after the German inquest into the incident. His death was a suicide. Allegedly.

The Chariots, thus far, were off to a dark and rocky start.

———

The debacle of Operation Title also brought to the forefront that the Brits and Americans had a difference in terminology. The British referred to the first day of any operation as "D.1" while the Americans used "D+1" for the day *after* the start. The subtle difference caused a "muddle" of some kind during Operation Title, and was partly blamed for the botched evacuation and forced overland trek. The trek, should it have proved to be necessary, had been planned to take advantage of a shared network of local Scandinavian agents, but it was foiled by incorrect timing.

To prevent the same confusion in the future, the British decided to switch to the American practice. The first day of any major military operation would henceforth be referred to by all Allied countries as "Day D."

———

The same day as the departure of the fake MFV *Arthur* out of Lunna Voe, Scotland, for Operation Title, Hans Kalmus's parents, Arnošt Kalmus and Elsa Kalmusová, were forced onto a train. It was the second train they had been ordered to board, and they shuffled alongside 1,865 other forcibly dehumanized prisoners, mostly Jewish. They had been confined in Prague until two days earlier, when the first transport train had brought them to their current location in the prison camp at Terezín, Czechoslovakia. The second train, the one they now boarded, was destined for Auschwitz.

———

Four days later, J. B. S. Haldane and Helen Spurway took turns in the chamber, traveling to a depth of 170 feet on an oxygen-rich gas mixture to see if they had seizures. Ursula Philip and Hans Kalmus, also still active divers, were risking their lives to oppose a country that would have taken them.

Mixed results and nascent status notwithstanding, the Admiralty decided to plow forward with both the Chariots and the COPP teams to see if they could get both or either to work for beach scouting.

The new Siebe Gorman factory was more than water tanks. On the same industrial block as the experimental diving laboratory, a neat, organized row of sharply roofed buildings housed manufacturing facilities capable of producing almost anything. One building contained a battery of milling machines. Another, drills. Lathes were lined up in a tidy row, like metal soldiers, in the lathe room. And one entire massive, endless warehouse was filled with field after field of tables topped with black sewing machines, as if black sewing machines had replaced ears of corn in a fertile field.

In wartime, most of this equipment was staffed by women, from the grinders to the drills. The Royal Navy personnel diving for the oxygen tests seemed to have thought of that gendered staffing as a perk to reward their test participation, with one diver recording after his first day of trials that "a couple of the lads soon fixed dates for themselves that night with a couple of girls in the factory."

More practically for the war effort, though, the factory also had a shop for making waterproof diving suits, referred to as diving dresses. Inside the white-painted walls and beneath the high, peaked roof of the "dress making shop" of Siebe Gorman, workers laid out patterns and created custom protective suits for the divers of the world. Thick canvas, sturdy rubber, and hardy rope were their tools. Quickly after the Admiralty's decision that beach reconnaissance was the way for future landings, the employees of Siebe Gorman rushed to cut patterns for a new thermally protective suit for swimming on the surface. Courtney and Clogstoun-Willmott had improvised their original thermal protection by soaking clothing in grease to waterproof it, which was a messy and imperfect solution. This suit would be better.

The new suits would be custom fitted to each handpicked, heavily trained COPP swimmer. They were bulky in appearance because of the insulation stuffed underneath them. Once fitted, they should not restrict the COPP personnel in their ability to swim or paddle a canoe. The suits had rubber seals at the wrists and around the perimeter of the face; the seals were fitted tight to keep out water. A waterproof diving slate could be tied around the left wrist. At least eleven cargo pockets of various sizes adorned the front and sides alone, allowing the COPPists to carry whatever mapping and charting equipment they needed. In addition, a special slot-shaped pocket held a massive knife. The workers glued coils of thick rope onto the shoe soles to provide grip on rocks and shores.

A COPP "surface swimsuit," with the hood pushed back off the head. *(DEFE 2/2084)*

The first models were slapped together hastily. The COPPists described them as "inefficient and uncomfortable but possible." At least some of the swimmers chose to pee in their bulky suits rather than deal with putting them on and taking them off again. In the same month the first new suits were pieced together, December 1942, Allied troops plucked from a captured *Maiale* operator one of the Italian Navy's strange inventions: flat blades that could be strapped around the heels, turning a diver's feet into fins. Slowly but surely, the Allies were turning their sailors into frogs.

On the dark morning of January 3, 1943, five two-person teams of divers straddling their five British Chariots, wearing new Siebe Gorman waterproof diving suits and breathing through oxygen rebreathers, worked their way quietly toward the Axis-held Mediterranean harbor of Palermo, for a mission dubbed Operation Principal. They knew they had to stay shallow or risk oxygen seizures, but they did not yet know how shallow. They were using the science to date, but the science to date was still not complete. Each team carried limpet mines to attach to the hulls of enemy ships. They had launched the Chariots out of host submarines because, even in warm waters, the limitations on battery life and the use of oxygen meant they were able to transit for only short distances. One of the host submarines was the formerly doomed HMS *Thetis*, refilled with air, raised from the seabed, renamed the HMS *Thunderbolt*, and scrubbed until its insides shone free of rotten flesh.

Host submarine *P.311* had been caught and blown up by an enemy patrol before the Chariots were launched. It had carried three Chariots; all the Charioteers died, in addition to the crew of the submarine. Of the five remaining Chariot teams launched from the two surviving subs now lurking toward the Palermo harbor, three lost control and scuttled their rides without placing charges. Two made it into the harbor. Both teams placed bombs, but one forgot to remove the pin from their limpet mine to activate it. In total, the eight teams of sixteen divers placed one successful bomb and sank only one ship. For that ship, fifteen died, and two were taken prisoner.

The results from the British Chariots remained, at best, mixed. The Italians had made such human torpedoes work at Alexandria, but they had started on the idea seven years before the Allies had. In the underwater war, the Allies were behind.

A few footsteps away from where the new suits were being made, J. B. S. Haldane swallowed ten milligrams of a bitter white drug before climbing into the horizontal dry chamber. The Benzedrine was similar in its chemical structure to pure powdered adrenaline, and it was the newest trend to keep military pilots, soldiers, and sailors awake for days on end. In tablet form, it had been tossed about as a miracle panacea by physicians since its discovery a few decades earlier, and British doctors zealously dosed and arguably overdosed patients with the powerful, addictive amphetamine to treat everything from narcolepsy to schizophrenia to radiation sickness to menstrual cramps. The German Army was consuming it in bulk.

According to Norwegian wireless operator R. Strand, after Operation Title's failure, the harrowing multiday journey on foot across Norway to Sweden through snow, starvation, and frostbite had been fueled by Benzedrine pills. The drug could sometimes induce a paranoid psychosis akin to schizophrenia, but it would keep anyone awake. It would keep them moving. Later that same month, November 1942, the Royal Air Force flipped from a total ban to approval of the drug for pilots, because stable brain chemistry is not relevant if someone is dead from other causes. "He got a good night's sleep before the Nazis caught him" has never been engraved as a compliment in any soldier's epitaph.

However, most critically to the UCL lab, the drug's effects were poorly characterized, especially if it was taken before diving. The impact it could have on Charioteers, midget submariners, and COPP swimmers was completely unknown. So, as usual, the lab went first.

After dosing himself, amphetamine-fueled J. B. S. Haldane sat in the white metal tube and breathed a mixture of 60 percent oxygen and 40 percent nitrogen through the square breathing bag of the Siebe Gorman Salvus rebreather strapped to his chest. Inside the dry, hot air of the chamber, his exhaled breath further warmed the chemical canister that scrubbed the carbon dioxide out of his exhalations. The canister felt hot to the touch from the chemical reaction happening inside. Spurway stared at Haldane intently, feeling slightly drunk but not problematically so from the effects of pressure at 110 feet. Haldane made it through thirty minutes on the bottom, and had itching and pain in his left thumb and ear on ascent. However, the times were similar to his normal tolerance times; the potent Benzedrine did not seem to increase his susceptibility to oxygen toxicity or to decompression sickness. Helen Spurway and

Martin Case had already been dosed and tested on a similar profile, as had several Royal Navy divers. Ken Donald thought the Benzedrine lengthened the time available before an oxygen seizure, but his conclusions were not based on statistics, and the UCL math wizards did not agree. However, they did agree that the drug did not make anything worse. Using Benzedrine before diving was fine, if it had to be done.

Elsewhere in the complex, factory workers feverishly cut materials to pattern for swimming and diving suits. They glued rope onto soles and sewed special cotton drawers and singlets. The COPP operators would be coming to the factory for at least a week apiece to learn to repair their equipment, footsteps away from the researchers doing the specialized science needed to keep them alive.

———

That science could not come quickly enough. Freshly formed COPP teams 3 and 4 were sent to Malta in early February 1943 "to assist in carrying out reconnaissance of the Sicilian beaches." Their highly trained leader, Lieutenant Commander Norman Teacher, "had had to leave the UK without proper fittings" of his swimsuit. He died on the mission, and his death was suspected to have been because his suit flooded through a poorly fitting should-have-been-watertight seal, drowning him. Without his navigational experience to guide them, the rest of the hurriedly assembled, undertrained team of COPP 3 disappeared four days later, having paddled about without compasses and without direction, in futility, into a "featureless" night. They had not been able to find the host submarine on their return from the beach; they had not been able to spot the infrared signal that had been lit to beckon them home. Their bodies were never found.

Only Lieutenants P. R. G. "Bob" Smith and David Brand of COPP 4 survived, and to do it, the determined duo paddled seventy-five miles with limited supplies over a horrific three and a half days all the way back to Malta, rather than riding back in comfort inside the host sub.

The debacle showed that, like the Chariots, the results from canoes were mixed at best, too. The Allies needed more ideas for beach scouting— ideas that put COPP operators onshore but without their having to paddle so far out to water depths manageable for full-sized submarines, which at those distances were harder to spot and return to.

An MI5 phone tap eavesdropped on a member of the Communist party talking about how Haldane had been knocked into a coma for a month as the result of a chamber experiment. Around the same time another UCL faculty member observed to a mutual colleague that "JBS Haldane has aged very much in the last two years." His test schedule shows no monthlong gaps, which suggests the coma story was a cover for his workload-induced absence from Communist party activities.

Vice Admiral (Submarines) Max Horton wrote a memo that the first X-craft midget submarines were supposedly going to be ready for their inaugural test run in safe, controlled "Home Waters" soon. These small subs would be able to approach closer to shore than full-sized submarines because of their tiny size. Horton sent the memo to First Lord of the Admiralty A. V. Alexander. Alexander scribbled a handwritten note in red ink and passed the optimistic timeline to Prime Minister Churchill.

The X-craft might work as an alternative to the canoes and the Chariots, if they could be built in time. If there was a way to provide the right mixture of breathing gas to keep people alive inside such a small, confined space. If the craft could overcome the problems of undersea survival that had been thrown into sharp focus at the start of the war when the submarine HMS *Thetis* (now the *Thunderbolt*) had gone down. If, if, if.

The Aliens War Service Department granted official permissions for Hans Kalmus and Ursula Philip to stay uninterned, and MI5 granted official permission for them to begin working on the same military research on which they had secretly already been working. Two days later, the lab completed experiments 367 and 368. For the second one, Haldane took a dose of sodium bromide, an anticonvulsant, to see if it helped prevent oxygen symptoms. It did not.

The lab group had to move out of the Piggery when a stray bomb missed the house but rattled it to its core, causing minor damage but enough to cut the utilities. The landlord had a difficult time separating what was damage from explosions and what was damage from the raucous inhabitants. The cracked sink was likely the bombs; the scratches to the carpet on the landing of the stairs were probably Spurway's cat. The members of the group scattered to wherever they could near Rothamsted.

Still missing from the lab work was Jim Rendel, whose lungs kept collapsing over and over again, no matter what the doctors tried. He gave up being able to return to scientific duty with UCL and took a job with the Ministry of Aircraft Production while Joan was pregnant with their second child. Haldane kept trying, but failing, to get the Admiralty to help with the medical bills, and he had to relay to Jim Rendel the unfortunate news that "everyone concerned says it is not their business."

Haldane tried to comfort Rendel by telling him that the dives were for a worthwhile purpose: "Our underwater work is going on, and a number of improvements in technique which Spurway and I worked out last Christmas are already in operational use. So I hope you will realize that your work, though of a preliminary character, has led on to results of real import—how important I hope to be able to tell you after the war."

The rest of the lab was hard at work, and perhaps waiting to see who would be injured next. On June 9, 1943, Ursula Philip's morning experiment went fine. Of course, "morning" was a relative term, since the lab's working hours started at lunchtime and ran through the night, and the experiment began at 11:53 a.m. "Fine" could be a relative term, too.

She sat patiently inside the dry chamber at ninety feet, her loose cloud of onyx curls only partially restrained by the straps of the rebreather's mouthpiece. Both J. B. S. Haldane and Helen Spurway stared at her lips. She felt her lips twitch at the same moment they saw it. "Slight wavey [*sic*] feelings" rippled the length of her leg muscles 5.5 minutes after she started breathing oxygen. Later, she explained in her thick German accent, after some degree of color had returned to her pale face, that the oxygen had also induced in her a feeling of distress.

The same group was going back in together that afternoon. Same chamber. Royal Navy diver Chief Petty Officer Smith would jam in alongside them as their attendant, because for this test, all three of the UCL teammates were to be guinea pigs.

Philip, Haldane, and Spurway each cracked open a pearl capsule filled with amyl nitrite and sniffed a massive huff of the escaping gas. Haldane's face turned red, he felt faint, and he started coughing. Philip blanched and could feel her heart thumping. Spurway's cheeks flushed, and her world spun more than normal.

Amyl nitrite was first described in medical literature as a vasodilating drug, meaning it expands and broadens the blood vessels of the body, allowing blood to circulate with less effort and more enthusiasm. Amyl nitrite drops the blood pressure of the user; it was originally studied alongside nitroglycerin as a possible treatment for patients mid–heart attack to help coax their bodies toward survival. It did not work as well as nitroglycerin, so it mostly faded into the background catalogue of unpopular chemicals.

In the early 1980s, however, amyl nitrite would bloom back into the world with a Technicolor neon vengeance, finding a new niche with its little pearlescent capsules as the heart-thumping party drug nicknamed "poppers." It was going to become so prevalent that it would be one of the first-guessed causes of the new, mysterious plague of Kaposi's sarcoma and opportunistic infections demolishing the communities of gay men among whom the drug was most popular.

But all of that was decades in the future. For now amyl nitrite was still the tool of science, at least ostensibly, even though it was sometimes handed around at chemistry lectures like a party favor. It was the topic of that day's investigation, data point number 433. Perhaps the vasodilator could help the ears clear more quickly during pressure changes, during rocketing urgent excursions up and down the water column to escape a bomb or a sunken sub. The ability to clear the ears was used to screen potential divers; therefore, if there was a drug to help it, that would increase the pool of eligible applicants.

Spurway and Philip both had ears that were difficult to clear, so they were the perfect test subjects. Therefore, in the name of science, in the name of diligence, in the name of maritime safety, and in the name of freedom and humanity, Spurway, Philip, and Haldane got stoned.

The drug did not help. It did not seem to make Haldane more sensitive to oxygen, as he was breathing pure O_2, but it did not help anyone's ears. Philip's nose still leaked a trickle of blood afterward as she came down from her buzzing high. The experiment had been worth a shot. Some sacrifices were greater than others.

After the disappearances of COPP teams 3 and 4 off Sicily, the Admiralty sent two more COPP teams to the Mediterranean in response to "a further urgent demand" from that locale's commander in chief. Apparently, before their brutal disappearance into the sea, the lost COPPs 3 and 4 had on their first few days done some mapping and gleaned beach data from Sicily without ruining the crucial element of surprise. The scouting work they had finished was enough military candy to tantalize the "C in C Mediterranean" and stitch his eyes permanently open to the value of the COPP teams despite the horrific deaths. That demand coming from him, despite the preceding disaster under his watch, provided the final argument in persuading the "higher authorities" that "COPP work was of real value in amphibious warfare, and was a necessity if seaborne assaults were to have any degree of success." Even though the teams still needed a better way to get onto beaches. Even though the rest of the work was still in progress.

C in C Mediterranean got his two teams. COPP 5 left the UK first, as soon as they got their custom-made canoes, and COPP 6 shortly afterward, even though they had to pillage storage and "borrow" supplies from COPP 7. The few short months that separated their departure from that of COPPs 3 and 4 had been advantageous, because these new crews had completed their full training and were "very much more adequately equipped." They missed the first recorded demonstration of fin swimming in the UK, but in any case, "proper fins [were] not yet available, even in prototype."

COPPs 5 and 6 were equipped. They were trained. They were ready. They brought some newfangled device invented by the Americans to try out. It was called the "Walky Talky." They also brought Chariots to see if the Chariots could get them onshore more efficiently than the folbots.

They loved the Walky Talky but hated the Chariots. They tried to make the ridable torpedoes work for a solid month while they struggled to survey the beaches of Sicily, but the COPPists argued that "the Chariot method was slow," the nights had limited hours of darkness, and the full-body rubber Chariot diving suits, with their narrow viewing panels and oxygen rebreathers, restricted their ability to see. Their surface swimming suits were easier to wear, even if they meant swimming on their backs into shore and even if being limited to canoes meant more work paddling than riding. On a night when they might have five hours of solid darkness to utilize in a canoe, the breathing apparatuses needed on board a Chariot gave only an hour and a half of oxygen. In the opinion of the COPP personnel, the Chariots were a waste of too many darkened hours of a proper cloudy night. Yet nonetheless they got their work done in Sicily.

Before dawn on July 10, 1943, the first small Allied ships kissed the southern shores of Sicily, and the wet soles of Allied boots came next. For the next three days, more than a hundred fifty thousand troops, supported by three thousand ships and four thousand aircraft, flooded the land scouted by COPPs 5 and 6 and their Chariots. The night before, the COPP teams had "succeeded in marking the beaches with human markers," so they could once again guide the troops to the correct spots on the morning of the invasion and help to eliminate uncertainty for at least one element that had previously relied on luck.

However, despite the clear and unbridled acrimony unleashed by the COPP leaders against the Chariots in the memos they wrote after the invasion, these vicious fans of their foldable canoes had to admit one crucial fact: "Bad weather causes the folbots to drift and tries the crews. Cold weather reduces endurance and lowers morale." Weather could be a problem. The canoes had been used only in the relatively protected, warm waters of the summertime Mediterranean and on pleasant tour-isty beaches to which people traveled from around the world to swim and sunbathe and build sandcastles in peacetime. Neither method was enough for the northern shores that the leaders knew would come.

The invading Allies crept across the island of Sicily like a purgative fog, and the instability of the Italian government reached its breaking point. A new government rose up in Rome and arrested fascist tyrant Benito Mussolini, reinforcing the Allies' hopes that Italy might be an unstable battlement in the Axis fortress.

German and Italian forces retreated across the Strait of Messina from Sicily to the Italian mainland. As they did so, they scattered a rich deposit of explosives, including hundreds of bombs in Messina's harbor itself, to foil or at least slow the obvious next step in the Allied plan.

Inside Messina's harbor, as the Germans left, they deposited for the first confirmed time a new and special kind of underwater bomb. A bomb with a booby trap. A bomb that exploded if you tried to lift it out of the water before disarming it, as the disposal methods for underwater bombs had prescribed to do thus far. Seven bomb-disposal experts dis-covered the booby trap the unfortunate way. Five died. The other two recovered in the hospital for weeks. Bomb-disposal expert John Bridge was sent to Messina to replace them.

Bridge's first act was to ask someone to row him around Messina's harbor in a rowboat. He described the water as "clear and quite warm" and said he "could in fact see down to a depth of forty feet." From above, he spotted the bombs, including several with the new kind of trap.

In order to clear the munitions without moving them, he demanded a diving suit. Over three days, Bridge spent twenty-seven hours inside a Siebe Gorman diving suit with a hard helmet, his air supplied to him from a crew in a boat on the surface. He plodded along the bottom and put his bomb-disposal skills to use among the fish without the need to jostle or lift the underwater munitions. He defused a bomb with the booby trap mechanism and extracted the device intact. Using his recov-ered example, he figured out how it worked.

Before the war John Bridge had been a grammar-school teacher, bouncing from temporary job to temporary job, rebuffed from permanent employment because he lacked a fancy degree from Oxford or Cambridge. When he volunteered for the service, the Royal Navy figured he would be good at puzzling out the electronics and trigger mechanisms of bombs because he had a background in physics. They figured right.

John Bridge and his team cleared Messina's harbor for use by the Allies in three days, breaking every rule about maximum time breathing air at depth for the sake of the urgency of the task. He somehow escaped injury, likely because the water was shallow.

Bridge reported their success on a Wednesday afternoon. The invasion of mainland Italy started two days later on September 3, and launched safely out of the harbor he had cleared.

The lesson of Messina was that the Germans were concocting new bomb types: bomb types that could not be lifted or moved; bomb types with delayed fuses that would set them off after days of sleep. John Bridge had been successful in his novel diving approach because he had been operating in a calm, controlled harbor where his surface ship could float happily in the quiet and the peace and pump the air he needed down to his helmet for as long as he needed it. But the same bombs might get placed in harbors where they could not be defused in advance, where they could not be defused before ships had to rush toward the land to off-load troops who would take the hostile shores.

The Chariots had been something of a bust, but they had led to the development of better breathing apparatuses and the accompanying UCL science so that those apparatuses could be used underwater, not just in emergencies but in planned, safe fashions. Deliberate fashions. Fashions, perhaps, that could enable a diver to work on a bomb in a slow, leisurely manner without the need for a surface ship and without fear of seizures or decompression sickness. The aborted human torpedoes did have a silver lining.

Over the summer, while troops massed in North Africa to launch for Sicily, Ken Donald submitted his first report about oxygen toxicity and safe diving depths to the Admiralty. He praised the military test subjects, sang their deserved honors, and said truthfully that "they are, without exception, men of 'iron nerves.'"

However, he missed one important thing: math. His report was a loose assembly of descriptions and time points organized by divers' names rather than in any useful fashion. He finished the report with a vague conclusion that most divers had symptoms—at some point— when breathing oxygen in sixty feet of water. He stopped short of prescribing how deep and for how long divers would actually be safe. He wrote a second report with more data points in it, and this time he plotted them on an informal graph, but he still stopped short of analysis, instead eyeballing a loose curve by hand for the graph and admitting that "the curves on the graphs are drawn by hand and are not mathematical ones." He misused and misunderstood basic terminology, substituting the word "medium" in a nonsensical fashion for the elementary mathematical term "median." He vaguely promised that "a further series of Time/Pressure curves are to be done in the near future after which a safety curve will be produced."

Each and every one of the UCL group members was a crack statistician forged in the bowels of a scientific lab where math was used—nay, insisted upon—to describe everything from the hatching success of duck eggs to the colorful fur splotches on feral cats. They thrived on numerical data the way a plant thrives on sunshine. They were fueled by numerical data the way an exhausted Charioteer plodding through the snow was fueled by Benzedrine. Ken Donald's data points were available to the same Admiralty-led working group to whom the UCL members had been presenting their own data and observations, such as the epiphany sparked by Elizabeth Jermyn's underwater seizure. The UCL group snorted the new data like it was amyl nitrite, and they got to work. They were going to make actual safety curves.

———

At the same time as Donald's data analysis, at the same time as the long-awaited invasion of Sicily, on July 9, 1943, Helen Spurway sat upright on a stool in Siebe Gorman's largest, most spacious vertical dry chamber for experiment number 461. Hans Kalmus was her attendant, and though Kalmus did not know it yet, he was working to battle his parents' murderers. J. B. S. Haldane puttered nearby, too, because of the excessive space available in the luxurious metal tube. A legless brownish reptile that looked like a snake but wasn't one rested coiled at his own little reptile station. He was scientifically titled *Anguis fragilis* but less creatively nicknamed "Anguis," and he had been invited along for the ride.

Kalmus tested Spurway's peripheral vision to establish a baseline measurement of where her eyes naturally could and could not see. They compressed the chamber. They hit ninety feet. Spurway thought Anguis might have looked uncomfortable, but after Kalmus had done some considered prodding, he disagreed. Anguis, despite not being a nautical creature, was unperturbed by the depth.

Spurway began breathing oxygen. Kalmus waited four minutes, then tested her peripheral vision as he had before. He was checking to see if the oxygen—which caused so much visual dazzle, so much eyesight waviness, so much tunnellike narrowing of the periphery—might cause problems with the vision of operators in the water. Operators scouting beaches, operators being attacked, or operators working on delicate, booby-trapped bombs—in short, operators who needed their eyes to work correctly. Kalmus used as a visual test the bowl of an old smoking pipe, which he stuck into a DSEA tube the same color as the background that he held up behind both objects. Spurway's job was to tell him what he was holding.

Spurway's pulse rate was sixty-eight. Then sixty. Respirations thirteen. Kalmus tested Spurway's vision again. Her field of view was contracting, especially on the sides, in the temporal direction. Kalmus noted a 10-degree change in all other directions, but 20 degrees there. Spurway's pupils dilated.

Her face and neck went tonic. All the muscles contracted. Kalmus jerked out her mouthpiece. She seized anyway. Their notes, presumably sketched down afterward, describe it as "not very severe. . . . Her jaw is dislocated."

Kalmus, technically a medical doctor but one who had never practiced,

reduced her jaw, popping it back into place. They laid her down on a pillow on the floor, where she began to struggle and retch. "She is more
agitated than last time, and repeatedly dislocates her jaw." Kalmus had
to pop her jaw back in four separate times before they reached the surface. Spurway vomited. She was taken to the sick bay. She vomited again.
It was the fifth time she had seized during an experiment. She seemed
unfazed. She was determined to amass enough data points for the military divers to be able to use their rebreathers safely.

The next month, on August 27, 1943, COPP cofounder Nigel Clogstoun-
Willmott gathered with several trusted fellow officers in a room inside
Norfolk House, a geometrically staid seven-story flat edifice of a brick
building in London that had become a hub for Allied military planning.
They had new beaches in their sights for a mission freshly titled Operation Overlord. Beaches along the northern rim of France. They laid out
the bullet points of what would be required to scout them.

"About 12,000 yards of beach required investigation," they wrote
when starting their calculations. They wanted to be ready for Operation Overlord by the desirable summer months when the water temper-
atures of Normandy would edge above their wintertime hypothermic
levels and make it possible for drenched troops to recover once on land
despite soaking-wet clothing and equipment. That would mean scouting
during the winter before. However, "no experience [was] available, except from exercises, as to continuous employments of COPP units in
really European winter weather." They had never done this work in winter before. It might take longer than expected. Starting early would be
crucial.

To complicate things further, "the use of proper S/M's [submarines]
is barred." They were too large, too likely to be detected to be used to get
onshore. Smaller motorized landing craft were a nonstarter, too, as far
as being used to execute the whole scouting plan, because they could not
"compete with anything except say 1 line of echo soundings per night,"
meaning they could not get enough done. Plus, they, too, would probably
be spotted as they dragged back and forth along the target shores, "ow-

ing to Radar, and patrols." In other words, the regular submarines were too big, the small landing craft were too small, and anything on the surface would be spotted by radar.

Clogstoun-Willmott and the other officers reached one final, inevitable conclusion: "The use of X-craft (4-man S/M's) as carriers for COPP...appears the only solution." The COPP swimmers and the X-craft submariners had thus far been separate entities, but to make Operation Overlord possible, they would need to learn to work together.

The Allies needed to get the midget submarines to work. If they did, they would be the first to succeed. But they were not the first to try.

CHAPTER 14

The Midget Subs

The plan was simple. Pilot their five midget submarines through the mouth of Pearl Harbor, Hawai'i, and drop down to the harbor floor. Wait there unnoticed until dawn. Attack with the rising sun, with the airplanes above. The date of the attack—the "D Day," to apply American parlance to a Japanese assault—would be December 7, 1941.

However, immediately after his vessel's release from the host submarine on the night of December 6, Ensign Kazuo Sakamaki struggled to level out the ballast inside the two-person, seventy-nine-foot, narrow, cigar-shaped midget sub christened *HA-19*. He and his crewmate continued to fail to stabilize their boat for hours, working away in the darkness until the dawn of December 7 broke to find them lost, still outside the harbor. Because their compass was broken, they were unable to navigate to their target. Finally they did stabilize themselves, but patrol boats were crisscrossing their desired entrance under the revealing spotlight of day.

"Learn the peculiarities of these harbors," their captain had lectured the submariners over a map two months before the attack. "Commit them to your memory."

They had tried. But even then, the maps were outdated, and the positions of the American ships drawn on them were all wrong. The maps had not shown any underwater obstacles during the approach, because the scouting before the attack had not been thorough.

Unable to navigate underwater because of the broken compass, Kazuo Sakamaki and his aide inside the minisub decided to switch to a surface attack instead. It would be a literal suicide mission, barreling through the patrol boats toward the now smoking row of damaged battleships inside Pearl Harbor. They would hurtle toward the American ships until their submarine got close enough to ram one, to cause some kind of carnage in exchange for the price of their own lives, to participate in the first attack of the war on American soil. With that plan in mind, they accelerated.

WHAM. They slammed into an unmapped coral reef. Death would have to wait. They were stuck. The batteries, jarred by the impact, began leaking some kind of noxious fumes. The compressed air, which was carried to provide the breathing atmosphere inside the subs, began leaking from its canisters at an uncontrolled rate, which also slowly increased the pressure level inside the watertight space.

The duo worked to free themselves from the unseen reef while "the air inside the submarine was reaching forty pounds," about equivalent to them being ninety feet below the ocean waves, but with loads of mystery chemicals swirled in because of the leaking batteries. As the submarine's internal pressure increased, the effect of each and every chemical component floating around inside the crew compartment became more severe. The air was "becoming dangerously foul." The two crew members began to feel weak, drowsy, "weary."

They worked their way off the coral reef by shifting the ballast back and forth inside the long, narrow space. The broken batteries delivered them repeated electric shocks while they crawled about in the confined tunnels that were the bow and stern. But they got free. They tried to aim—a second time—for carnage. They began to move.

WHAM. Another unmapped reef, another natural obstacle.

Sakamaki's account of the rest of the day gets muddled after the second reef impact. He describes working hard to get free a second time, he describes sweating heavily while crawling about in the confined space to move around ballast, but he also openly acknowledges that he could "not remember the rest of the afternoon," except that at some point he aimed the midget sub toward their rendezvous point with their host

Ensign Kazuo Sakamaki's midget submarine, *HA-19*, where it washed up on the shore of O'ahu, Hawai'i, and was photographed by unnamed US Naval personnel on or about December 8, 1941. *(Image Naval History and Heritage Command NH 54302)*

submarine and then "collapsed." His aide in the sub passed out, too: "After a while my aide fell asleep, perhaps unconscious." When they both woke up, they discovered that their submarine had "sailed all night in the moonlight with the hatch wide open." He sounds as if he had been surprised to find it so.

Sakamaki and his aide, beyond exhaustion, attempted to scuttle their submarine and themselves with it, to sink both the boat and their own corpses beneath the warm camouflage of the deep blue Pacific. But the bomb they used failed. The aide drowned in the process. A profoundly fatigued yet alive Sakamaki drifted up onto one of the sandy, humid, mountain-backed, windswept beaches of O'ahu, where he was found by American troops and became the first prisoner of war taken by the United States of America in World War II. His midget submarine drifted up after him. All nine of the other Japanese midget submariners who entered Pearl Harbor died.

Japanese airplanes were infamously successful on December 7, 1941, but the five midget submarines of Pearl Harbor have been largely forgotten because of their failure, lost, sunk, or captured. However, the *HA-19* was found intact, sitting in the sand, waiting to be reverse engineered. Waiting for the paper documents inside it to be examined, dissected to

American military personnel tow Sakamaki's submarine from the sea after its discovery on or about December 8, 1941. *(US National Archives ARC ID 133891335)*

see what the Japanese submariners had known beforehand and, perhaps more important, what they had not.

The idea of small, stealthy submarines was an old one. The single-person *Turtle* puttered unsuccessfully about New York Harbor in 1776 during the American Revolutionary War. In France in the early 1800s, Robert Fulton made a small prototype named *Nautilus* that worked but had an unfortunate tendency to leak. The Germans built the *Brand-taucher*, which got stuck in a harbor bottom in 1851. The eight-person, hand-cranked Confederate *H. L. Hunley* met with debatable success in 1864 during the American Civil War, but killed its entire crew in the process. The Japanese were not the first to build midget submarines, but they were arguably the first to build modern ones, miniature models that seemed to have the potential to work from the beginning to the end of a mission . . . if only they had figured out the problems of breathing gas and scouting beforehand. It is probable that some of the finer points of Sakamaki's war memoir were lost in the translation from its original Japanese, careful as that translation was; however, his story reeks of contaminated breathing gas, finally let out by the open hatch door.

People like Vice Admiral (Submarines) Max Horton and Commander Cromwell Hanford Varley received the bulk of the well-earned credit for pushing the same concept through the Royal Navy until their dreams became a reality called the X-craft. However, internal Royal Navy documents acknowledged where they drew inspiration for that particular war: "As the Italians first developed the Chariot and the Japanese the midget S/M [submarine], it is probable that in planning harbour defences, attack by Special Service Craft has always been a consideration in Axis minds."

English-language histories have largely passed over Ensign Kazuo Sakamaki and the nine other Japanese midget submariners of Pearl Harbor, but they and their tiny vessels were at the forefronts of the minds of the Allied naval community after the discovery of the wrecked *HA-19* on that Oʻahu beach in December of 1941. After Churchill's bellowed order to copy the Japanese and the Italians, they were at the forefronts of the minds of those who made and used the first X-craft and Chariot prototypes.

Over the course of the war Germany, Japan, and Italy built more than two thousand midget submarines. The Allies, however, were going to use science to make midget submarines work better. By the time the Admiralty called with increased urgency in the fall of 1943, the X-craft and their crews had already been drilling away in western Scotland, partly

Lieutenant J. E. Smart rides on top of an X-craft in the waters of Scotland. *(IWM 22899)*

enabled by creativity, partly enabled by patriotic, work-all-night grit from the United Kingdom's shipbuilders, and partly enabled by the UCL lab's first few years of wartime respiratory science.

Each three-person X-craft was a squat black "ugly duckling" of a submarine, with so little room inside that from the command station, the captain could reach out and jab both of the other two crew members with a probing forefinger. The boats had been designed so they could travel with ease over the tops of underwater minefields. They had a maximum height of ten feet, which let them glide above mines when covered by high tide, and they were fifty-one feet long, a bit longer than a standard school bus. Those dimensions on the outside left not enough room for a full-grown adult to stand up on the inside, except at one special spot in the center of the vessel, which must have been a popular spot with the crew. Also piled inside the frame were a diesel engine for use when traveling on the surface, batteries for use when traveling submerged, electronics, hydraulics, life-support equipment, navigational equipment, and a trunk beneath one hatch for entry and egress.

Sublieutenant K. C. J. Robinson sits at the controls inside an X-craft. *(IWM 26933)*

The X-craft could travel at a maximum of six knots, about one-fifth the speed of a typical battleship but double the pace of the Chariots. They could support their crews for five to seven days across up to thirty-five hundred miles in a cocoon of some degree of comfort compared to the Charioteers' exposed and meager five hours and fifteen miles in the water. It was not a pleasant ride for the X-craft crews who spent multiple days inside that cramped, dark little industrial cave. It smelled constantly of fuel oil. Myriad leaks and spurts of water through tiny holes in the hull made the atmosphere so humid, the crew often used a ladies' hair dryer on the damp electronics. Fatigue from survivable but chronic levels of carbon dioxide was a constant issue. However, the X-craft crews could stay alive on multiday missions, whereas the Charioteers could not. In addition, unlike on a Chariot, the X-craft crews' snack options were broader than paste squeezed out of waterproof tubes smuggled inside pockets.

The X-crafts' bombs were bigger, too. They were positioned like saddlebags on either side of the vessels, augmenting their smashed-looking appearance. The X-craft could lug not one but two detachable charges, with each bomb full of a hefty four thousand pounds of gray explosive granules that looked disturbingly like instant gravy. The Chariots, in contrast, loaded all their hopes into just one charge containing six hundred pounds.

The pressure-proof X-craft could dive down to three hundred feet to hide. The Chariots would have lost their riders to seizures long before that.

It was clear which craft gave its crew the better chance of survival. The X-craft could also be released 150 miles from their destination, lessening the odds that a full-sized submarine would be spotted too close to shore and ruin the surprise of the coming attack. Full-sized submarines could carry and release folbots, but it is hard to kayak 150 miles, and with the X-craft's ability to travel farther toward land, the folbots were no longer needed at all.

X-3 and *X-4** were prototypes, welded-together manifestations of a concept that was still in the exploratory stages of thought, and unsur-

---

X-1 was a full-sized submarine and *X-2* was a captured Italian submarine, so the names were started for the experimental boats at *X-3*. *X-3* was made by Varley Marine, and *X-4* was made by Portsmouth RDY.

prisingly in training they proved to have some flaws. But by *X-5* through *X-10*, the little, painstakingly manufactured boats were starting to impress. A crack team of welders, machinists, and engineers was piecing them together one widget at a time at the firm of Vickers-Armstrongs. The Admiralty clamored for still more boats beyond that, but Vickers-Armstrongs was already overloaded building regular-sized submarines for the war effort, so they shared the designs with three more shipbuilders. The four firms started assembling raw materials and piecing together the first few inchoate parts of what would become, eventually, someday, twenty-three more tiny subs.

Young Doris Wright was a crane operator at one construction site, and she said of the frenzied efforts to build the X-craft that "we worked hard and played hard in those days but it was all worthwhile and very rewarding to watch the wonderful subs taking shape." X-craft shipbuilder Marjorie Warwick called it a time of "long working hours and anxieties."

The Italian *Maiale* had been constructed using specialized engines made by Alfa Romeo. The name alone conveys an impression of sleek, dynamic, powerful engineering; of the *Maiale* prowling like a sports car toward mission success; of the subtle sound effect of an underwater growl from the vessel's high-performance machinery, although in practice it was silent. The X-craft, on the other hand, were powered by the sturdy and reliable engines of London buses. Less glamorous, maybe, but certainly more endearing.

The potential power of the X-craft as a tactical tool meant that they were kept so secret that in 1943 when a telegram mentioning them was finally sent out to the British fleet, at least one commander in chief had to write back to ask if the unfamiliar term was a code or a mistake. The Admiralty had to send out clarification that, no, it was not a mistake; the X-craft were submarines that did in fact exist. They not only existed but, at that point, crews had been practicing with them in secret for almost a year. They had been hidden in Scotland.

———

Loch a' Chàirn Bhàin in Gaelic or Loch Cairnbawn in English, a sea-loch with one narrow side cracked open to the ocean, is a remote, devastatingly beautiful natural hiding place to practice secretive naval activities. Packed rows of rocky almost mountains form a fortress of land around the three closed sides. Behind that choir of rounded peaks, two visible, jagged full-mountain summits emphasize the little loch's isolation from all living things, except the tenacious shrubberies clinging to the rocks and the Atlantic porpoises playing in the cold water below. Ripping winds have carved these encroaching mountains and hills into a foreboding, Picasso-like landscape of blunt geometric edges, and yet despite those frequent winds, somehow a mixture of thick, chilling mist blended with low, murky clouds settles there. Little additional human security is needed to form a safe perimeter for this gorgeous, eerie, poetic, humbling place.

The nearby island Gruinard was determined in 1942 to be so desolate, so far from any habitation, that it could be used as a test site to seed the land with experimental anthrax bombs. The bombs had been developed in case biological warfare became necessary against the Germans, and after their use, the island became untouchable by any living animal. The same year as those tests, the Admiralty established a secret X-craft practice site in the winds and waves of Loch Cairnbawn, a full day's mountainous hike farther north than that.

Two hundred fifty miles south of that awe-inspiring place, closer to the major cities of Scotland, the Admiralty had also taken over the small, Swiss-inspired row of buildings of a therapeutic waterfront "hydro resort" retreat carved into the side of another sea-loch. They repopulated it with submariners as well as a bevy of WRNS, Women's Royal Naval Service members, pronounced out loud as *Wrens*. They called the newly occupied resort "HMS *Varbel*" and made it the official base for the freshly formed Twelfth Submarine Flotilla, whose members were to be the masters of the midget submarines. From there, the X-craft and their attendant vessels would travel up to Loch Cairnbawn as needed. Over the course of the war, according to some accounts, Joseph Goebbels and his propaganda machine would several times trumpet that HMS *Varbel* had been sunk, despite the fact that it was actually the code name for a row of buildings safe on land.

The masters of the midget submarines dubbed the northern practice site in Loch Cairnbawn "Port HHZ." They placed to drift in HHZ's narrow waters HMS *Bonaventure*, a sturdy but otherwise forgettable depot ship that had been conscripted from a shipping company. The *Bonaventure* was there to be X-craft support, but it was also used for X-craft target practice. Loch Cairnbawn's job was to mimic a Norwegian fjord for training before another run at the *Tirpitz*. The *Tirpitz* and its guns, after all, still threatened all Allied ships that braved the waters off northern Europe. *Tirpitz* was a gatekeeper, and the gatekeeper needed to be felled to clear the passage before any northern beach landing could occur.

The Wrens of HMS *Varbel* performed typical WRNS duties to support the X-craft, such as hauling munitions and working "cheek by jowl" with able seamen "in the bowels of corvettes and sloops" for training scenarios, transporting casualties after drills, acting as crew for the military transport boats, and conducting engine maintenance on their own vehicles. HMS *Varbel* also became the site of an X-craft excursion that made a different kind of history when Wren Angela Cooper was asked to step outside her normal tasking to become the first woman documented to pilot a submarine.

An impatient midshipman "badly wanted to get out to the trot"—a trot being a type of floating jetty connected to land where submarines could dock. However, his first lieutenant had not shown up, so he tapped Angela Cooper to drive the little boat with him instead. He stood on the casing of the X-craft, above the water, at the special rails that had been installed on top so officers could navigate while tethered to the boat to avoid being swept overboard. She took the helm below. She was used to working on navy boats in gale-force winds, but unused to the fully enclosed protection of a submarine's pressure-proof hull. She glued her eyes to the black pointer that indicated the ship's heading, followed the mid's navigational cues from above, and piloted the sub without mishap.

The meeting where the Royal Navy officers first floated the thought of "the use of X-craft" for reconnaissance for Operation Overlord took place in Norfolk House in London on August 21, 1943. A week later COPP master Nigel Clogstoun-Willmott sat down with a few other officers to discuss again those beaches, the X-craft, and how such a concept might work in practice. That was the meeting where they decided the X-craft were their only option to survey their twelve thousand yards of beach.

"By COPP methods . . . a successful beach reconnaissance requires a minimum of five hours [sic] darkness" to make it worth the trip to shore. Those continuous hours of darkness were key.

Clogstoun-Willmott and his compatriot Logan Scott-Bowden provided a margin for error in their calculations because the tasks were physically demanding, and they made the assumption that each COPP team could cover about three hundred yards of beach in one night. Based on the requirement for five hours of darkness, there were only five sufficiently moonless and lengthy nights per month, and then only in the winter. If they razed the length of beach that needed scouting down to three thousand yards from twelve thousand yards, then two COPP teams working every usable night would need to operate for six months straight. Assuming there were zero casualties. On top of that, March 23, 1944, would be the last possible night for the teams to do reconnaissance before summer 1944. So, the teams would need to start scouting by October of 1943, in slightly over a month from the meeting.

That timeline seemed theoretically plausible, at least at first. However, the number of teams available to do the work was an issue. COPP 1 was the name for the instructors at the depot; it was not an actual team. COPP 2 was a name that had been set aside for a specific leader, who then got sidetracked by other "special work" and never assembled a team. COPPs 3 and 4 were dead, except for two survivors, who had been reshuffled. COPP 5 was in Algiers on their second back-to-back assignment with no break, and COPP 6 was in Malta on the same kind of exhausting, relentless schedule. COPP 7 was in India, where the commander in chief was pushing and clamoring for two more teams, so it seemed a safe assumption he was not about to send the personnel of COPP 7 back. The members of COPP 8 and 10 had recently been selected, had no training at all, and were promised to help in India. COPP 9 was the name of another nebulous group that had not yet been formed. And that was the full roster of COPP personnel. So, in short, there was no one to go.

Conducting a COPP beach survey was a "cold-blooded and unpleasant duty," physically brutal and mentally exhausting, and around the same time as Operation Overlord planning began, the few COPP officers who were in the field had also begun submitting letters stating they had all "applied for a reversion to the General Service, owing to the conditions under which they are working." In other words, the Admiralty had no spare teams, and what few teams they had were threatening to resign from sheer fatigue.

The technical documents and memoranda of the time did not describe the exact scene in the room when the leaders had that realization, only its results. However, it is easy to imagine Nigel Clogstoun-Willmott and Logan Scott-Bowden hunched over a conference table in Norfolk House, maps and papers and pencil-scratched sheets of calculations splayed between them. To prevent another Dieppe, they needed fully trained personnel who did not exist—and who would not exist unless they prolonged for another full year the bombings and the deaths and the shortages and the horrors not yet uncovered on the mainland, all because of the inexorability of the schedules of the moon and sun.

It was the end of August, so they wanted to start the scouting in a month. Five months were necessary for training in beach analysis, scouting, hand-to-hand combat, et cetera. Six weeks to two months on top of that were required for training inside the X-craft. In order to make Operation Overlord possible, they needed to have started training new COPP crews five months before they had made their plan. It is easy to imagine Clogstoun-Willmott and Scott-Bowden staring into each other's grim, pallid faces in that moment—the moment when they both understood that they would be the ones to go. There was no one else.

"COPP 1 will become operational, and COPP 9 will form round nucleus and become operational, both in 'X' Craft. This will leave a bare skeleton to man depot, with no training staff." Development of new COPP teams would have to wait. The original members of Party Inhuman, the COPP instructors, would start their X-craft training in November, as soon as they could get to Scotland.

The week of August 21, 1943, the idea of X-craft for scouting had been a mere thought. On the twenty-seventh, Clogstoun-Willmott and Scott-Bowden did the math. On August 29, the thought became a plan.

Two days later, J. B. S. Haldane and Helen Spurway took a break from their oxygen-testing protocols to sit together with Ken Donald and Royal Australian Naval Volunteer Reserve officer John Stuart Mould for a test inside the locked chamber, but with no added gas pressure. Donald did light physical work to simulate mechanical operations. They stayed inside the chamber for 298 minutes, almost five hours. They ended with skull-splitting 6.71 percent carbon dioxide and dizzyingly low 13.17 percent oxygen. Haldane and Spurway were repeating their earlier work for the small-space carbon dioxide tests with Martin Case. They were making sure that, with a crew of four inside the same kind of small, enclosed space—or midget submarine—if all else went awry, they could survive without refreshing the breathing gas for at least five hours. It was experiment number 509.

On September 1, the next day, Port HHZ was sealed off for training. The one rural road somewhat nearby was blocked, even to locals. From that day until the next X-craft mission was complete, nobody would go in or out. Nobody would be provided the opportunity to leak any insight that might sink one of the tiny subs. Training began, first for the *Tirpitz*, then for Normandy.

That same day, COPP team 5 guided troops onto the thoroughly scouted beaches of Salerno, Italy.

Also that same day, Helen Spurway and Ursula Philip climbed into the dry chamber at Siebe Gorman, carrying rebreathers, to get more data points about using gas mixtures with elevated percentages of oxygen.

Their previous work was being used in the X-craft as the scientists focused on the next step: the beaches.

––––––

Inside a space as small as an X-craft, crowded further by diving equipment, beds, navigation equipment, food, supplies, and the special hair dryer that kept the electronics working, breathing gas is always the limiting factor in survival. The two circular entry hatches up top could be opened whenever the boat was on the surface, or new gas could be pumped in through a snorkel to refresh the oxygen supply and remove the carbon dioxide. However, when they were submerged, the laws of respiration ruled, and the walls of the submarine imposed a firm clock on survival. An X-craft crew at the edge of their gas supply could keep breathing and suffer dizziness, pain, and vomiting down below. Or they could come up and get spotted. The choices were certain death below or chance of death above, because the laws of breathing gas do not negotiate.

When submerged and counting the minutes of usable atmosphere, the three-person crews inside the X-craft would slowly release oxygen from compressed gas bottles to replenish what they breathed. One record indicated that each minisub carried twelve Luftwaffe bottles for that purpose, with each precious aluminum canister having been snatched from a downed German aircraft. One diver referred to these containers as "marvelous jobs" and described how the submariners usually did not even bother to remove the German-language markings before using them.

To remove the carbon dioxide, the crew members periodically opened boxes of material called Protosorb. The crumbly tiny rocks of Protosorb were chalky chunks of CO_2-absorbing chemicals, and they looked like bleached kitty litter. They bonded to the gas and pulled it right out of the air, no help needed, just an optional small fan to keep the air mixing and moving through the pebbly crumbs. The sealed boxes of Protosorb, which would have made a tidy stack somewhere inside the X-craft, were the descendants of the same material brought on board submarines by John Scott Haldane with young child JBS-then-Jack in tow as his assistant. They were the same material tested by Haldane and Case in the

dangling metal compartment in the waters off Portsmouth at the request
of Vice Admiral (Submarines) Max Horton.

The X-craft crews' bodies were the only portable carbon dioxide detectors in existence, so the crewmembers opened new boxes as they experienced the symptoms they had been warned about from the results
of the experiments. The number of Protosorb boxes and oxygen bottles
needed and the fact that it was possible to live underwater while waiting
out the duration of the revealing sunlight hours before heading back to
the surface for fresh air were the products of careful scientific work.
Siebe Gorman packed the boxes. The UCL researchers determined how
many to stow on each sub.

The crews inside the boats knew where the boxes had come from but
little else. They barely knew what they had signed up for at all until it was
too late to back out, until they found themselves scrambling through a
diminutive circular door in the top of a tiny sub.

The X-craft recruiters had honestly but mysteriously limited their ad-
vertising descriptions to phrases like "special and hazardous duty" when
they asked for volunteers from the general fleet. One recruit was asked if
he was interested in "hazardous underwater work" with no other clarification. In order to qualify, the volunteers had to know how to swim, and
they had to be single; nobody was allowed who had a spouse or children.
Some were asked if they were claustrophobic. From there, they were sent
to the Admiralty's submarine training escape tower in the south of England, but even there, they were given no further information.

Hugh Michael Irwin, shortly after being recruited via those mysterious means, crouched huddled inside the enclosed, narrow, upright metal
cylinder built against the bottom right angle of the also cylindrical but
much larger tank of the submarine training escape tower. This bigger
tank of the escape tower was an open-topped, white-walled container
with water forty feet deep, housed inside a brick submarine training facility in Gosport, in the south of England. Irwin was in air inside the
smaller metal construction at the bottom, but he was nonetheless

breathing through an oxygen-filled Siebe Gorman rebreather positioned squarely on his chest. It was his only apparel besides goggles and tight swim trunks. He was in the air for now, but he was about to get wet.

The last of Irwin's three training partners climbed upward to enter the still dripping practice hatch, leaving only Irwin and the instructor. One of these three partners, a man named Cleary, would die of oxygen toxicity in the year to come on one of these very excursions for which he was training. But for today, they were practicing how to exit a submarine in the clear, controlled water tank of a home naval facility. The ocean was for later. The third recruit to leave the tank closed the hatch behind him.

Normally, the tower was used purely to practice escapes to let shiny, newly minted submariners think they had a glimpse of hope, a chance of swimming to the surface should their vessels go down at sea. In truth, four submariners had made it out of the HMS *Thetis* alive using their DSEA oxygen rebreathers after it sank, so the odds were not zero, but they were small. The normal use of the tank was for the submariners to practice leaving in a hurry, racing for the surface, racing for daylight and the hope of rescue by a surface ship.

Jim Rendel was the unseen harbinger of the dangers of the tower.

A trainee wearing a DSEA swims upward from the escape hatch at the submarine training tower. *(IWM A 13873)*

Rendel had suffered an air embolism during the depth changes in the dry chamber. A small portion of his lung, isolated by some unknown pathology, had refused to release the air inside it in a proper fashion. Instead, the lung had burst, sending the air careening into Rendel's body. Rendel had survived his accident with no more than a collapsed lung, but a disturbing fraction of the time, such emboli were fatal.

The same thing could happen during a submarine escape. If a submariner panicked and did not exhale, or even if they did not panic and they still forgot to exhale, the trapped air inside their lungs would expand as they rose to the surface. The lungs are weak, squishy structures that cannot contain the laws of physics. The laws of physics will win every time. An ascent from three feet of water is enough to rupture a lung, hence the need for practice in advance.

The Admiralty had built the escape tower for such practice, and then the utility of midget submarines and divers had dawned on the Royal Navy. That same tower was available for troops to practice leaving X-craft more intentionally, for less desperate, more clandestine diving purposes. As a result, Hugh Irwin sat and waited.

After his third training partner disappeared into the training hatch above him, Irwin could hear the flow of water and air shifting around in the piping, out of sight behind the thick, isolating metal door. After some time and some more sounds, the door in the roof reopened, again dripping water, but now with nobody in the chamber. Irwin was the only one left in the cramped, air-filled space besides the instructor. It was his turn to go up.

Once Irwin was inside the small white upper space of the lock, the water began to fill slowly around his toes, flooding upward around his crouching body. It covered his feet, his ankles, his hips. It began to encroach on the bottom edge of the leather breathing bag of his rebreather. The air volume narrowed in kind, retreating upward past his waist, his shoulders, his neck, his chin, his lips. Before, he had been breathing into the rebreather because the instructors had told him to. Now he needed it. Now he had either to use the oxygen rebreather or suck down the cool, clear water of the training tank. It was oxygen rebreather or drown.

Irwin was supposed to sit still and let the piping of the escape hatch work its pumping process until the air was purged from the space. He was supposed to sit still, be patient, keep breathing until there was no air left in the hatch at all. The training was practice for leaving the X-craft, and if there was air when he exited, the gush of released air from the hatch hitting the surface would provide a gigantic, fizzing, bubbling red flag to any enemies above. Bubbles would signal to the enemy where to drop their bombs. The Axis had built midget subs first, so they would not be sufficiently mystified by the concept to provide time for a diver or crew to escape, should a diver open the hatch early.

But Hugh Irwin got impatient. It was only training, he figured while crouched inside the cramped white metal chamber. He rationalized to himself that he would be diving exclusively at night, so who cared if there were bubbles during his practice run? No watchful German eyes would be able to see them for lack of sunlight even in the real event when he exited the real X-craft. So why the hell would he wait? He "wanted to get out quick" and finish the day. He opened the upper lock door. There was still air in the hatch, but he sprang the upper door open anyway and climbed out into the clear friendly water. His rogue bolus of illegal bubbles hit the surface with a jubilant splash, and Irwin swam up at a controlled pace behind it.

When he reached the top Irwin was surprised to find a circle of officers glaring down at him from around the rim of the tank, all with disapproval etched into the lines of their faces. They had come there to watch the demonstration, to watch the Royal Navy's new, stealthy rebreather divers climb out of a practice X-craft hatch smoothly, without bubbles, in a way that would confound and dismay the enemy. The officers wanted to watch the divers escape the practice hatch and proceed to the surface to wreak mayhem as they would wreak it undetected against the hulls of enemy craft, beneath the Germans' very noses, in secret. They wanted a show. Hugh Irwin ruined their fun.

Irwin got into a bit of trouble, but the rest of the training went well, and the crew was declared officially ready to attempt the same maneuver to exit the hatch of a real X-craft in a Scottish loch. This was no small task; submarine escape training remains one of the most common sources of fatal air embolisms for navies around the world.

After they learned what they had signed up for, Irwin and the others were sent to Scotland and Port HHZ to practice in more realistic conditions, against HMS *Bonaventure*, until enough of the X-craft had been churned out by busy shipyards for a small number of eager crews to go on a real mission.

It was time to take down the gatekeeper, to clear the Atlantic. It was time for another run at the *Tirpitz*. The X-craft, the divers, and the science lined up.

———

"No nation had ever towed another submarine underwater before. The Germans said it couldn't be done. The Italians said it couldn't be done. We decided to have a go," submariner Roy Broome said of the reason he and the rest of the crew of the submarine HMS *Thrasher* were summoned to Loch Cairnbawn. Amid the mist and the porpoises of the lonely sea-loch, they practiced using their full-sized sub to tow the midget submarine *X-5*. The scheme worked. It was a feasible way to get the tiny boats to Norway with less stress on the X-craft crews inside.

Six new X-craft had been completed so far, and they were numbered uncreatively *X-5* through *X-10*. All six would go on the mission and divide up to aim for three different targets. The crews of *X-5* through *X-7* were assigned to the big show and sent to aim for the critical *Tirpitz*. Those in *X-8* through *X-10* were to go after the less critical German battleships *Scharnhorst* and *Lützow*. The parent submarines would tow the smaller subs from Scotland to Norway, where their targets lay in harbors.

One crew per sub would ride in each X-craft during the North Sea passage as a precaution against a repeat of the incident in which the Chariots had broken free and been lost. When it was time for the attacks, fresh, rested crews would leave the comparative comfort of the full-sized submarines, fold into the midget subs, and execute the missions.

Each midget's crew had a diver who could lock in and out through the wet and dry "W. and D." of the forward hatch, like they had practiced in the submarine escape training tower, to execute tasks in the water. Much of the German fleet was protected by skirts of steel netting to ward off small boats and torpedoes. The divers had spent plenty of time faux-attacking HMS *Bonaventure* by locking out of the X-craft to use special underwater tools to cut away such nets and allow their subs quiet, fully submerged passage. The divers were also in charge of attaching the mines directly to the undersides of the target enemy vessels because closer proximity between the charge and the hull would do more damage. The amount of time the divers had to breathe on pure oxygen at any depth remained uncertain, and there had been oxygen casualties in training. The crews were all wary of their oxygen rebreathers as a result, and loath to use them. Haldane and Spurway were under strict quarantine and unable to experiment because of an outbreak of chicken pox. HMS *Truculent* with the *X-6* in tow slid out of the narrow opening of Loch Cairnbawn and began the eight continuous days and nights of the journey to Norway.

———

After Haldane was sprung from quarantine while the X-craft were under tow, Ken Donald, underwater munitions specialist John Stuart Mould, and he climbed for a second time into the dry space of hyperbaric chamber number three. They had long wrapped up the CO_2 exposure tests for the midget subs, but in mid-September, as the X-craft mission went on, with no known directive describing why, the researchers of Siebe Gorman decided to shove one more stray test into the completed X-craft series.

Chamber number three was the small, horizontal one at Siebe Gorman, only a hundred cubic feet in volume, which happened to be roughly the same size as the internal gas space of an X-craft. Haldane, Donald, and Mould opened cylinders of CO_2 to bring the internal percentage up quickly without their having to wait to exhale it naturally.

They stayed at one atmosphere of pressure, just like in a midget submarine. The crew of three kept themselves sealed inside at those elevated carbon dioxide levels. They panted and huffed inside the oppressive levels of toxic gas for one hour and thirty-one minutes before switching to oxygen rebreathers. Haldane vomited into his own mouth, but he swallowed it. The situation was unpleasant, but survivable. It seemed that an X-craft crew could escape, if need be, so they would not be lost like the crew of the *Thetis*.

The towing got off to an inauspicious start. The submariners on the mission began with six big-sub-little-sub teams traveling on the surface, one for each X-craft. *X-9*'s tow rope broke in transit, and the tiny submarine and its passage crew snapped loose from their parent sub during particularly vicious waves. The parent ship simply could not find the small midget sub again in the open ocean, and the little boats did not have radios that could transmit. The *X-9* and its crew were lost forever at sea.

X-8's rope snapped, too, during a brief but steep dip after a smack from a wave. With his boat still functioning, however, that captain decided "to plough along on his own" toward Norway, in the hopes that the crew of the parent sub might come back and find them. They did, but the *X-8* had to be scuttled anyway after the side charges flooded and partially exploded, causing damage to the hull. Even before the passage to Norway was complete, there were only four of the six X-craft left.

The line of a floating mine snagged on the towing hawser of *X-7* and the mine "became impaled on the bows." The *X-7*'s captain, Godfrey Place, climbed up on the casing and kicked the deadly high-explosive charge free, taking care to avoid its pressure-sensitive horns. Still four X-craft left.

The newly minted *X-10* experienced mechanical problems, including a leak and a malfunctioning pump in the W and D compartment. Submarines do not work as well when filled with water, so the crew had to return to Scotland. Later analysis of German documents would show that their target ship the *Scharnhorst* was out on exercises anyway, so their target would not have been in place even if the boat had worked perfectly. Their convoy was down to three—the three aiming for the *Tirpitz*.

The waves had been "mountainous" in the days before the crew of HMS *Thrasher* finally let slip the hawser connecting their larger vessel to the smaller *X-5* behind it, but as the submariners worked through that process, the brutal North Atlantic winds died down with a serendipity that some of the crew interpreted as a portent. The attack crews had piled into the midgets with "high spirits," joy, and eagerness, even though the odds of success inside the small, risky vessels were so low that, according to the other submariners, the X-craft crews "basically knew they were not going to come back." The final three X-craft puttered off into the darkness, away from their host subs, just after midnight on the new day of September 22, 1943.

The new day did not bring new luck. The *X-6*'s periscope flooded. The crew had to surface because they could not see. The outer, anti-submarine netting of their target harbor was open to allow a small ship in, so they followed the ship rather than risk an oxygen dive to cut the net. They drove skillfully in the wake to render their own disturbance invisible. The Germans swung the gate closed behind them.

"So now we've had it as far as changing our mind," joked *X-6* crewman Edmund Goddard after the gate closed behind them. Once the *X-6* was within the harbor, one of the explosive charges flooded, and the small craft took on a decided list. By daybreak the crew found a small, pre-made gap in the second layer of steel netting closest to *Tirpitz*; the gap had been opened on purpose to allow smaller German surface craft to travel to and from the larger vessel. They had to stay on the surface because they could not navigate with their damaged periscope, but like Ensign Sakamaki before them, they decided to attack anyway.

In the growing light of the morning, the diminutive *X-6* aimed for the hull of that gray behemoth the *Tirpitz*. When German soldiers saw something intrude inside their protective netting, they started attacking the shape, firing rifles, throwing grenades, using basically whatever they could find, so the crew of the *X-6* could not deploy divers to execute the ideal careful attachment of their last remaining bomb. Grenades and bombs in the water might have killed a diver.

The X-craft got caught inside another set of nets. The crew "floundered around inside the nets" until, at 7:15 a.m., they set their bombs with a time delay of one hour and simply dropped them on the seafloor as close beneath the *Tirpitz* as they could. They bailed out of their entangled midget sub. Then they scuttled the craft on top of the ticking site of its own bombs, in the hopes that the explosion would damage it too much to provide good information about its construction. Afterward, they trod water on the surface while waiting to be taken prisoner. They hoped for the best.

The *X-7* got stuck sooner, first in the outer steel netting around four in the morning. Its commander, Godfrey Place, did not want to risk sending out a diver "unless it was really necessary." The water was too deep, too cold, too dangerous and the oxygen too risky. They were not trying to cut through a net, he figured; they were trying to extricate themselves from an entangled one, and who knew what the diver might face upon peeling open the forward hatch door? It would not be like practice. After maneuvering the little boat back and forth, back and forth as much as they could budge in the imprisoning meshwork of steel, they managed to shake themselves free.

Once in the harbor, they got stuck again, that time in the netting around the *Tirpitz* itself. While the crew of the *X-6* was actively working to scuttle their own craft, the crew of the *X-7* was nearby trying to shake themselves free from that second set of nets. They succeeded. But then they got caught a third time.

At 8:12 in the morning, "there was a tremendous explosion." The charges placed by the *X-6* on the seafloor went off on time, and they managed to blow off hearty chunks of the *Tirpitz*. After waves shook free the struggling *X-7*, the little boat surfaced. The crew thought they had been damaged by the friendly fire of the *X-6*'s unwittingly close blast, so their leader, Godfrey Place, climbed onto the casing to signal surrender. But before he could, German gunfire sank their vessel, sending the *X-7* once more to the bottom without control. Godfrey Place, in heavy boots and a thick wool sweater, stepped from the casing onto a nearby German gunnery target, from which he was plucked and taken prisoner.

Sublieutenant Bob Aitken, the sub's diver, plunged to the bottom with the *X-7*, as did the two other submariners of the crew. For that mission, they had packed four into the diminutive space, just like in the final chamber tests: three regular crew members and a specialist diver for cutting the nets and placing the charges. Because Aitken was the diver, he felt responsible, now that they were underwater, for helping the other two men escape. Crewmate Bill Whittam cut their DSEA sets down from an overhead storage rack. He passed one forward to Aitken and one to crewmate W. M. Whitley. Aitken gave the other two a brief refresher on their escape drill and how the hatches worked.

They opened vents to let in the sea. Letting in the water would relieve the pressure holding the hatches shut and allow them to swing open, so the crew could make their way to freedom. There was no way to get out without first letting the ocean in.

The icy cold water of Norway's fjords trickled in around the crew's ankles, sapping heat from their waiting bodies. Some of the water-inlet vents were stuck, so the process took even longer than it should have. Like Hugh Irwin in the escape trunk, except with no instructor monitoring their safety, the three felt the deadly dark water creep around their ankles. Then their hips. When the water rose to the level of the midget sub's batteries, the batteries shorted, spurting noxious fumes into the shrinking wedge of breathable atmosphere that remained. The crew had to start breathing the pure oxygen from the DSEAs to avoid the poisonous gas. Their boat sank to a depth of 120 feet. Water continued to seep through the vents. The pressure inside the submarine continued to increase as the water crept in slowly, too slowly. The broken vents meant Aitken, Whittam, and Whitley experienced the pressure for longer.

When the boat seemed to be flooded sufficiently for them to be able to move the hatches, Aitken tried to prize open the forward one—the one for the divers—but it would not budge yet. To check on the others, he pulled himself toward the stern of the boat through a flooded maze of freezing water. Whitley lay propped at an unnatural angle at his post by the flooded space of the periscope. He was dead. Aitken probed farther inside the dark frigid grave, which had previously felt too small but now was too cavernous to search. He was looking for his other crewmate, Whittam.

Aitken felt his breath stop short. The DSEAs contained only enough gas for shallow dives. Nobody was supposed to be breathing oxygen at 120 feet. Certainly not for the extended period of time it had taken for the remaining unbroken vents to fill the submarine. One hundred twenty feet was too far down. The deeper divers traveled, the more gas they used for each breath, and the bottles had not been sized for that depth. Aitken had run out of gas. He had nothing left but the breath in his lungs. He went for the hatch.

He blacked out on the ascent. He remembered regaining his vision sometime later as he shot upward through the water column, and being thrilled to find his body had executed the routine properly. Somehow, without dying, without seizing, without suffering an air embolism, he made it from the bottom to the surface. The Germans were impressed. Aitken, Godfrey Place, and the surviving crew of the *X-6* were treated as heroes by the Germans out of respect for the difficulty of what they had done. They were toasted and given hot drinks.

The fate of the remaining midget sub, the *X-5*, remains a mystery. It was glimpsed once by the Germans before they began firing at it, but it is not known whether the crew dropped their charges and contributed to the explosion beneath the *Tirpitz* or whether they were sunk before they could do so. The *X-5* and its crew were never seen again.

Of the six original X-craft, only the *X-10* returned to base, but the overall mission was nonetheless considered a success, because even though the *Tirpitz* had not been sunk, it had been hobbled. While the X-craft submariners were being interrogated and toasted in turn by their German captors, the damage to the mechanical leviathan became clear. *Tirpitz* would need extensive repairs before it could sail again. And after some surgically placed follow-up bombs dropped by the Royal Air Force while the *Tirpitz* was in dock getting those repairs, its destruction would be complete. The gatekeeper of the North Atlantic was dead, and the beaches were one step closer.

CHAPTER 15

Airs of Overlord

I t had taken all of Randy Lovelace's charm and finesse to talk his way into his current situation. The young aerospace physician stood poised 40,200 feet above the crust of the earth, and unrestrained, he stared down through the open hole in the airplane's metal belly. The symmetrical rectangular doors of the bomb bay of the B-17E Flying Fortress had been peeled open, creating a chasm in the bottom of the ribbed, industrial-looking compartment; savage winds whipped throughout. Lovelace had a dazzling, unprotected view of the mountainous Washington landscape through this glass-less picture window, and the breadth of the space between him and the land below was wide enough that the very color of the air itself tinctured the ground with a hazy gray-blue. The picture window beckoned him.

Lovelace was in his mid-thirties; he had a broad mouth and close-set dark eyes. Two oxygen bottles were strapped to his legs. Over his nose and mouth, he wore an oxygen mask of his own creation. Without it, at that height, he would have been unconscious in seconds. He was dressed for thorough protection against the unforgiving chill of that extreme altitude, which was almost half again higher than Mount Everest, and he wore two gloves on each hand in addition to his thick, full-body thermal protection.

He jumped. Into the open square, into the brutal gray-blue of the cavernous, windy nothing. The B-17E continued the slow, spiraling descent that had prompted his departure, and Lovelace plummeted feetfirst with much less control. His body hurtled first through the bomb bay doors and then through the sparseness of high altitude, the comparative heft of his human flesh forcing its way through the puny resistive efforts of the thin atmosphere. He reached a remarkable velocity remarkably quickly. The tether connecting him to the airplane pulled taut. From the other end, the tether's tension peeled loose his carefully packed parachute, a glorious twenty-eight-foot circle of fluttering nylon that snapped with abrupt attention into its fully expanded shape.

With the expansion of his parachute, Lovelace's hurtling body snapped, too. The deployment of the massive source of drag jolted him to a relative near stop. The change in speed was sudden. It was focused on his torso. His body buckled like his chest was being ripped backward by the unseen claw of a giant hand. His head and neck whipped forward. Then backward. Then forward again. His limbs whipped the same. Both gloves tore free from one hand from the force of the ricochet. He was knocked unconscious. He dangled limply, unmoving, like a neglected puppet suspended from a floating nylon circle in the sky.

Thanks to good engineering, however, his oxygen-mask contraption worked. Over the course of his ambling twenty-three-minute-and-fifty-one-second float back down to earth, the mask he had built gave him lifesaving oxygen, and he regained consciousness as he drifted. He settled to the ground awake and no worse for the experience—mostly—with his only reported injury being frostbite on the degloved hand.

It was his first-ever skydive. He had jumped not for the military, not directly at least, but instead for science. William Randolph "Randy" Lovelace II also believed in testing on himself first. Since the first murmurs that war might come, he and the other scientists of America's aeromedical research departments had been hard at work, trying to do for those in the sky what the UCL lab was doing for those underwater. Most of them in the United States, like their counterparts across the pond, used themselves as the guinea pigs.

Fighter pilots can get decompression sickness, too. Any too rapid decrease in pressure is sufficient to allow dissolved nitrogen to spring forth from the hidden chemical crevices of the body and wreak havoc as bubbles in the tissues. A quick ascent from the depths of the ocean is the most obvious scenario for decreasing pressure, but pilots rocketing from low levels to high altitudes are subject to it, too. If pressure decreases fast enough, they can get just as sick as divers. Paralysis is rare but possible. However, previous wars had been limited to sputtering biplanes and slow-rising dirigibles, which hardly met the criteria for biologically quick ascents. So the problem had not yet been studied. Lovelace and the others in his clan of aerospace medical experts had discovered and proved that breathing oxygen before a flight was the optimal way for fighter pilots to avoid DCS. Breathing pure oxygen in advance was like studying for a test; the "prebreathe" allowed the nitrogen to work its way out before the real-deal ascent.

Lovelace's rogue, unsanctioned jump demonstrated that aviators could survive bailing out of airplanes at high altitudes, altitudes without enough oxygen to sustain life, provided they were given the right kind of breathing equipment. His breathing equipment. Although, he also recommended, there should be a redesign to make parachutes open slower at heights with such thin air.

Lovelace theoretically should have gotten in trouble for his pseudo-piracy of the aircraft and his self-risk, if the proper policies had been followed, the recommended paperwork filed. Instead, the military awarded him a medal, because he had solved a problem.

While the British were hard at work developing tools and tactics for survival underwater, the Americans focused on the same in the air. The two teammate countries traded insights. The scientific problems of survival undersea and in the air are shockingly similar. People need to bring their own breathing gas, because neither world has enough oxygen to sustain us. Ascend or descend too fast and you risk decompression sickness. Both worlds are cold enough to chill you to the bone. Breathing in both environments is a scientifically purchased bargain with the clock before the environment manages to kill you.

Air support had been lacking before the landings at Dieppe, and as a result too many of the German armaments onshore had remained intact. Air support could not be lacking before Normandy. Like the members of the UCL lab still pressurizing themselves to depth on a near daily basis, the Yankee scientists were climbing into chambers to make the military needs possible. Except their chambers went to altitude. As did the Germans'.

German physician Theodor Benzinger climbed into the wheeled metal pressure chamber he and his compatriots had built at the *Erpro-bungsstelle Rechlin* experimental aeronautical base northwest of Berlin. Roughly six feet tall and eight feet in diameter, their sturdy upright cylinder spoke to the universality of pressure-resistant designs through its similarity to the circular shapes of the equipment at Siebe Gorman. That one had casters attached to its four gangly support legs so it could be rolled to new positions; it also had horizontal rows of metal-reinforced, slatted windows sliced into its girth.

The entry door was a vertical, rounded rectangle in the side, and it was ringed by a perimeter of heavy-duty restraining bolts. After Benzinger—with his buzz-cut hair, striking high cheekbones, and straight, strong nose—climbed through, someone pushed that door shut behind him. He took his seat on the rounded white metal basin that served as a chair and settled in. Nearby, there was another chamber attached to his through a set of pipes and valves. Personnel on the outside worked the valves and levers to depressurize the other chamber to an altitude of fifteen thousand meters, or 49,200 feet, which was higher than Lovelace's jump. They depressurized Benzinger's chamber to three thousand meters, or 9,800 feet. He cleared his ears and waited.

Once he reached his assigned altitude, Benzinger yanked backward on the long white handle near his right hand; that handle opened the valves that separated the two chambers. Gas burst out of his chamber through the newly opened pipes into its lower-pressure neighbor. It was a laboratory simulation of what would happen when a bullet or a bomb or really any sturdy object tore through the pressure-proof canopy of a military aircraft in flight and the pressure got released: explosive decompression. The roar of the traveling gas reverberated through the metal cylinders as the disparity in pressure between the two spaces sucked the air and oxygen from Benzinger's chamber and from Benzinger's lungs.

He knew to breathe normally and exhale, and he avoided an embolism. His chamber was fully depressurized to the dizzying height of fifteen thousand meters in less than six seconds. He gasped for breath.

Questions of physiology know no political boundaries. The Germans, too, had new airplanes; they, too, had new submarines. They had their own underwater oxygen breathing apparatus, the *Tauchretter*, so the Germans, too, had the same questions of physiology. The war was a race of technology and science: a race to develop the first radar against submarines; a race to develop planes that could go higher, faster, longer; a race to turn human beings into amphibious beasts. As the Americans worked on aviation and the British worked on diving, the Germans worked on both.

"As this problem [of explosive decompression] could not be solved by animal experiments," Benzinger wrote, with echoes of J. B. S. Haldane, "the investigators tested the effects of explosive decompression on themselves."

Benzinger was shocked to survive his own chamber test. He wrote, many decades later, after Operation Paperclip brought him and many other German aerospace scientists to American shores: "The American authors probably were as surprised as their German colleagues when experience showed that the physical phenomena of explosive decompression have no harmful effect on man." They had all expected to die in the testing. Benzinger's use of the word "colleagues" to describe the scientists of warring factions was a bit of a stretch, especially since many Nazi physicians and scientists are rightfully lodged in the halls of infamy for their forcible and evil experiments on unwilling, usually Jewish subjects. Even Benzinger's compilation book of aerospace research from other German scientists has suspicious gaps in some of the methods.

The Americans and the Germans reached similar conclusions. Above about thirteen thousand feet, pilots need oxygen. Below about thirty thousand feet, pilots could survive if they were in an unpressurized cabin with oxygen. Above that limit, however, not even pure oxygen was enough. The cabins needed to be pressurized as well. The good news was that, should explosive decompression occur, the aviators could survive.

Should they need to bail out at 40,200 feet after a sudden reduction in pressure, they had Lovelace's breathing system, which would get them down safely. With oxygen, and with an oxygen prebreathe beforehand to reduce their chances of decompression sickness, they had "several minutes" to drift their leisurely way down to a lower altitude. Otherwise, they would have less than fifteen seconds.

Around the same time as Benzinger, Hans Kalmus also climbed into a chamber, but a high-pressure one. Had Benzinger's political leaders been less prone to abusing minority groups as easy scapegoats for their political problems, the two men might have been true colleagues. Instead, they raced. Theodor Benzinger for the glory of his homeland. Hans Kalmus for his family's survival.

Kalmus and his family had just returned from a holiday on which they visited Naomi Mitchison at her estate in Scotland. The immigrant refugees had previously been under government-ordered travel restrictions that confined them to the areas of the bombings for so long that Hans, Nussy, and their boys, Peter and George, had all gone a bit stir-crazy. So the moment the government loosened the travel restrictions, the Kalmuses jumped at the chance to take a break, and Mitchison was happy to play host. She had been hosting friends and strangers since the bombings began, in addition to organizing community dinners to alleviate the workloads of the local women, many of whom were faced with both raising families and working full-time.

The Kalmuses got a blissful if temporary reprieve from the war. They rowed small boats and waded in the ocean with Naomi to cast nets to catch iridescent fish at night. Peter and George were thrilled by the dragons that Naomi had let another guest paint on the walls of a bedroom. The happy family even found a camera with a few spare frames of heavily rationed film to capture the moments.

But the break was over. Now it was time to see if Hans Kalmus's vacation from the testing had affected his oxygen tolerance, either by increasing or decreasing it. The goal was to take advantage of his idyllic reprieve to see if there was a way to enhance the oxygen tolerance of military divers, either by pressurizing them or giving them a break from diving before their missions. Helen Spurway stared into Kalmus's face while he breathed through the rubber mouthpiece of a Salvus oxygen rebreather at a depth of sixty-six feet. He made it 11.5 minutes before he began to twitch.

Immediately after his removal from the chamber, lanky athlete Hans Kalmus folded himself onto a bench in the office of Siebe Gorman's tank room. With a light illuminating his angular face from the right-hand side, he stared straight ahead at a paper chart marked with circles and angles, his nose aligned with the center line. He carefully indicated where he could and could not see. Either he or Helen Spurway made small dashes on the chart to mark what was beyond his field of vision. They were testing to see if the oxygen had any lasting effects, to see if the divers would have any blind spots when they crept from the ocean.

Six days later, Kalmus repeated the experiment with a blend of 81.5 percent oxygen at a depth of 105 feet, to make sure the two gas mixtures acted the same. If the Admiralty gave oxygen-rich gas blends to divers, might divers experience blind spots during critical moments in combat? Kalmus lasted 9.5 minutes before twitching. The mixtures both narrowed his peripheral vision, but neither his vacation nor the type of gas blend seemed to make a difference as to how much.

Ursula Philip performed the same test a few weeks later. She breathed on the Salvus rebreather. Helen Spurway struggled to take her weak pulse. Philip reported she did "not feel right." She began "shivering violently." She began giggling. The personnel began to decompress. Philip was "laughing uncontrollably" as bloodstained watery discharge leaked from her nose. Hans Kalmus monitored her, watching her pale skin for twitches. Philip stared at the same paper printout of an eye chart. She plotted where her deep brown eyes could and could not see. Across the North Sea, different Germans did similar work.

The tests were a success in the scientific sense of the word, in that the mixtures were working as designed. The oxygen-rich blends could be used to dive deeper than pure oxygen would allow, but with less risk of decompression sickness than air. They were a working compromise with the sea.

There were not enough trained Combined Operations Pilotage Parties swimmers to do the necessary beach scouting for Operation Overlord. There were not enough X-craft submariners to send any to get COPP training. Even if there had been, the COPP trainers would not have been at the depot because the extreme shortage meant they were being deployed. COPP 1, the instructor unit, was picked for the Operation Overlord mission because it was "the only complete unit in the country at the moment." The COPP training personnel would need to learn to be X-craft submariners, too, and fast.

"In connection with beach reconnaissance for Operation Overlord, it was decided that there was a requirement for up to four X-craft for use with COPP units for reconnaissance," one planning meeting concluded. To complicate things, though, all but one X-craft had recently been sunk or blown up in a hostile-held Norwegian fjord where their metal carcasses could not be recovered. The American partners across the ocean had also been promised an X-craft, and they were tapping their fingers on the proverbial table, waiting to receive it.

The Overlord planners whittled the number of teams required from four to three. The Italians had thirty operational *Maiale* roaming the ocean, and the Allies struggled to prepare their first X-craft–COPP team. However, three was the lowest they could force the number. "These midget submarines were the only carrier craft that would be expected to work in the area without fear of detection by the enemy," they recorded.

Meanwhile, elsewhere in Norfolk House, other Overlord planners in the Admiralty discussed the problem of underwater obstacles. Not just natural obstacles like the reefs encountered by Ensign Kazuo Sakamaki but also manufactured obstacles such as concrete and metal constructions, some even bedecked with explosives, that could be hidden beneath the surface of the water. Before Dieppe, manufactured underwater obstacles had lingered in the backs of the minds of the more prescient tactical planners as a theoretical possibility. Few had thought such obstacles would be placed on defensive beaches, though, so they were not a concern—not really. There was some evidence the Germans had tried placing barbed wire underwater, but it seemed to rust away too fast to be effective. Some personnel had tried to make feeble efforts to plan for underwater obstacles in advance of Dieppe, to figure out how to dismantle them or render them harmless, just in case. But it seems as if so long as such obstacles remained mere theory, those personnel were shouted down.

"There is no user demand for this method, so why not concentrate on land methods, improving our underwater technique where necessary," wrote one officer in charge of combined training methods, brushing off the concept and refusing to allocate resources to the experiments.

By the month after Dieppe, however, the perspectives had switched. "The evolution of a technique for destroying underwater obstacles is a most urgent matter, particularly in view of their discovery for the first time at Dieppe," the problem statements suddenly read. Often in the military, especially during wartime, nothing is considered a problem until it becomes an emergency. The big brains went to work tossing around fun ideas for obstacle disposal such as throwing depth charges at them with launchers from above. However, in true government fashion, the biggest hurdle would turn out to be deciding who was responsible.

The obstacles were underwater or mostly underwater at high tide, which suggested the navy. But they were exposed at low tide, which suggested the army. The conflict dragged on for months of debate, even though the new tactics had been declared an "urgent" need. The level of hostility in the written notes of the discussions, which hinged on the surprisingly fuzzy definitions of dry land and shallow water, seemed sufficiently contentious that a suggestion of teamwork might have exploded the entire conference room in which the conversations were held. Should the demarcation line be high water? Or should it be two feet of depth at high water? What about using the line of the tide at the time of the planned landing? What was the predicted depth at which landing craft would ground, and would it make sense for the army to be in charge of water shallower than that, with the navy in charge of deeper? Or should the army be in charge of taking their people from start to finish?

On top of that, which personnel were sufficiently trained in explosives that they could be redirected toward such obstacle work and retrained for doing it underwater in time?

The answer to the last question, at least, was easy. The navy had already formed an experimental group called the Boom Commando Party, with a tasking slightly less exciting than the name. They were in charge of swimming on the surface and using a variety of methods, including but not limited to explosives, to dismantle the protective booms guarding the mouths of hostile harbors.

Keith May Briggs ended up in that group sort of by accident. He had been an apprentice plumber before he joined the Royal Marines at age seventeen to get out of some "family troubles," but he then found himself constantly seasick when on boats. His original job was to serve food in the wardrooms, but he often heaved over the questionable vittles while nauseated by the shipboard motion alone. So, to escape shipboard service, he took and passed a test to become a plumber on a shore outfit, which ended up getting dissolved to have all its personnel funneled into the Royal Marine engineers instead. He learned "all work to do with explosives" and went through the same bomb and demolition courses as the sappers from the Royal Engineers. Then, one day, a recruiter asked if anyone in the group knew how to swim. He raised his hand. He became a not-quite-volunteer member of the Boom Commando Party—in the sea, rather than on it.

After more months of tiring memos debating the exact name, the appropriateness of the words "commando" and "party," and the order the words should appear in, the Boom Commando Party would eventually be given the far-less-appealing moniker "Landing Craft Obstruction Clearance Units," or LCOCU. However, thankfully, their work got started before the government agreed on their title. A series of reconstructed replicas of German underwater obstacles had already been built at a naval facility called Appledore, on a north-facing beach on a western shore of southern England. In the loosely constructed Appledore "camp," which was really a series of low-cost, prefabricated corrugated-steel structures called Nissen huts, Keith Briggs and the other swimmers of LCOCU were charged with using their explosives skills to figure out how to destroy the obstacles—how to render them impotent in the fastest, most efficient way possible.

Firing bundles of detonating cord out of their pistols at the structures did not make an impact. Dragging long, thick tubes of explosives over the top before setting them off blew holes in the sand but made a "negligible" impact on metal and concrete. Mines thrown downward at the obstacles from launchers often missed and caused less damage than desired. Essentially, the LCOCU experimenters found that unless the explosives were placed with "close contact between the charge and the obstacle, and unless very large charges are used," the bombs did nothing useful. They blew giant holes in the sand, since sand is more pliable and easily shifted than concrete and metal, but adding giant holes to the landing pathways of troops created more of a barrier to progress, not less of one. The expression about "almost being good enough for horseshoes and hand grenades," it turned out, did not apply underwater.

However, also in the arsenal of the LCOCU personnel were the DSEA oxygen rebreathers. Their Nissen hut camp had a pump to fill the oxygen bottles. So these expert swimmers donned thin rubber suits for protection and walked into the water from the sandy, gradually sloping shores. They used the rebreathers to spend careful time underwater, placing explosives at strategic points on the obstacles that had been "copied from photographs of Normandy." Across the ocean, the Americans held their breath to dive but worked on the same concepts, with matching replicas in the waters off Fort Pierce, Florida. Both groups assembled lengthy catalogues of techniques for use against underwater obstacles of every conceivable shape and design.

The divers did not know what they were training for, because their unit name had not yet been switched to the more obvious Landing Craft Obstacle Clearance Units, but Keith Briggs had his suspicions. They had three main obstacles to learn to target. Element C, also known as a Belgian gate, was one of them. It was a massive, ten-foot-tall, ten-foot-wide wedge-shaped scaffold welded together out of five-inch-wide steel angle iron. The German troops would position the narrow end of the wedge facing toward land, and the broad, flat side facing the sea. The Element C had spikes of steel projecting from the top of the flat ocean-facing side, and those spikes were often topped with explosive charges. Landing vessels would get blocked by the strong wedge shapes, torn apart by the spikes, blown up by the charges, or all three.

The divers were also expecting tetrahedra, angled bars of steel or concrete joined into a sturdy, sharp, four-sided pyramid, also often topped with explosives. Finally, they needed to know how to dismantle hedgehogs. The hedgehogs were bundles of three straight bars of steel attached in the middle so that they looked like jacks from the children's game. No matter how the hedgehogs were rolled, some of their jagged corners always pointed upward toward the bellies of ships. They, too, were often bedecked with bombs.

Blowing the steel frames to bits was a futile effort, but if the charges were placed accurately enough, the frames could be collapsed. Working in the chilly waters off Appledore, the LCOCU divers figured out that the Element C could be collapsed from its prohibitive height of ten feet down to an impotent pile no thicker than a few inches. If a diver could get down there to place the explosives in exactly the right place.

LCOCU frogmen leaving the water at Appledore following training. *(IWM 28997)*

A US Navy frogman works to dismantle a partially submerged hedgehog during training.

"Placing charges accurately enough from landing craft is not easy," the LCOCU experimenters wrote after one attempt to perform the same precision blasting from a boat on the surface. In contrast, though, a swimmer "underwater is comparatively safe from swell." In other words, the ocean's swell and surface waves were less able to throw around those who were fully underwater. The risk was greater, but so was the accuracy.

The COPP scouts were the key to finding where the obstacles and explosives would be. The X-craft were the key to getting those COPP scouts ashore. The oxygen rebreathers would be the key to LCOCU dismantling the obstacles. The UCL lab's work had provided the key to using the oxygen rebreathers.

By November 18, 1943, COPP 1 was in Loch Cairnbawn learning about submarines and experimenting with methods of deploying dinghies

A row of Element C "Belgian gates" guards a beach. Additional obstacles farther up the beach are tetrahedra and solitary log posts topped with explosives.

from X-craft. LCOCU diver Keith Briggs was hard at work training against Element C and hedgehogs at Appledore, using an oxygen re-breather to dive beneath the waves. At 1:26 p.m. that same day, blood-stained fluids oozed out of the nose of Ursula Philip as her chamber reached the surface after yet another oxygen test.

One hour after Philip's test, at 2:30 p.m. on that otherwise innocuous Thursday, seven members of the Admiralty's Sub-Committee on Under-water Physiology, including Ken Donald, held a secret meeting "in the absence of Professor Haldane." Something was up.

The notes and accompanying memos of that secret meeting dance around naming who called it, but that truth is evident through process of elimination among the speaking characters. The notes are clear that *someone* on the Sub-Committee on Underwater Physiology had complained about Haldane, specifically about his personality. That someone was present at the meeting to discuss kicking him out. There were only seven people present. The others present weighed in to agree with the original complainant.

"Having had personal experience in dealing with Professor Haldane on other occasions, [the representative] can fully endorse the suggestion of the Sub-Committee that he is a most difficult person to work with."

"The individual experimental work carried out by Professor Haldane and his assistants has become to some extent a hindrance to the programme of work by service personnel."

"It is understood that difficulties of this kind have arisen elsewhere, and are not confined to [Haldane's] association with the navy," one member said in the context of talking about Haldane as a difficult person.

"The personality of Professor Haldane and the presence of his assistants at Messrs. Siebe Gorman's compromises [*sic*] the smooth working of the Admiralty Experimental Diving Unit."

Young and brash, Ken Donald was the kind of person who would later take full credit for starting the Admiralty Experimental Diving Unit even though the real officer in charge was William O. Shelford. Donald, based on the incompleteness of his own scientific reports, was seemingly unable to perform the basic math needed to analyze the results of the research that he claimed he was in charge of. It is clear from the pattern of the speakers in the notes that Ken Donald requested the meeting, and he requested it because he did not like J. B. S. Haldane.

Few did. Haldane was always convinced he was right, and he lacked basic social graces. At one point, years later, he would live with Helen Spurway in a cat-saturated apartment a few floors above Elizabeth Jermyn; her then husband, Nariman Bamji; and their young son, Andrew. When Haldane was accidentally delivered a telegram meant for Bamji, Haldane never bothered to pass it along, even though it had news of the death of Bamji's brother. When Bamji confronted Haldane about the telegram after finding out about it weeks later, Haldane responded with the explosive and breathtakingly tactless declaration, "What does it matter? He's dead anyway!"

However, sometimes it is the pushiest, most tactless leaders who make the most progress, especially in wartime. Haldane and his group had, at that point, conducted 562 of the most novel, trailblazing, and informative tests in diving history. All on themselves.

But Ken Donald worked for the government, which meant he held power. Also, because he could order military divers into the tanks, he had access to more young, breathing bodies than Haldane could muster.

The subcommittee sided with Donald. They decided to tell Haldane that all experiments on civilians, in all fields of science, were going to be stopped. Their strategy was to hide from Haldane's righteous fury behind the badge of the government: "It is almost certain that a hornet's nest will be stirred up and therefore, I think it is better for you to shield behind the Admiralty organization." However, they still wanted to be able to ask Haldane questions about physiology and to keep him in the working group, so they would have access to him when they wanted him. So they lied and pretended their personal dislike was a new policy.

Plus, UCL was starting to ask that the government pay Haldane's salary, since he was working for them full-time. The government repeatedly, for months, had been dodging the issue. Kicking Haldane out would get them out of paying him.

Haldane bought their explanation. He accepted their decision gracefully. They told him he had until January 31, 1944, at which point he would be asked to stop experiments at Siebe Gorman. But they hoped after that he would still be at the disposal of the subcommittee.

He replied with class: "I am, of course, very willing to do so, and am already attempting to arrange for part of the work with which I am concerned to be carried out elsewhere." Haldane did not get it. The bullies of his youth were back, trapping him once again under a table loaded with sandbags as punishment for his inability to emotionally connect with their group. In the midst of the discussions, Haldane demonstrated his ability to cooperate productively as he traded results and analyses with Commander Oscar D. Yarborough of the Experimental Diving Unit in

Washington, DC. Yarborough was asking him for more data points about oxygen diving. Their Experimental Diving Unit had done some work, but not enough. Haldane sent his data.

A month after the decision to exclude him, physiologist George Lindor Brown, the chairman of the subcommittee and the person who had presided over the eviction meeting, wrote to Haldane a secret letter asking him to verify a report about diving written by Ken Donald. Brown wanted Haldane to check Donald's work to make sure that Donald had correctly interpreted how to use oxygen-rich gas blends for bomb-disposal divers. A copy of that letter never made it into the official subcommittee files.

CHAPTER 16

The Operations

The crews of the Admiralty's smaller boats agreed that "slight" on the wave-grading scale was rough "from the point of view of a very small craft." On New Year's Eve, Friday, December 31, 1943, much of England took the day to muster what level of celebratory glitz and glam they could before that night's blackout restrictions. The crews of two little Allied boats, however, stayed on duty. They worked their way through the wintry afternoon waters off the northern coast of Normandy. Their craft were very small. The seas were "slight."

Lieutenant Commander Nigel Clogstoun-Willmott hunkered down low between the gunwales of the gray landing craft, personnel (large) (LCP[L]), otherwise known as a Higgins boat. The crew transport vessel was modest in size despite its name, and as it had been intended as a short-distance shuttle for personnel, it had no top or protective covering to shelter its passengers from the steady rain now pouring down around them. The boat was based on a design born out of the oil fields of Louisiana, and it was at odds with the mightier forces of the English Channel. Clogstoun-Willmott and his fellow COPP 1 instructors, Major Logan Scott-Bowden and Sergeant Bruce Ogden-Smith, got rocked by every surge of water hitting the tiny craft from the "slight lumpy cross swell" out of the northwest.

Scott-Bowden was an officer with the Royal Engineers on the edge of turning twenty-four, with heavy, thoughtful eyebrows and the noble cleft chin of an illustrated cartoon warrior. Ogden-Smith had been born at the tail end of the First World War and had almost two full years on Scott-Bowden, although his full cheeks, dense, wavy hair, and sparse mustache gave him the look of someone much younger. Clogstoun-Willmott was the head of the new, crack team, and they had come a long way from improvised foldable canoes and rogue scouting missions from submarines.

Those three COPP instructors, at least, were wearing rain gear in the form of their surface swimsuits. The thick, warm insulating suits covered them from head to toe in a medium tinge of robust khaki designed to blur into any European beach. The rope soles of the integrated boots worked equally well to gain traction on a wet boat deck as they did across a rocky ocean floor. The suits even had pockets specially sized for brandy flasks and combat knives. The rest of the crew on board LCP(L) 290 for this mission—Lieutenants Glen, Wild, and Richards—lacked the same protection. As did the crew of LCP(L) 292, the other small craft driving through the rain as a standby. To add to the misery, everyone was seasick.

Clogstoun-Willmott and the others on board 290 had spent too much time trying to get their new QH radar system working. The passage across the Channel had gone smoothly, according to plan, with large escort ships towing the LCP(L)s and everyone riding on board those bigger, more stable vessels. But then they fiddled with the QH systems for too long. They lost all the time they had gained in the efficient passage. To stay on schedule, to maximize the cover of the darkness of the coming night, they now had to plow the smaller craft through the swell and the waves at full speed, resulting in their seasickness.

Scott-Bowden and Ogden-Smith had been practicing inside X-craft and were now proficient in them. Ogden-Smith had even needed to place a rush order for a new surface swimming suit because of "the high death rate of suits" and his number of hours inside the tight, snag-filled quarters of the midget subs. The newly constructed *X-20* had received the royal treatment: a special transport train for a pampered, cautiously plodding four-day journey to the X-craft's departure point off the southern coast in Portsmouth.

The COPP personnel did not want to waste the precious moments of the waxing-moon period before and after New Year's Eve 1943. If they did, their beach scouting for Operation Overlord would be even further behind, and the beach landing might not be possible in the coming new year. But the waves would not comply.

The X-craft were high risk and in short supply. COPP 1 had been on only one full training exercise with the crew of *X-20*. Clogstoun-Willmott had barely recovered from the flu. They were still testing out how few COPP personnel they could manage with for the operation and, conversely, how many they could pack into an X-craft and keep the breathing gas survivable for five full days.

"It had not, at this time," explained one formal report, "been fully established that two X-craft personnel only were an adequate crew for a 5-day operation, particularly when five people (as against the usual three) were breathing the atmosphere in the boat."

It was too high-risk to use the miniature submarines, the C in C had decided. The weather was foul, and the submarines too precious. He changed the plan the day before the mission. The COPP scouts would have to try to deploy from the LCP(L) instead and hope the surface craft did not get spotted by sentries onshore. The "policy with regard to enemy forces" was one simple word: "evasion."

By 2200 hours Friday, the radar conditions had improved, and the newfound signal let the QH help Clogstoun-Willmott correct their location. They were seven miles offshore. He shifted their heading and pointed them directly toward the target beach. The swell now propelled them from behind. They pushed the LCP(L) to maximum speed while the last reflected glow of the winter sun disappeared beyond the western horizon. The beam of the Pointe de Ver lighthouse onshore emerged before them through the rain and the dark mist, three distinctive flashes of light grouped and recurring every fifteen seconds, confirming their new course heading was correct. The sight of the beam came as a relief to Clogstoun-Willmott.

An hour before midnight, the crew of LCP(L) 292 lowered a hurriedly adapted dinghy anchor until they felt it lock into the bottom, seven feet below the surface. They had lost their own anchor to the rough seas in tow. The stern of their boat pivoted around the firmly set point and swung to stream. On LCP(L) 290, Scott-Bowden and Ogden-Smith grabbed the depth-measuring pole that Clogstoun-Willmott and they had improvised en route. It was annoyingly short. They had tied fluttering handkerchiefs and other miscellanea around it at measured increments. Their original sixteen-foot-long black-and-white measuring pole had been lost to the waves, too.

The rain continued to pour. Scott-Bowden and Ogden-Smith slipped into the water. They had 170 yards to swim to reach an estimated water depth of three feet, where their legs would comfortably reach bottom, and 150 yards to wade after that. The incoming waves swept them toward the blackout-darkened French beach town of La Rivière.

Once in the shallows, while wading, Scott-Bowden thought he felt soft mud beneath the soles of his boots, or perhaps peat. It was firm but slippery. He tried to hold his breath, dive down, and take a sample with an auger, but he got hammered by the surf. The dark glop ran out of the auger tube as soon as he pulled it out of the water. Instead, as a compromise, he reached down with one hand and grabbed a fistful of the underwater mystery terrain. He rubbed the muck across his face, using the sensitive, exposed skin of his cheeks and other facial features to examine its texture, to assess if it would hold a tank. Ogden-Smith and he had reached the depth ranges where that might matter, depending on the state of the tide.

By the stroke of midnight, they reached the dry shore. The blackout in the French town was complete. Neither scout saw any sign of life onshore. They stayed hunkered down anyway, as they had been trained.

The slowly spinning beacon of the lighthouse above them onshore backlit the long, unfurled coils of razor-sharp concertina wire the Germans had used to crown the exits. The lighthouse was like a mechanical spinning spy, a friendly French presence on the occupied shore, and with each sweeping flash, it revealed to the scouts where Nazi defenses had—but also had not—been placed. "A gap in the wire exists 5 yards west of Groyne B," Scott-Bowden and Logan-Smith noted. The gap could be a clean egress point for landing troops. The beach had groynes built outward into the water to limit beach erosion, and those low seawalls were proving useful for mapping.

A sliver of light broke through the darkness up by the lighthouse. It was someone opening a door, presumably a German trooper, given the hour was past the curfew for regular citizens. Scott-Bowden and Ogden-Smith froze in silence. The light disappeared. They continued to work westward over the sands. Their current beach was "a belt of clean sand on top of rock" about eighty yards across. They took a sample of sand and placed it in container number one, one of the many small tubes wrapped around them in belts like bandoliers. They probed the depth of the sand. It varied between three and nine inches, but it was never more than nine inches deep above solid rock. The duo continued to take samples as they moved down the length of the shore.

Ogden-Smith huddled on Scott-Bowden's right, looking inland, and he thought he detected movement at the back of the beach a mere forty yards away from their own flattened shapes. Ogden-Smith saw a light blink twice across his vision. He thought it might have been a flashlight. He signaled Scott-Bowden. They froze again. Thankfully, they never found out for sure what Scott-Bowden had seen. The rain worked to their advantage, and "the visibility was very bad." They waited, and the light disappeared on its own.

Another section of beach was covered with long stones "worn smooth and flat by the sea." The largest were about eighteen inches long and less than three inches thick. The duo pocketed more rocks to bring back as examples.

The two scouts had been briefed before their mission with aerial photographs and interpretive intelligence from the 21st Army Group. The aerial photographs were spectacular but needed correcting in places. "We both observed the difference in appearance of Groyne A from that indicated and studied previously on Aerial photographs," Scott-Bowden wrote.

The clock counted down to the new year of 1944 with the two men flattened against the beach, placing soil and sand in sample tubes and unrolled condoms. A few minutes past midnight, Logan Scott-Bowden and Bruce Ogden-Smith edged their way back into the vigorously break-ing surf to get off the beach that would one day be code-named "Gold." The waves battered them, treating their bulky insulated swimsuits and equipment like aquatic sails to hurl them back onto the shore, and it took them half an hour to swim the 250 yards out to the small craft wait-ing for them at anchor. In the process they lost a fighting knife and both of their sampling augers.

The tools, they hoped, would stay buried without detection. They watched new waves roll in and knew their footsteps onshore "would be washed out by the flooding tide." They had timed the mission that way. By the time the German troops woke up the next day, all visible traces of Operation KJH would be gone.

Scott-Bowden and Ogden-Smith drew their detailed measurements up into a map and summarized their assessment in a report that they presented along with the samples of soil, rock, sand, and seaweed. They highlighted the places where the German defenses were weak. They re-turned their information to the 21st Army Group, who had determined the simple priority for this operation: "The important object was the ex-amination of certain beaches." Examined the beaches the two had.

Clogstoun-Willmott, Scott-Bowden, and Ogden-Smith were aware of the import of their acts. "On these operations depends to a very great extent the final success of Operation Overlord," Clogstoun-Willmott wrote in his final report about the scouting of beaches Juno and Gold.

There were many more beaches to be covered. Scott-Bowden and Ogden-Smith had spent a mere forty-five minutes on land so far, but they needed to cover all of the terrain that would be swarmed by the armies of many nations. They needed more information, and to get it, they needed the *X-20*.

The crew of Operation KJH returned to England on January 2, 1944. They met with the photography scouting masters of the 21st Army Group by noon the next day. There was no time for a break. They immediately started working on modifying their techniques to finalize how they would operate out of an X-craft. The moon would once again slip into an acceptable phase of darkness on January 17. They had two weeks to pre-pare.

In between the COPP missions, the UCL lab conducted eleven more experiments on themselves in their efforts to finalize their scientific answers about using oxygen underwater before Ken Donald's manufactured
eviction deadline. The reports of Operation KJH are mum on the topic of
underwater obstacles, but as of December 31, 1943, the Admiralty's understanding was that "at present there are no known underwater obstacles in the area of the Overlord assault." The divers of the Boom
Commando Party, soon to be renamed LCOCU, practiced their disposal
techniques anyway. The Germans might sink obstacles at any time. The
divers needed to be ready.

On January 14, 1944, Haldane started breathing oxygen at sixty-six
feet of pressure at 12:25 p.m. At 1:37, he recorded "vomiting." Then, at 1:41
"trying not to vomit." A description of his vomit from 1:42 said, "not much
actual fluid brought up." By 1:54, the chamber had depressurized to
reach the surface, and he was no longer alone in his gastrointestinal
fireworks. Ursula Philip dashed out within minutes of the hatch opening. In his sloppy, jagged handwriting, her boss scratched a penciled note
that she was off "vomiting in [the] lavatory." More data points were
added to the pile for the oxygen rebreathers. Avoiding vomit was helpful
for divers, but it was not mandatory.

HM Trawler *Darthema* towed the "ugly duckling" submarine *X-20* south
through the final protective boom off the English coast at a hair past
noon on Monday, January 17, 1944. The X-craft's London bus engines were
finely tuned and ready to go. Its bombs were prepared and firmly attached on either side. Its belly was full of food, fresh water, and equipment
for the crew. Two ML motor launch vessels escorted the towing pair.

The operation was titled Postage Able, and there was no pomp and
circumstance in its simple launch command. At 0600 that morning, the
commander in chief at Portsmouth had spoken the top secret signal to
start, and that signal was: "Carry out Operation Postage Able." The
weather was better that time.

Clogstoun-Willmott was on board again, as were Scott-Bowden and Logan-Smith, because there was no point in using inexperienced personnel for that most important mission. The two X-craft crew members who would squeeze themselves into the limited remaining space were also seasoned veterans: Lieutenant Kenneth Robert "Ken" Hudspeth and Sublieutenant Bruce Edward Enzer had been in charge of the *X-10* when they'd had to abort the mission en route to the *Tirpitz* raid in Norway. That raid had been some arguable version of successful, but Hudspeth and Enzer had been the only X-craft crew to make it back with their craft, and they were therefore the only X-craft crew with operational experience.

Nobody had much hope for the crew's survival, but they believed that they could glean enough battle-critical information to make the risks and sacrifice worthwhile. Five lives risked now could save thousands during the landings.

The team of five's command had allowed them a maximum of five days for Postage Able. If Scott-Bowden and Ogden-Smith were able to get onto the beaches three nights in a row, they were ordered to take a break on night four. Aircraft would be flying recon lines, scanning for five miles on either side of the submarine's planned transit path. Those flights were not for scouting; they were for safety. The planes would be looking for a midget submarine that needed rescue. Packed inside the *X-20* was an aircraft rubber dinghy and a yellow fluorescent marker block, in case the crew should sink. The *X-20* had no wireless communication except for a simple radio receiver, so the five submariners could listen for codes snuck into radio programs, but that was it. They would not be able to signal home for help.

Their mission that time was to scout "the American sector." Those were the beaches near the towns of Les Moulins (Saint-Lauren-sur-Mer), Colleville-sur-Mer, and Vierville-sur-Mer. Those beaches would one day be labeled with the code name "Omaha." The *X-20* was also packed with as much "abundant intelligence material" as MI9 was able to shove inside the cramped space of the limited hull.

At 2158, the *X-20*, freed from the tow that had brought it this far, slipped into the final protective Allied minefield between England and

France. By 0007 in the starry early hours of January 18, the crew was clear of all friendly Channel obstacles, and Clogstoun-Willmott plotted a new orientation of 215 degrees magnetic compass heading. Hudspeth made the turn. They headed straight for the crescent-shaped cove of the Normandy beaches intended as the site of future Allied landings.

Sublieutenant Enzer slowed their journey but not intentionally. The weather picked up and with it the seas, and Enzer's position was to be the officer on watch. On board a normal ship, that role would have included binoculars, rain gear, and a vantage point somewhere as high up as possible. On an X-craft, there was no such thing as "high up."

The upper casing of an X-craft contained a small platform large enough for one person, although it sometimes held more when the conditions were good; around the rim of that casing ran a metal bar. Enzer's job was to stand aboard the casing, his body strapped to the induction pipe, and hang on to that bar. Under normal, cold but pleasant conditions in a protected Scottish sea-loch, Enzer would have been able to ride there like a celebrity in a parade and wave to the fish and porpoises he passed. In the English Channel, however, with the submarine traveling on the surface at 5.2 knots, Enzer had to cling to the bar "with fervour, while floating on his front like a paper streamer on the bosom of the ocean." The rest of the craft was submerged out of view beneath him, and he got dragged as if he were seaweed caught on the casing. Hudspeth, inside the small *X-20*, picked the maximum engine speed at which the waves allowed Enzer to "lift his head above water for breathing purposes." . . . Breathing purposes.

Clogstoun-Willmott wished for "an intelligent merman to fill this role," writing that "legs are liable to considerable injury."

At 0316 they sped up. They had crossed inside the protective hook of a crescent of land, the seas had calmed as a result, and the same Pointe de Ver lighthouse from Operation KJH came back into view ahead, blinking its comforting, familiar groups of three. However, that time the lighthouse was quickly extinguished by someone onshore for unknown reasons, and it was not relit. Hudspeth, Clogstoun-Willmott, and Enzer had to pilot the *X-20* using landmarks and headings alone.

A gentle haze and drizzle hovered over the darkened shoreline, so the group of five decided to take the risk of approaching slowly on the surface, hidden by the cloak of the misty weather. By 0810 they were close to shore. Once Enzer climbed safely inside, they dove the submarine and settled on the sea bottom at twenty-seven feet, which is just over eight meters. Deep enough to hide. Deep enough to rest. The first shift took watch. The rest tried to sleep.

By 1340, still huddled on the bottom, the crew had just eaten, and it was time to work again. They raised the periscope, a slim optical tube "about the thickness of a moderately thick walking stick." They poked it six inches above the water, occasionally a foot, in intermittent bursts. Each time they left it up no longer than a few minutes, and they kept it moving to make sure it was not an obvious target for spotters onshore. They wanted it to look like ocean flotsam.

For two hours they "slithered" the miniature submarine back and forth along the sandy bottom to maintain periscope depths during the changing waves and tides. They would have been long past the limits set by the gas supply of their small volume had they not had Protosorb containers and Luftwaffe oxygen bottles with them. Especially with a five-man crew packed inside. No more than four people had been thoroughly tested, and five was a gamble. Case and Haldane had advised not to exceed a maximum of thirteen to sixteen hours for that gas volume, and the crew was getting close.

Without warning "the whip crack of Mauser bullets" echoed throughout the water. The Germans onshore were firing their weapons, and accurately enough that the sounds of the bullets pinged throughout the X-20's hull. The five were alarmed. The gunfire might have been aimed at random true flotsam, shot at by bored soldiers on watch, but it also might have been aimed at the periscope. The crew of the X-20 lowered the periscope and "gently slithered away" into deeper waters.

The Germans kept firing into the water even after the submarine was gone. They never escalated to machine guns. They never used heavy artillery. The gunfire that felt like it had been directed at the midget submarine, it seemed, was a fluke. Practice. Maybe a false alarm. Or maybe the shooters simply talked themselves out of believing what they had in truth spotted. Regardless, the lack of escalation was a good sign.

The clock continued to tick with each new box of Protosorb Ken Hudspeth peeled open to refresh the increasingly foul submarine air. Despite the gunfire scare, the three COPP passengers decided that their mission time was too short for excessive caution. They would go ashore despite their uncertainty about having been spotted. "Time was short . . . risk of hunting was therefore accepted."

Hudspeth and Enzer brought the small vessel to the surface five minutes after the disappearance of the last lingering fingers of nautical twilight, and they maneuvered it closer to shore. Scott-Bowden and Ogden-Smith donned their protective khaki swimsuits. They strapped on their bandoliers, each with twelve sample tubes to hold bits of the beach's sand and muck. They pocketed a trowel. They clipped a reel and stake to their belts for measurements. They also secured their .45 automatic firearms and two spare magazines apiece. They were scouts, but they knew that, as written in the COPP training manual, "success depends on ruthlessness. Kick a man as hard as you can when he is down. There are no rules—there is no referee."

They moved forward to the X-craft's wet and dry compartment and crawled upward into the hatch. The first door sealed below them. They opened the outer door to the night sky and crawled onto the pitch-black casing.

The submarine was three hundred yards from the edge of the water and four hundred eighty yards from the back of the target beach. It was positioned across the water from the end of a road stretching inland. Two sloppily blacked-out houses twinkled glimpses of light around the edges of their windows from shore, where rows of reinforced former homes with knocked-out upper-story windows had been converted to military-esque fortresses. Clogstoun-Willmott, Enzer, and Hudspeth stayed inside the submarine and ventilated their breathing gas for the first time in twelve hours, while Scott-Bowden and Ogden-Smith crept through the quiet, waveless water.

When the swimming duo reached a water depth of 3.5 feet, they froze. A German sentry stood at the back of their target beach, about two hundred yards away. He was within eyesight. He was within firing distance. The surface of the water was smooth, and the surf was negligible, so their dark forms, camouflaged against the sand as they might have been, would nonetheless cause a visible ripple where any water was forced to flow around them. The sentry swept his flashlight across the water. He shone it directly on them. They kept still. For at least five minutes, they waited for the sentry to leave or move.

The sentry had not fired his weapon. However, a second sentry was approaching, and this one wielded a flashlight, too. Scott-Bowden and Logan-Smith opted for the risk of motion. They "edged" off silently, slowly, cautiously creeping backward into the water from which they had emerged. They swam northwest, letting themselves sweep out to sea before coming back to shore again, west of where the sentries still loitered.

The beach seemed clear. Scott-Bowden and Logan-Smith dug into the seafloor with the auger, carefully filling the sample tubes in their bandoliers one at a time. They saw a single line of footprints perforating the sand; they noted it on their waterproof wrist slates as an indicator that the area was not mined.

By 2150 they had made it back on board the *X-20*. Their fingers were numb from pulling through the frigid Atlantic water, but they had determined that the section of beach was suitable for tanks. It was "hard well compacted sand." For the next five hours, they cleaned their weapons, maintained their equipment, and kept their turns on watch before they could get some rest. They also wrote down everything they had learned and made sure all their samples were labeled, in case they did not survive the next sortie.

They repeated the task the next night, the nineteenth, once again dodging sentries, flashlights, and "slow deliberate" gunfire that thankfully never escalated beyond stray shots from rifles. Each day, lurking beneath the waves, the crew used the periscope to survey the land first. When not moving, they tethered the submarine using a hemp cable that they could cut to break free if under attack. They marked new landmarks to supplement the scouting photographs and charts and assessed where Germans and farm animals were allowed to walk and therefore where there should not be mines. Each night, Scott-Bowden and Ogden-Smith crawled ashore to fill in more details, to note the locations of pillboxes and earthworks projects under way. To update the level of demolition of the beachside buildings. To note the fine differences in soil color that marked freshly dug minefields and booby traps.

The lines of the mission report are peppered with references to the misery of living with the carbon dioxide, which was kept at moderate but not complete bay by the Siebe Gorman Protosorb boxes.

"Withdrew. Air was getting very poor."

"Came awash 1½ miles offshore for ventilating."

"Surfaced. Ventilated by engine for a short while. . . . The Major [Scott-Bowden] amongst others, had been feeling very sick owing to bad air, but improved on the casing."

They had to surface at least every twelve hours or less. They sometimes drugged themselves with Benzedrine to keep going. They reported drinking far less water than normal, despite carrying a plentiful supply, a possible sign of nausea caused by their sucking in toxic gas levels. They wrote that, to achieve a less painful level, "it would appear desirable that more trials be carried out" regarding the prescribed use of oxygen and Protosorb for five people. The environment was uncomfortable but survivable. They could have thanked the hundreds of tests done by the UCL lab for that, had they known about the work.

"Withdrew from shore. Air becoming foul in spite of use of oxygen and CO_2 absorbent, so decided to proceed offshore to ventilate."

"After 11 hours submerged the atmosphere was foul, and the Major showed signs of distress."

By the night of January 20, they had had enough, and Clogstoun-Willmott, officially the leader of the mission, made the decision to head home. Scott-Bowden and Ogden-Smith had crawled their way westward over the bulk of the remaining beach in the American sector, scouting most of it, gathering more than enough information to consider their mission successful "for a first attempt." They had been mandated to take a break on the fourth night anyway, and the weather outside was worsening, implying through the sky's looming darkness that to loiter for a fifth night would have been "abortive in any case."

So, just after suppertime, after Clogstoun-Willmott made one final note reading that "air became very bad (none fresh for 11 hours) and everyone showed signs of distress," Hudspeth and Enzer brought the *X-20* to the surface. The weather was calm, and Enzer crawled back up to his casing, lashing himself to the bar yet again. The crew headed north, home, back through the mapped gap in the friendly minefield, and they reached berth under their own power almost twenty-four hours later. Once safe on English soil, every member of the crew collapsed both physically and mentally, freed from the strain of "one of the most cold-blooded jobs in the war."

Professors, government scientists, and military analysts immediately got to work processing their results. The sand, the rocks, the measured angles, the mapped divots and gullies would all be used to find beaches in England that matched those in France. Then the landing teams would use those beaches to practice. To make sure the tanks would run. To test the engineers' solutions for traversing gullies. To practice climbing the barriers and circumventing the sand walls that might trap them and cause another massacre like the one at Blue Beach in Dieppe.

Operation KJH and Operation Postage Able would result in medals for the crews, and those medals were about to rain down in glittering abundance. However, the missions had been so secret that no paperwork describing what the men had done could be filed without jeopardizing their results. The officers and the others in the know had to stand before the medal awards tribunals and describe what had happened. Nothing was written down except vague, gallant adjectives like "intrepidity," "skill," and "leadership" and one comment remarking on how far submarine technology had progressed.

At the end of January 1943, while the unspecific awards paperwork was being filed, stamped, and sent up the bureaucratic chain for approval, Ursula Philip and Helen Spurway became the last two members of the UCL group to crawl out of a hyperbaric chamber. The science, the Admiralty had decided, had reached a sufficient state that experimentation was no longer a priority. The oxygen rebreathers were ready for military use. Ursula Philip's twitching lip would be the last civilian one to do so for that research. William O. Shelford would lead some additional experiments with military divers to verify that carbon dioxide levels stayed low in breathing loops for divers using additional custom types of breathing apparatuses, but for the most part, all that remained was for the scientific data points to be compiled into usable guidance.

The usable guidance was needed by LCOCU divers, who would attempt to dismantle and flatten underwater obstacles in advance of landing vessels. It was needed for the newly forming Naval Party or "P-party" divers like John Bridge, who, in the harbor of Messina, Sicily, after the battle's end, scrutinized every inch of watery harbor floor and dismantled all the lurking underwater bombs. Breathing apparatuses were being given not just to submariners but also to tank personnel, should their tanks burble sadly to the seafloor as so many had done during Dieppe. The action was moving from experiment to practice, and from practice to reality.

Most of the divers would never know where their oxygen and DCS guidance came from or at what risk it had been obtained. A leading officer of the P-parties was John Stuart Mould, who had participated in at least three chamber experiments, including some of the final assessments of carbon dioxide for the X-craft. Mould had sat in the smallest chamber at Siebe Gorman, sweating alongside Haldane, Spurway, and Donald, trying to fight the inevitable razor headache that came with rebreathing high-carbon-dioxide air for too long. Mould and his burgeoning P-parties drove the work with oxygen-rich gas blends; as an ordnance expert, he had known what to expect, what to fear, and what the divers might have been asked to do to find and defuse bombs. Aside from Mould, however, it is not clear whether any of the divers knew where their information came from, or whether they simply accepted the paperwork and numbers handed to them.

Ken Donald compiled his data points of young male divers into a report titled "Variation in the Oxygen Tolerance of Human Subjects in the Wet and in the Dry" and submitted it to all the members of the Underwater Physiology Sub-Committee, Haldane included. All of Donald's data points were in tables again, and he again stopped short of providing any mathematical analysis or of extending his findings to provide a useful conclusion to divers. Helen Spurway, master statistician that she was, got to work.

After the success of Operations KJH and Postage Able, the COPP teams were not permitted to go back to Normandy for more scouting, for fear that they would be caught and jeopardize the landing plan. Instead, the X-craft were reassigned. They would practice serving as navigational beacons, as marker buoys for incoming troops aiming for land. The team of Operation Postage Able had remarked on how difficult it was to spot several of the beaches using landmarks alone, especially the beach to be named Omaha, which had been earmarked for the Americans. At Omaha, small mistakes in guidance could end up with vessels stuck on sandbars, with troops floundering and drowning in the unexpected and invisible underwater gullies that the COPP teams had mapped. Each planned beach location threaded a needle between German defenses, and misguided troops could land directly in front of hostile batteries. Rather than risk it, the X-craft would serve as advance crews, post themselves on the beach edges, and guide the landing craft in on the day of the attack, the D-Day. At least, they would for the British. The Americans declined the service, even though it was offered.

On February 25, 1944, the Allied, Japanese, and scientific worlds collided one final time when water intrusion into the hull of the practicing *X-20* caused a battery explosion resulting in the internal release of clouds of toxic chlorine gas. Ensign Kazuo Sakamaki and his shipmate had suffered and choked on the same during the invasion of Pearl Harbor, and J. B. S. Haldane and his father, John Scott, had inhaled the stuff willingly during WWI. They had "repeatedly risked their lives, and sustained a certain amount of injury, in experiments" to solve the riddle of breathing masks that could function and keep chlorine out of the lungs.

Because of the previous science, the *X-20*'s crew was able to don their oxygen-breathing apparatuses, and they survived. Instead of yet another horrific submarine sinking, the battery explosion became a footnote to history, remarkable only for being a nonevent, for marking another problem solved by science.

CHAPTER 17

The Longest Days

OUISTREHAM, FRANCE
JUNE 4, 1944
TWO DAYS UNTIL D-DAY

German troops were on the beaches of Normandy. But these Wehrmacht soldiers had come to the sands to play, not to fight. They frolicked back and forth, some knocking a ball about, others enjoying their pleasant summer Sunday afternoon by jumping into the chilly, welcoming waves.

Watching over the festivities from behind the broad stripe of shoreline were the last standing buildings of the formerly bustling tourist location. Those few remaining churches and homes were isolated, their human occupants long displaced, but they stood nonetheless, stalwart and tall and damaged, behind the picturesque swath of hard-packed, German-occupied sand. If the streets of London were smiles with piles of brick marking the occasional tooth knocked out by trauma, then the beaches in front of the French town of Ouistreham were mostly empty, screaming maws holding a few lonely broken canines clinging to existence in a mouth long lost to rot.

Resilient bushes, trees, and grass now worked to gain a foothold in the rubble and sand between the beach and the town, and enough had succeeded over the past several years of war to give the warped impression of a picturesque green French countryside—if you did not look closely enough to see the guns hidden among the leaves. Behind the encroaching shoreline wilderness, vehicles, presumably German military ones, rushed through the rest of the small town.

George Butler Honour watched the view from a mile and a quarter offshore through the stalklike periscope of the brand-new midget submarine *X-23*. To him, that beach was code-named "Sword." He hoped none of the gamboling, jubilant German troops were strong enough swimmers to discover his tiny lurking boat with its crew of five. Honour watched the Wehrmacht soldiers play, and he thought of how little they knew about what was coming for them.

———

Meanwhile, across the Channel in England, the people of Portsmouth pretended nothing was happening. However, radar operator Albert Bovill would never forget what he saw when he arrived there, because hundreds upon hundreds of barges, ships, and aircraft were assembling in Portsmouth Harbor. Tanks. Tank-landing craft. Gliders. Ship after ship after ship. The harbor was packed to the brim. Amid the brewing bad weather, Bovill remarked to himself that "you could have walked across Portsmouth Harbor without getting your feet wet." Hundreds of thousands of troops from every Allied country had assembled there, too.

It had become something of a British household custom for the residents to offer passing troops whatever they could. Tea, as long as the pot held out. Biscuits, if there were any. Sandwiches and desserts, if soldiers were lucky enough to pass the right house. The custom had become so ingrained that when the military ran training exercises in the more isolated parts of Great Britain—the parts selected to provide realistic preparation for surviving in the bombed-out French countryside or the wilderness of Nordic countries—they had to warn the citizens in advance not to give the troops food. Otherwise, it would not be much of a wilderness-training exercise.

In contrast, in Portsmouth, when Combined Operations officer Edward Findlay Gueritz marched through the streets beside his compatriots in uniform, nobody looked at him. Nobody acknowledged him. Nobody asked about the portable Mulberry harbors being assembled. Nobody wondered out loud about the innumerable ships. According to Gueritz, "Everyone just ignored it all." There were rules in place about keeping military movements quiet, about not spreading information that could trickle across the water and inform Hitler. And so, the citizens forced themselves not to see things happening. The soldiers passed by as ghosts.

The ghosts were young. Many were teenagers. Too many had lied at the recruiting office to sign up at the age of sixteen or seventeen, barely old enough for facial hair, much less death. They had been selected for the operation for their youth and inexperience, because most had never been under enemy fire before, and so they did not know how afraid they should have been.

Edward Findlay Gueritz was twenty-four when he mustered in Portsmouth, making him the old man among the brewing crowd. He was not unscarred. He was nineteen and in a movie theater when the HMS *Thetis* had gone down, and he had been told to leave by a message flashed across the cinema screen. He had boarded one of the left-behind vessels, the ones that lingered to collect the stragglers, and he had arrived at the wreck site in time to see "the melancholy spectacle of this submarine's stern sticking up out of the water, and a circle of ships stopped, gazing, wondering what was to be done."

Gueritz had survived the worst of the Blitz in Portsmouth. He had listened to the "splatter of splinters" falling onto corrugated iron roofs after each bomb blast. He had been on board HMS *Jersey* when it was blown in half. He and the other survivors had stood together on the steps of a building after reaching land; they had been covered in fuel oil but tried to light cigarettes anyway.

Edward Findlay Gueritz had spent time ramming against practice waves in a blunt-nosed landing craft in a Scottish loch before coming back to Portsmouth that one last time. He had spent a recent month doing nothing but beach training. There had been eight full-scale practice operations on friendly beaches. At an age when most are emerging from school to experience the world for the first time, Edward Findlay Gueritz already felt in his bones the "perpetual fatigue" and weariness of war, of "day after day, day after day, with no foreseeable end." He had grown up in that war.

He knew about the beach massacres at Gallipoli in WWI and at Dieppe a fresh two years earlier. For the upcoming beach attack, he was the wizened one. He was going to be in charge. He had been designated a beachmaster, more particularly a beachmaster for Sword. Edward Findlay Gueritz would be one of the early troops ashore, a member of the first few groups onto the sands now being watched over by George Butler Honour and the *X-23*. Once the weather cleared.

George Butler Honour and the rest of the crew—Jimmy H. Hodges, George B. Vause, Jim Booth, and Lionel Geoffrey Lyne—surfaced their X-craft just before midnight on Sunday, June 4, to listen to Niton Radio, an English coastal radio station. The storm interfered with the reception, and as the choppy broadcast crackled across the airwaves, the crew had to turn off a piece of equipment called a gyro repeater in order to get any intelligible words out of their set. Despite listening intently, they could not hear the secret message labeled with their code name of "Padfoot." But at 0100 on Monday, they heard it. Turning off the gyro had helped. The weather was too bad for the operation to proceed the next day, the message read after decoding. It was delayed.

Honour worried about the oxygen supply in their small, confined space. They had a strict, prescribed rate at which they needed to let the oxygen hiss out from the lightweight aluminum Luftwaffe airplane bottles when submerged, and they had to stay submerged for eighteen out of every twenty-four hours, except the darkest hours of night, or risk being spotted. Honour did not know how long the weather delay would last; he knew how long the oxygen could.

They had food for ten days, but oxygen for only a few. If the oxygen ran out, they "would have been in a bit of a predicament." They had false passports with them in case they needed to scuttle their tiny boat and make their way inland, but Honour harbored dark nightmares about a nameless yet massive Nazi rifleman pressing his gleaming, sharpened bayonet into Honour's back and demanding information about the coming beach landings. Honour wished he knew less about the Allies' plan.

Lionel Geoffrey Lyne was a specialized, highly trained navigator from COPP team 9. He had homed them in to the exactly correct place on the outer edge of Sword, the easternmost beach of the Allied plan. As they faced the land from the ocean, Juno was the next site to their right, one step west. Gold was third. Ken Hudspeth, the pilot of the *X-20* for the beach-scouting missions, waited with the rest of his X-craft crew in a similar state next to the western boundary of Gold. Those three—Sword, Juno, and Gold—were the beaches marked for the British and Canadian troops. The *X-20* and the *X-23* bracketed them.

Each beach was nominally referred to by the nationality of the majority of the troops that would storm it, but in truth, the Allied troops were intermingled, and the attack was unified. Ships and crews from Norway and Poland as well as troops from France were mixed among the assembling armada in Portsmouth, too.

The X-craft were waiting to guide their brethren in along that long, sloping, beautiful but largely featureless piece of coastline where navigating to the proper sites was modestly described as "difficult" and where the enemy now roamed. The Americans were destined for Omaha and Utah, even farther west. The beaches formed a swath, but none of the landing points on those beaches were directly connected to one another. Rather, each landing point was at a targeted gap in the Germans' onshore defenses. In between lay pillboxes with artillery. Machine guns. Flamethrowers. Death.

To the left of Sword, outside the intended beachhead, the shoreline was backed by low cliffs, ominous reminders of the fates of those who had landed on Blue at Dieppe two years before. Gun batteries pockmarked the cliff region. To the right of Sword, in between it and Juno, the scouts had determined that the picturesque sandbanks there were flat but soft, and marked with row after row of runnels, which were deep gouges perfect for trapping any vessel trying to land. The landing point for Sword beach itself was just under a mile long between the big guns and the bad terrain. Edward Findlay Gueritz's job was to organize the landings at that sweet spot.

The team of the *X-23* dropped anchor to maintain their exact location as determined by Lyne's "absolutely bang on," careful navigation. They drifted downward thirty-five feet to the smooth, sandy bottom and filled their ballast tanks with water in order to dampen the swell of the coming storm.

On the night of June 5, just before midnight, the secret codes over Niton Radio said the invasion was on for the next day. The crew of the *X-23* resubmerged but did not rest. They started rigging equipment, their lights, sound signals, and flags. Meanwhile, a massive convoy of Allied ships began to move across the English Channel.

Some of the soldiers on board shook hands with one another, trading quips like "See you in Paris" or "Before the war I paid to visit France, and now they're paying me to go." One group of the Fighting French Commandos belted out rousing songs on their way out to sea. For some, their last vanishing sight of England was of searchlight crews onshore waving them off into the darkness. Once under way, they were told where in France they were going.

———

Everyone's goal, the goal of all of it, was to get troops onto the beaches, then off. Every ship, every gun, every airplane and diver that participated in the plan, they were all there to get the troops out of the water, onto the sand, and then swiftly inland beyond the deadly open shores. As many troops as possible. As fast as possible.

There were no secret priorities that time, no calls to spare buildings. There would be no smoke screen blocking the battle from the view of command, no officers hiding behind it. The single-minded goal of the past few years for the scientists was now the single-minded goal blazing inside the head of every one of the hundreds of thousands of Allies advancing on Normandy: troops onto the beaches. Then immediately off, immediately inland.

———

When the moon reached its zenith on the newly assigned D-Day of June 6, 1944, slightly over an hour past midnight, the first airborne troops began to jump out of planes on the landward side of the selected French beachfront towns. While these soldiers drifted, suspended from their circular parachutes, through the skies under the bright, nearly full moon, their counterpart personnel on the ships of the Allied armada blazed ever closer through the waves.

The smallest of the troop-carrying vessels had been secured on the decks of the larger ships to be launched once they were close to shore. The troops aboard other small but self-driving vessels, however, learned why the scouts of Operation KJH had complained about the violent effects of moderate seas on small craft. Moderate seas for those on a small craft are rough.

Someone vomited on Edward Findlay Gueritz. It seemed like everyone was vomiting. Gueritz went up on deck to try to scrape some of it off.

The ship crews had been ordered not to use lights until dawn and to keep radio silence until the surprise had ended; they were isolated from one another on the journey. So when Gueritz emerged on deck from his cocoon deep inside the enclosed command ship LCH 185 in the early-morning hours well before sunrise, he looked out with fresh respect at the extent of the silent, blackened armada of which he was now one tiny part. In every direction, he saw nothing but dark ships and craft stretching across the hostile waters. All of them were crewed by thousands of others with the same unified goal: troops onto the beach, then off.

Gueritz stared ahead. His uniform was soggy with the vomit of one of the people with whom he had spent so much time training on friendly practice beaches. After his feeble attempts to clean himself, he witnessed the first sign that this beach landing would be different.

At 0340 that morning, the Royal Air Force arrived. Gueritz watched with awe as "the whole horizon seemed to lift up in the air" in front of him. A massive cloud of smoke and soil and fury released by the first dusting of munitions onshore kicked the entire visual definition of land higher up into the starlit sky.

Local French citizen Michele Grimond thought the bottomless rumble was the thunder of a coming storm. When his trinkets began to move and clatter on the mantelpiece, he realized it was an invasion.

While the bombers strafed the land, each ship in the convoy aimed for a specific point at one of the five beaches to start the invasion from sea. The Sword-bound ships of the convoy had a designated position on their maps of Sword called the lowering position, where they were supposed to go before lowering smaller craft off their decks and into the water. No more trying to unload tanks onshore and under fire like at Dieppe, not this time, at least not for the first wave. The tank-carrying vessels—the landing craft, tank, or LCTs—of Group I would go to the lowering position, and then between there and the shore, those ships would unleash about three dozen waterproof tanks into the water. Tanks that had their own propellers. The odd models were from a cluster of inventions dubbed "Hobart's Funnies" built specifically for beach assault. They were officially called DD tanks for Duplex Drive, a nod to their amphibious talents, but many people referred to them as Donald Ducks. Many of the vessels in the first group were led by COPP scouts, who were all navigational experts and sure to keep focus on the correct locations ashore.

The DD tanks would form a swarm. The swarm would merge and organize into a line parallel to the beach as the tanks were piloted ashore under their own power, fully loaded, strange-looking, and ready to fire. The vessels carrying the first troops to proceed on foot would follow. If the tank-launching process failed, the captains of the LCTs would beach their ships and let the tank crews fire from the decks. In short—that time—no troops would be landed without heavy fire support in place and roaring. No more trying to unload tanks in the moment.

At seven minutes past five a.m., with the armada on its final approach, George Butler Honour and the crew of the X-23 began their final task. Their gyro compass had started acting oddly after all of the switching on and off to get the radio signals, so Lionel Geoffrey Lyne of COPP 9 did not feel confident enough in its readings to raise the amber beacon that would have meant precision in their location to the exact foot. Instead, he picked the green light to signal that they were close. The official name for their location was position DD.

One of the X-men clambered up through the sub's wet and dry, and the slosh of the vicious storm hurled more water than expected down into the tiny compartment below. He aimed the green beacon seaward, and the light began to flash in a pattern of one long, then two short. It was Morse code for the letter "D," their location. The crew of the *X-23* did not document who took the first shift on the stormy casing, holding the light, but each of them seems to have taken a turn. At some point, each of them got washed off, saved only because they had tied themselves to the boat.

Before the time that the X-craft beacon was lit, the return fire from the Wehrmacht onshore was not dramatic. The British and American pilots overhead, breathing through new, advanced oxygen-supply systems, continued their bombings from air. In the water, the Allied minesweeper crews cleared the first channel up to the lowering point and buoyed it for safe navigation. Two main headquarters ships and six massive troop-carrying landing craft, infantry (LCIs), of the armada were in their prescribed tidy line at their correct appointed places on the attack map, with said tidy line in a column perpendicular to the beach. From there, the troops on each of those massive ships could file into and launch the tiny flat-bottomed landing boats off either side. The first three of the bombarding ships, which in total were eight battleships, had started to form their own row parallel to the shore, and their munitions experts were starting to launch their explosives onto the beach, too.

Before the X-craft beacon was lit, those in the gun batteries onshore, it seemed, had hunkered down in response to the surprise and overwhelming fire. No one on the Allied side had yet reported a casualty. Things were running on schedule. Everything was neat. Everyone was in place. But then again, no Allies were on the beach before the beacon was lit.

Group I's convoy of LCTs was about to approach the lowering position. At that point, in the words of one officer, "events started to move swiftly."

Allied aircraft "streaked low across the eastern flank" on the left, along the outer edge of the landing pathway. The pilots laid a protective smoke screen hiding the path from outside view. The smoke screen was to stop the soldiers from being caught in the crossfire of batteries outside the landing channel. It worked. But it also provided cover for German torpedo boats.

A B-25 bomber laying a protective smoke screen on the water. *(https://catalog.archives .gov/id/204950052)*

Three snuck in. They sent two torpedoes rocketing through the horizontal row of bombarding ships, in between the battleships HMS *Warspite* and HMS *Ramillies,* and a third torpedo skimmed within a few feet of the farther-out major command ship HMS *Largs.* They missed. A fourth torpedo hit its target, and the boiler room of the Norwegian ship HNoMS *Svenner* burst into the sky in a roiling cloud of steam and fire and broken shards. The whole ship seemed to lift out of the water, and the funnel ricocheted aft. The back of the ship was broken, split into two halves. The halves started to sink. The crew jumped into the water.

The freshly mangled pieces of the *Svenner* had barely begun to float away when the cloud-covered sun first cracked the horizon. The ship crews lowered the rear ramps of the LCTs into the salt ocean to vomit out their odd, amphibious cargo. The tanks rolled smoothly out from the protection of the high, solid gray metallic walls that had encased them on their trip over. They floated, exposed. The tank crews were almost ready to go. At least, as ready as they could get. The DD tanks of Group I began to assemble into their swarm and headed toward shore. Troops organized to follow.

Return fire flashed bright from the muzzle of an enemy soldier hidden inside the windows of a large house onshore. The support gunships "set [the house] well alight." When day broke fully behind the clouds, the crew of the *X-23* replaced their flashing beacon with a flag bearing the capital letter "D" on a massive eighteen-foot pole. They hoisted it high to be sure nobody mistook them for the enemy. The X-craft was trimmed so that the bulk of its squat, awkward shape was hidden beneath the waterline, which meant it was invisible except for the extra-large flag on the extra-long mast and whichever officer was lashed to the casing.

Royal Marine Donald B. King of the 30th Assault Unit had a belly full of bacon, eggs, and four shots of rum for courage when he first saw the midget submarine. King questioned his eyes. To him it looked as if "a man with a flag was walking on the water." Fellow Marine Bill Powell thought the same. He questioned whether the man on the water was a mirage or a vision induced by rum.

At 0633, while those at Sword were still assembling, working on an earlier schedule, the first troops set foot on Omaha beach. Or they almost did. They missed their target. Their DDs and flat-bottomed boats had been pushed awry by the wind and waves, and with no anchored X-craft to guide them, many either foundered or landed directly in front of the gun batteries.

At 0655, Sword beach's first line of swimming DD tanks passed by the X-craft on their way in, followed by soldiers on ten flat-bottomed, troop-carrying boats. The troop-carrying boats, called Higgins boats, were the same model LCP(L) that had been battered about during Operation KJH, but now, instead of a skeleton crew of scouts, each was filled with troops. Brimming with troops. Troops clinging to the outer gunwales and practically dangling over the sides. However, that time, the first wave of troops brought with them their own heavy guns in the form of tanks. And those heavy guns were ready to fire.

"Many of you will be in the first assault. Many of you will be killed," one senior officer had lectured the night before. "But don't worry. A second assault wave will pass over you, and if that fails the third one will pass over you also until we have gained a foothold on that beach."

The words were probably shallow comfort. Gueritz rode behind the early lines on LCH 185. Two navigational ships took forward positions to lead ashore the swimming tanks. A high-speed ML motor launch craft flanked everyone on the left to keep anyone from veering toward the cliffs. Six gunboats of various types went, too, with crews ready to provide heavy artillery fire. Three of the gunboats would run aground on purpose, so their gunners could fire from the front line. The troops were riding into a nightmare, but they had a heavily armed team beside them.

The tanks landed first. The first DD line was supposed to land at Sword at 0717 and thirty seconds, exactly seven minutes and thirty seconds before the official operational start time for Sword—the "H-hour"—of 0725. After the tanks touched down, the support vessels offshore were supposed to lift their fire and focus their wrath farther up into the hills to avoid their own but still hammer the opposition. Results, however, as happens often in war, were mixed.

The tanks yawed heavily in the treacherous waters, enough that the crews had to slow their approach. They fell behind schedule. Brutal, choppy waves battered against the canvas waterproof coatings. Thirty-four of the DDs had successfully launched from their crafts, but two sank en route. The crews of those two were saved because the tanks had been equipped with dinghies. Many of the tank crews had been given DSEAs, just in case. Thirty-one tanks made it to the beach.

By 0730 the beach was fully obscured by smoke. The vessels offshore shifted their supporting fire to the flanks and upward as planned, and the tanks and gunships onshore were now wholly, completely responsible for the troops they accompanied. More German gun emplacements opened fire in the chaos. Many of those gun sites had not been visible from the positions of the ships offshore, and some could not be reached by the guns of those ships because of the angles. Bright muzzle flashes popped deadly betrayal of their existence between the leaves of beachside greenery. Group I beached and unloaded quickly. The crews of their carrying vessels turned around to grab another set of troops.

To Gueritz, riding on the final approach of LCH 185, the view struck him as gray. Gray clouds above. Gray smoke from firearms ahead. Gray mushrooms of smoke expanding upward from detonated rockets and bombs.

The chief hazards for the smaller landing craft were the beach obstacles. As of Operation Postage Able, no obstacles had been present, but they were there now. The minesweepers had cleared the deeper waters, but the shallows held ten-foot wooden ramps with anti-tank Teller mines strapped on top. They held wooden stakes made out of young fir trees, each eight to ten feet high and covered in more Teller mines or shells. Hedgehogs with shells. Tetrahedra with more shells. Thankfully Element C was scarce on Sword, if there was any there at all. Sword beach had clusters of massive ramps farther out to sea than any of the other beaches on which the Allies were landing. Those ramps were backed by rows of bomb-laden stakes, which were then backed by staggered rows of sharpened hedgehogs. Most of the obstacles were exposed by the low tide, but not all. The enemy defenses were what the Allies had expected, thanks to the most recent scouting reports. But they still formed a tight labyrinth of steel, wood, and explosives.

The pilot of Gueritz's landing craft dodged the ramps and stakes of the deadly maze while the troops on board got battered about by the "nasty waves." Gueritz and his commander had been assigned bodyguards to watch their backs while they focused on organizing, and Gueritz's bodyguard, Kenneth George Oakley, stared at the stakes and hedgehogs and bombs with horror while they zigzagged toward the sand through the heavy mortar fire raining down.

———

Some of the captains of the smaller vessels got cocky after the relative ease of their first landing, which took place amid what would be the lightest volleys of the morning, from not-yet-alert Germans' fire. Temporary Lieutenant J. O. Thomas had his motor launch vessel start blasting over their loudspeaker the rousing big-band polka song "Roll Out the Barrel," complete with the lyrics "the gang's all here" and "we'll have a barrel of fun." A bugler on another craft trumpeted back at him a peppy song from the Spanish Civil War. Bill Millin, the personal bagpiper of Lord Lovat from Scotland, sent the martial song of his bagpipes out across the water.

Group II was hot on their tails and finished unloading within minutes of Group I. More troops and tanks and ships poured in. The beach became a swarm. Group II included landing craft with more kinds of specialized tanks, tanks that needed to be deposited and unloaded close to shore, if not on the sand itself. One of those landing craft was LCT 947.

The LCT that had been assigned number 947 was packed with Royal Engineers who were charged with blowing up mines and obstacles. It was supposed to beach itself early and off-load more of Hobart's Funnies, including specialized tanks, then let off the engineers. Crews sat inside their tanks for the ride to shore, ready to drive onto the sand. Before the departure from England, tank-riding Lieutenant Colonel Arthur Cocks had cheerfully encouraged the loading crews of LCT 947 to work past a mechanical problem with getting the tanks on board. He had reassured them "bad start, good finish," referring to his optimism about their arrival in Normandy.

Only one tank crew made it off LCT 947. The next tank was hit by a mortar. It was hit in the turret, the turret from which Colonel Cocks was leaning . . . leaning and waving the other crews on, as before, always leading, always encouraging. The explosion took off his head.

His tank was carrying Bangalore torpedoes, which detonated. A scream rang out from within. The explosion killed three, and LCT 947 went up in flames. Two other specialized Funnies tanks were put out of commission by the blast, jeopardizing the engineers' ability to make usable beach exits.

———

Edward Gueritz splashed off the ramp of his craft near the flaming LCT, as did his commander. Tall Gueritz grabbed his shorter bodyguard, Oakley, and helped him through the water. The tide was beginning to come in, and minute by minute, the obstacles were becoming submerged. Gueritz and Oakley ran past the fiery tanks and deceased Cocks, past the beached and beaching crafts, between the mine-laden obstacles that were still in the dry, and between the thickening splatter of their friends' bodies lying on the dark, wet sand. Gueritz found the sand and the slope of the beach to be exactly like the sands where he had practiced. Thanks to the scouting, and the ensuing practice, the shore felt almost familiar.

Partway up the shoreline, Oakley yelled, "We down here!" They fell to the ground. They scraped away a small trench, a modest, temporary shelter to give them a moment to regroup. The mortar fire was heavy but seemed to be mostly above them. If they stayed down, the Wehrmacht might not be able to lower their guns enough for a hit.

One of the DD tanks crawled up to their position in the sand, and a soldier popped open the turret from inside. "Where's the fire?" the soldier bellowed. Oakley pointed and shouted the compass orientation he had taken to a gun battery. The turret hatch clanged shut. The tank fired a projectile in the direction given. The gun battery went silent.

Oakley and Gueritz hustled farther up the shore past the high-water line. They established a position for Gueritz to start sending out signals to organize the incoming mayhem of vessels and people.

Early troops were supposed to place flags on the edges of the landing zone to help channel incoming traffic, and Oakley's friend Sid Coxton had been tasked with placing one. Oakley found Sid lying flat on the beach with a deep, open gash oozing scarlet across his back. Sid's kidneys were hanging out. Oakley pressed the kidneys inside, applied a makeshift bandage, grabbed Sid's flag, and placed it in its location.

"The horror of it was you couldn't stop to help your buddy." If someone stopped, they would be the next one to get it. Their only goal was to "get off the bloody beach."

Frederic Ashcroft was moving full speed about two-thirds of the way up the high-water mark of Sword when he was knocked sideways by the splinter of an 88mm round slashing across his right arm. He kept running. He ran until he made it to the back of the beach to a shallow but sheltered sandbank where he could apply a field dressing. While he was working, the colonel sheltering next to him rolled sideways. He had been shot in the chest. He was dead.

On unloading, Private Bill Gray saw a man with a thick, prominent red mustache who looked as if he were sleeping on the shore. On the reverse, behind his intact face, all of his brains had been shot out. Gray kept going. As new troops ran past the same body, they all knew that "the important thing [was] to keep the assault going." The rich smell of earth, freshly turned by the blasts, mixed with the metallic aromas of torn flesh and blood, which mixed again with the acrid stench of detonated cordite. The morbid perfume filled the air and the nostrils of the young troops.

As landing craft continued to arrive, wave after wave, group after group, congestion on Sword became a problem. The scheduled touch-down time for the LCOCU divers, laden with their explosives and diving equipment, was supposed to be 0745; they were to arrive with Group V a mere twenty minutes after H-hour on Sword. The Royal Engineers and the divers had been distributed among the fleet that time, divided up among many landing craft to maximize the odds that at least some would make it onshore still breathing. The divers were the ones in charge of the menacing spikes, tetrahedra, ramps, and hedgehogs, which as the tide rose were quickly becoming more immersed. Thankfully, their equipment was designed to let them both fight onshore and swim underwater.

Some of the vessels carrying the explosives landed late by about twenty minutes. The LCOCU bolted ashore at Sword, not at 0745 but at 0805, during the heaviest period of fire and casualties, with the water rising at about a foot a minute, and around the same time that a separate officer made the logical but confused observation that obstacle clearance seemed to be going slowly.

The LCOCU divers landing at Juno and Gold did not fare better, but they fared differently. Keith May Briggs, the seasick British not-quite-volunteer recruited because he could swim, had been assigned a landing craft headed for Juno, the beach designated for the Canadians. Once the density of the obstacles became apparent, the officers at some of the beaches reconfigured their plans and sent the obstacle experts in earlier waves. Briggs landed earlier in the assault and was supposed to clear the beach "before the major invasion took place." He and his crew were to make a six-hundred-yard gap using their supply of three to four tons of explosives. He was one of the first ashore.

Briggs's landing craft collided with a hedgehog in its final approach. The metal beast ripped a hole down the side, and the craft sank to the seabed. It was in only a few feet of water, and the crew clambered out. However, their explosives were lost.

Briggs and the other LCOCU divers made it ashore and curled up behind a low seawall. They hunkered down to avoid German snipers above. Someone put a tin helmet on a stick and poked it above their cover to see if the snipers would fire at it. They did. Briggs waited for the engineers of later waves to arrive with fresh blasting supplies and plenty of tanks. They did.

Once the engineers and heavy equipment joined Briggs and the other LCOCU at Juno, the LCOCU crews started to work with them to clear obstacles through "mechanical means." The hedgehogs could not be rolled, because no matter which way they spun, they had jagged-pointed tips aimed up at the skies and at the ships' underbellies. But they could be slid. Briggs tied hedgehog after hedgehog to whatever heavy equipment was available onshore, and the industrial vehicles pulled them aside one at a time amid the sniper fire and bombs to clear a landing path. Briggs noticed as he worked that the hedgehog that had torn apart his own vessel looked to be the only one without a mine on top of it.

Juno had Element C. For them, Briggs and the other divers swam underwater using their oxygen rebreathers. They tied ropes to the massive angle iron obstacles so that the bulldozers on land could pull them aside. The plan was not quite the original one of blowing them apart, but it worked. And, because of the careful scientific groundwork, there were no reported cases of oxygen toxicity in the process.

Lieutenant Harold Hargreaves, who had been a cotton salesman before the war, rode a vessel aiming for Gold beach. With him were bank clerks, carpenters, and engineering apprentices, all now LCOCU divers. When their little craft came in sight of the sands, they pushed it to top speed, going "hell-for-leather on the beach." Hargreaves hoped he was lucky enough that the obstacles would be on the water's edge, not in the water itself. It would save him some diving. He wasn't. They weren't.

The divers barreled out of the craft and started wrapping lines of explosive charges at strategic points around the posts, ramps, and hedgehogs, all of which were frosted with mines. Shells and bombs rained down, and heavy surf battered them. But "the obstacles were being slowly but systematically destroyed." The divers made a gap. They widened that gap. By the end of the invasion, by Hargreaves's math, they would have cleared twenty-five hundred obstacles off Gold alone.

The kapok blast jackets developed by Horace Cameron Wright and his team saved lives in the process. The LCOCU divers wore them under their rubber swimsuits. One petty officer was working in six feet of water when the detonation of a nearby bomb paralyzed him, as the most severe blasts had paralyzed Wright during testing. But he survived. He came to. It is impossible to say for certain, but it is likely that the kapok reduced the risk of trauma to his lungs and gut and saved his life.

Hugh Michael Irwin, the midget submariner who had gotten bored in the training escape tower and exited early, had allegedly been bumped from the X-craft marking crew in exchange for a bribe of cigarettes. Instead he found himself in command of landing craft 591, aiming straight for Gold, barreling down the channel made by Hargreaves and the other LCOCU.

Infantryman George Nicholson unloaded from a landing craft that had lodged itself slightly too far out in the water, and he had to weave his way four hundred yards through a chessboard of poles with mines on top. Through the haze of his furious sprint, he saw the LCOCU frogmen on the beach waving him in along their cleared path.

The tides chased the troops and coated the hard-packed sand, and by high tide, most of the obstacles were submerged again. The LCOCU divers were not able to operate safely underwater because of the huge volume of vessels still swarming the area and jostling for the shore, with their exposed propellers spinning and ready to shred. So instead the divers turned to help the Royal Engineers dispose of the mines and charges on land. When the tides went back out, they would turn again to clearing the seas.

———

With the landing positions on the British and Canadian beaches established and clear, the X-craft crew was finished with their duties. Too exhausted for extra physical work, the crew of the *X-23* cut their anchor line instead of reeling it in. They rendezvoused with their towing trawler.

British Number 4 Commando during their approach to Sword beach on D-Day.
(IWM BU 1181)

The battered, drained crew of five pulled themselves on board and swapped out with a fresh passage crew who would pilot the little boat home. The captain of the towing trawler gave his own cabin to George Butler Honour so Honour could finally get some sleep.

By 0900, the rate of enemy fire on Sword, Juno, and Gold had started to slow. The Allies were beginning to silence the batteries and the snipers one at a time. The British Number 4 Commando, many of whom were veterans of Dieppe, had landed on Sword and started to work their way inward, as had other commando units. The foothold was theirs. By 0906, the congested beach was providing clusters of targets and an organizational disaster for Gueritz. However, a minimum of two exits inland had been secured on Sword alone. The troops were getting off the sand.

On Omaha, over in the American sector, things did not go as well. Operation Postage Able had come back with several insights about that part of the shoreline, and they turned out to be portents.

The scouts had emphasized that beneath the waters of Omaha "runnels exist; near low water mark they are sufficiently deep to cause drowning of vehicles." Also, "shingle exists at the back of the beach" and "it is a natural anti-tank and anti-wheeled vehicle obstacle." They had mapped out a concave cliff at the top of Omaha's backing ridge and reported that the cliff did not appear passable. The scouts had specifically outlined where, away from the cliff, beach exits could be had.

They had described at length the sandbars offshore, and they had warned of the deep runnels between. Scott-Bowden and Ogden-Smith reported having to crawl over these sandbars and swim in between them, and they recommended that the engineers at that beach make blocking or filling these channels their top priority or soldiers would drown. They also wrote that that particular piece of coast was "featureless to a stranger at night" and that "competent assault markers would be necessary" to guide the landing troops. Slight deviations from the target would make the sandbars, runnels, gunners, and cliffs all worse.

The troops at Omaha had deviated from the target, pushed by the winds and waves, and they were unable to correct their course without sufficient marking vessels to guide them. Crews beached on the sandbars and let their troops out too soon. Troops unloaded into the runnels in water too deep to stand, much less run. Soldiers drowned, weighed down by their equipment. Their misdirection put them partially in front of the heavy artillery, in front of the fortifications, below the concave sloping cliff with no exits.

The scheduled bombing onshore also did not happen in advance of the landing. The navigational disarray in combination with the storm scrambled the vessels in the water, and "the order of landing was somewhat mixed." The order was supposed to be DD tanks first, to provide cover, then a wave of infantry to provide more cover, then a wave of Naval Combat Demolition Units (NCDUs) paired with their Army demolition counterparts to start blasting the obstacles. The DDs almost all sank. The infantry got scrambled. The NCDUs landed with no cover. They cleared the obstacles while being picked off, sometimes one at a time, sometimes groups at a time. They refused to stop.

Edward Findlay Gueritz made it across the sands of Sword among the early waves of D-Day, as did his bodyguard, Kenneth George Oakley. When the troops continued to land and add to the congestion onshore behind them, Gueritz and the other beachmasters did their best to smooth exit pathways under a gradually decreasing load of enemy fire. The German troops in the batteries above them were killed or taken prisoner one at a time, sometimes through hand-to-hand fighting inside the concrete strongholds. Their guns slowly went silent.

By 1300 that day, British commandos moving inland from Sword linked up with the airborne forces behind the towns. Others connected with Americans moving inland from Omaha. By 1555, Sword beach was sufficiently cleared for people to start unloading stores and equipment from the landing vessels directly onto the sand. Flail tanks with rollers covered in flying heavy chains whipped their way down the shores, looking for mines, and Royal Engineers and LCOCU divers searched with them. Sappers crawled their way inland on their hands and knees, using skewers to probe for the little wooden mines that could not be picked up

Royal Navy Commandos of the LCOCU clear obstacles on Gold beach at low tide. (IWM A 23993)

by metal detectors. When the tide ebbed again, the engineers and LCOCU divers walked back out across the wet, hard-packed sand to continue to demolish and collapse obstacles. To widen safe channels. To let more and more Allies flood in.

The American frogmen, the NCDUs, had planned from the start to abandon their heavily practiced underwater methods for the operation and to attack the exposed obstacles in the dry at low tide only, without risk from overhead ships. They wore high lace-up paratrooper boots because their Army counterparts had determined those worked best to keep out the sand. They brought their diving equipment with them, with their amphibious masks nestled in black rubber bags and their oxygen tanks in the landing craft, in case the plan went awry, but they knew that the rapid influx of the fearsome tides in the American sector would limit their efficiency underwater. On Omaha and Utah, the NCDUs and Army beach units continued to clear obstacles each time the waters allowed. On Utah, the same navigational difficulties as at Omaha occurred, but with the fortunate effect of landing troops at a better location, where enemy fire and obstacles were "far lighter than in the intended area."

Around 2250, the Allies released clouds of smoke to obscure their swath of beaches to protect themselves from dusk air attacks by defending Luftwaffe pilots taking advantage of the cover of night. Some bombs did fall through the smoke. Unfortunately, Allied planes arrived at the same time with reinforcements, and at least two were shot down by friendly fire.

At 0730 the next morning, June 7, the X-craft motored into port in England under the power of their own reliable London bus engines. The invasion of Normandy had been publicly announced and there was no more need for secrecy, so the people of Portsmouth, released from their vow of silence, flooded the streets with tea, cakes, and whatever food they could carry to celebrate with any troops they could find.

A B-26 Marauder gives cover to waterborne vessels at Sword beach on D-Day.

Within the span of a few days after the taking of Sicily, John Bridge had cleared Messina's harbor using a diving hard hat with gas supplied by an umbilical. Now that the five beaches of D-Day were secure, he would see what he could do in Normandy when backed by a horde of fellow bomb-disposal divers—and free-swimming, rebreather-using ones at that.

The P-party divers could swim under the waves with confidence, searching for additional bombs and hazards that might menace incoming ships and troops. They could do it while wearing the same rubber suits and breathing gear developed for the Charioteers, and while breathing high-oxygen mixtures of oxygen and air that would let them rocket to the surface if needed, without risk of DCS, to avoid an impending blast, but also without risk of an oxygen seizure while down below. They carried two gas mixtures to choose from based on the depths to which they would need to dive, and the 75 percent oxygen blend let them dive down to the maximum forty-foot depth of European docks without oxygen toxicity. They had all trained for this task in a secret rush in the intervening months between the final science and the first launch. Young divers had been picked for their stamina and energy, and surface tenders too old for combat were recruited from shipyards and factories to monitor the divers from the surface and help with their equipment. The teams developed father-son friendships.

Ever so slightly offshore from the freshly taken sands but in waters shallower than the ranges patrolled by minesweepers, the P-party divers laid out taut lines in the muck and water and moved down them with care one row at a time. They sifted through the featureless, waterlogged sand with their hands, pushing their way inch by inch. They were isolated from the world above, unable to hear gunfire or blasts. They breathed and re-breathed the same gas. Above the water, those onshore would see only markers to reveal the divers' existence; buoyant rubber floats tracked back and forth, back and forth above the divers' positions.

The divers used the curves for oxygen risk given to them by Ken Donald, who in all contemporary records of P-party actions gave himself credit for their novel oxygen-rich breathing mixtures. But despite rigorous searches, no records have been found showing that Donald ever experimented with the high-oxygen blends or had a hand in their creation. Only the UCL lab did. And Helen Spurway made their risk curves.

By early spring of 1944, Donald had finished his data collection about oxygen toxicity in general and handed it to the Admiralty in a disordered heap of tables and numbers with no clear guidance for divers in the field. Haldane had given the pile of numbers to Helen Spurway. Spurway and Haldane had plotted them, fitted them, and made equations and graphs to translate them in a usable way for divers in the field. Spurway did the mathematical heavy lifting. Her conclusions were eventually tacked on to Donald's data as an appendix to his official report. But by the time Donald sent prescriptions to the fleet, her name and Haldane's had been deleted, and Donald took the credit.

John Bridge of Messina's harbor landed at Gold beach on D+1, the day after the attack, with Naval Party 1500. He strode down the open rectangular ramp of a square-shaped landing craft, mechanized (LCM) into thigh-deep water. From there he waded ashore. The Normandy beaches would be his first time dealing with anti-tank Teller mines, which still

salted the ocean in abundance, but he was a bomb-disposal officer
who had already cleared a harbor and who had helped invent a new type
of diving. Once, while he was driving away from a disposal site, a fuse that
he had extracted from an enemy bomb ignited inside the glove compartment of his car, indicating that had he not gotten that fuse out quickly, he
would have had only minutes before the full bomb exploded in his face.
The former schoolteacher now bore scars down the length of his left arm
and leg from splinters of shrapnel. He could handle the disc-shaped Teller
mines, too. He and his party would get the ports open safely.

LCOCU divers on Gold beach, outside of La Rivière, France. *(IWM 23994)*

John Bridge and Naval Party 1500 swept Gold beach for mines for
about two to three weeks. During that time, he worked in the ocean and
on the dry sand. At one point, as the crew was walking across a soft area,
Bridge noticed that they had wandered into the characteristic stippling
pattern of a minefield. They backed out by retracing their own footsteps,
and then they disposed of the field of small wooden boxes that held
enough charge to tear off a foot.

After a while, Bridge and the party ran out of tasks to do. The ordnance had been disposed of. The beaches were open. Not a single diver

A sampling of the various land mines, booby traps, and bomb parts used to train the human minesweepers at Normandy.

on any of the five beaches had suffered from even a minor injury, much less from oxygen toxicity or decompression sickness. Once again, careful scientific research used by skillful operators turned what could have been a devastatingly fatal job into a heroic, successful nonincident.

Minesweeping activities at either Omaha beach or Utah beach, with a smoke screen still visible in the background. Such activities were crucial to making sure new waves of incoming troops could land safely.

An overview photograph
of the landings on one of
the beaches.

Two photographs taken from a similar vantage point on board an LCM. The photograph on the
left, taken on the journey toward the beaches, was originally captioned "Standing room only." The
photograph on the right shows the same model vessel used after the landings to transport
casualties on stretchers back out to sea for hospitalization in England.

CHAPTER 18

The Unremembered

Acting Temporary Lieutenant Ronald Henry Saull of the Royal Navy sauntered into Ouistreham, France, on D+1 after the majority, but not all, of the fracas had ended. Saull was a gangly, homely fellow with big ears, spectacles, an easy, broad smile, and the sort of disarming dimples that made him look approachable and friendly despite his penchant for high explosives. He had been an electrical engineer before the war, and he was recruited into bombs presumably for his ability to understand their circuitry. He was quiet, bright, and shy. His partner-in-munitions, Sublieutenant Donald Kirkland, sauntered with him. They had brought their rebreathers and fins.

Ronald Saull and Donald Kirkland were members of P-party 1502C, then later Party 1747. They were charged with getting the waters of the strategically positioned river town not only open but safe and open. Small boats were scattered throughout the Ouistreham harbors and the delta of the River Orne, which empties into the ocean next to Sword beach. The river also had a bridge and a series of mechanical locks to allow ships passage inland. Booby traps and timed charges could have been lurking in any of them. Or all of them.

Saull was an expert in bending explosive charges so they hugged the graceful semicircle of a C-shaped piece of steel. Each such partial hoop, about the size of a broken dinner plate, could then be slipped around the steel bars or wooden stakes of an obstacle blocking the waterways. The explosives would cut right through the obstacles, whose horns would pop off with ease. Or if he found a regular old bomb, Saull could disassemble it while breathing oxygen underwater in the near-blind muck. Because he was a leading officer, the dirty work of the hands-on disposal was considered his responsibility. He was one of so many who had been given "acting temporary" promotions to allow them to take leadership roles on the beaches.

Saull and Kirkland would grope their way through the dark waters near the mechanical locks at Ouistreham until they felt the hard cold edges of yet another demolition charge. Breathing their recycled oxygen-rich gas, they disassembled the volatile components that made the charges eager to blow. They saved the locks and a bridge, and kept multiple booby-trapped ships from being blown up and sunk to the bottom, where they would have become obstacles to river travel. Saull and Kirkland helped make the River Orne an access point rather than a barrier for Allied vessels.

An admiral summarized their work by noting that "the saving of the lock gates at Ouistreham was an outstanding contribution to the buildup," and both divers would be on the receiving end of colorful ribbons attached to hefty medals for the achievement. Ouistreham was one of the first mainland continental ports taken by the Allies. It would not be the last.

"As each port was occupied . . . large risks were run continuously and willingly so that clearances should be quick. Much diving of an arduous and dangerous kind has always been involved, first to discover mines, booby traps and other deadly devices left behind by the enemy, and then to render them safe." In other words, after they opened Ouistreham, the P-parties kept hopping down the coasts of Europe to dispose of bombs and booby traps as port after port, harbor after harbor yielded to the accelerating Allies. The risk was constant.

The awards for Kirkland and Saull's dangerous work—an Order of the British Empire for Kirkland and a George Medal for Saull—were announced on May 15, 1945, but three days before the announcement, possibly without ever hearing the news, they died together in a minefield.

They had survived the war in Europe, technically. Germany surrendered unconditionally on May 8, 1945, but the quiet, unexploded bombs that both sides had sown did not get the message. The spattered mine-fields and ordnance would continue to hibernate, uninformed of the cease-fire, and blow up to take limbs and lives for decades. On May 12, 1945, four days after the war in Europe had officially ended, Saull and Kirkland entered a minefield in the Netherlands. They did not make it out.

The duo had worked their way through the harbors of Europe. They had figured out the necessary disposal method for a massive, smooth underwater Type C mine with a new type of magnetic-acoustic trigger, which nobody had seen before. They had been an inseparable team through all of it. Ronald Henry Saull and Donald Kirkland are buried together in the Rotterdam (Crooswijk) General Cemetery in the Nether-lands. In their plot and row, Saull occupies space number sixty-six and Kirkland number sixty-five. Nothing is known of the specifics of how they met their end.

So many of the stories of D-Day and the years before and after will stay untold because the participants, like Saull and Kirkland, did not live to tell them. No historian can discern what was absent in the world for their losses. No book can tell the stories of the lives that did not get to be lived. Ronald Saull and Donald Kirkland are in this one as representa-tives of those whose stories will stay forever missing.

———

The troops storming the beaches on D-Day did so without body armor because of German Jewish refugee Hans Grüneberg, and that was a good thing. After he left the UCL research lab to sign up for the army, Grüne-berg applied the statistics of genetics to the spread patterns of ballistic projectiles instead. The math to predict distributions of eye colors among flies became the math to predict distributions of grenade frag-ments or blast shrapnel sprayed across an area. Grüneberg shot weap-onry into London telephone directories as a cheap and effective way of comparing firepower, and he used British Army case reports from gre-nade accidents to calculate which parts of a soldier's body were most likely to get hit.

After he heard the plan to supply the D-Day troops with body armor or some kind of armored clothing, Grüneberg sat down and did the math. Bullets and shrapnel that hit parts of the body where such injuries caused immediate fatality were not readily stopped by WWII-era armor. It was a time before Kevlar, back when armor meant steel. Bullets and shrapnel that hit other parts of the body could still cause fatalities, but steel armor did not cover those regions.

"The net effect of the armor was almost nil," Grüneberg reported. But "the extra weight was far from nil." "For many pounds of weight" the number of lives actually saved would be "ludicrously small." So Grüneberg handed in his math, and the "idea of the body armor was stopped." The soldiers ran ashore faster because they were less loaded down and less likely to get pinned facedown in the cold, unforgiving ocean.

During the rest of his military career, Grüneberg served by taking blood donations to deliver to the front lines, and he earned himself the affectionate nickname "Super Leech" by collecting more than fifteen thousand pints of blood destined to go to soldiers.

Elsbeth Grüneberg, who had married Hans shortly before his flight to England, had originally been denied refugee status based on her own merits as a dental surgeon. Thankfully, she had been able to come over with his. They had both been exempted from internment because of the length of time they had spent in the country before actual war broke out, so during the war, she ran her own dental surgery practice. Unfortunately, Elsbeth's struggle with what her family described as untreated bipolar disorder led to her suicide in April 1944, two months before the landings that would lead to Hitler's downfall. On June 26, 1944, only a few weeks later, with Hans and their two sons still reeling from her loss, a German V-1 rocket destroyed the Grüneberg family home. Hans Grüneberg was granted compassionate leave from the army to resettle his children.

Hans Grüneberg, ever the thorough, meticulous pragmatist, grew a sharply trimmed beard after the war, wrote several books about his substantial achievements in genetics, helped establish the humble mouse as the universal animal to use for laboratory experiments, and then shortly after his seventy-fifth birthday announced to his friends and family that he intended to have no more. The former refugee, now a thorough and proud Englishman, died five months later of a heart attack, right on his own schedule.

His sons, Reuben and Daniel, both went on to be extraordinary scientists in their own right. Reuben remembers taking long walks with his father on Sunday mornings, strolling down the streets of London near the UCL campus. His father talked to him as an equal even from the age of ten, and often discussed with the boys his latest research problems in the lab. Their extended family rose to high levels of achievement at the locations where they were scattered all over the world, wherever they had been able to go in their flights in the 1930s. For some of the family, Grüneberg became Gruneberg. None considered a return to Germany.

————

Both of Ursula Philip's brothers and her parents had escaped to America, thanks to the German police officer who had tipped them off and to the English fundraising efforts for their travel. Her younger brother Manfred was part of the Allied march inland after D-Day. Before the landings in France, Manfred waited in England with the US Army, and he stayed with his sister, Ulla, while he was there. The youngest Philip sibling, Georg, continued to grow and flourish throughout the war until, as feared by the Nazis, he reached combat age. Georg "George" Orkin Philip turned eighteen in late December 1942 and, as expected, promptly filled out his draft card, leading to service with the Allies in the Pacific. Both of Ursula's brothers were bilingual and native German speakers, which undoubtedly came in handy.

Ursula Philip—the warm, friendly, staunchly feminist, and ever-modest genius with the slight stoop and kind, sparkling brown eyes—decided she wanted a child. She and a man named Leslie Minchin were an unabashed "pair of hippies" with no plans to marry and firm beliefs in freedom, glory, and unfettered lives. They met in 1944, and Ursula had little in common with the suave, womanizing, pro-nudist, free-love-advocating Minchin, but by VE Day on May 8, 1945, at the age of thirty-six, she was "very clear she wanted a child." To celebrate the end of the war in Europe, they went to a bed-and-breakfast somewhere in Wales; there the owner expected them to be thrilled at being given the used, lipstick-marked bedsheets of some unknown Allied major and his wife. They were not. But on the trip, they made a daughter. The next year, the same year Catherine was born, Ursula Philip became an English citizen.

Haldane reacted with rage at the news of Ursula's motherhood. Scientists were not supposed to be burdened with the pragmatic chores of home life, and mothers must single-handedly assume those chores, he seemed to think. His condescending responses suggested that he thought Philip's genius being applied to raising a child was a waste. Despite the fact that she struggled with serious pregnancy-related medical issues before and after her daughter's birth, he wrote her off, calling her lazy, downgrading her position relative to Grüneberg, Kalmus, and Spurway, and sending to other employees of UCL nasty letters about her saying he thought they should hold off on giving her any kind of pay raise or promotion until she proved whether she could still manage to work. In contrast, Haldane had given Grüneberg a raise in 1934 when he first found out the German scientist had gotten married, and he fought for Grüneberg to get a UCL post with an elevated rate of pay after the war, because he was a single father.

Despite Haldane's sexist inanity, Ursula Philip and her young daughter, Cathie, led a beautiful, happy life together, with Philip legally adopting her own biological child to try to use paperwork to plaster over some of the social stigma of being a single mother. Philip became a scientific lecturer at Newcastle University and escaped Haldane's difficult reach, and she approached teaching her students with the same kind, level mentorship that anyone who knew her would have expected. She was known for making up her own fun English words like "giggilation," which she came up with to describe group outbursts of giggling, and her sweet nature made her grow to appreciate the less brusque mannerisms of British culture over those of her Germanic heritage. She raised Cathie within a cultural community that called themselves "refu-Jews," World War II refugees of the Jewish religion, and she remained lifelong friends with the Kalmus family.

Ursula Philip was not one to brag about herself. Before the war, she came up with a statistical test that proved another scientist was making extravagant claims based on fabricated data; her methods are still used to detect data fraud. During the war, she pressurized herself in a chamber at severe risk for the sake of the Allied effort. She was a single mother whose daughter justifiably brags about her. But Ursula "Ulla" Philip never revealed the purpose of her secret war work, even to her family. The work stayed classified until long after her death, and in the words of her daughter, "It's typical of children that they never ask the right questions of their parents."

The inland march of the Allied troops slowly revealed layer upon layer of inhumane depravity until the discovery of concentration camps revealed a level of horrific acts so unprecedented that the word "genocide" had to be coined to describe it. Humankind had committed psychotic acts against its own members for millennia—slavery, colonization, the treatment of women as property—but never had such a systematic, regulated machine been built for the efficient and intentional extermination of human beings en masse. Worse, the massacre had been justified by the lie that it would benefit humanity.

The concentration camps forced discussions about the practice of applying the theories of eugenics. The people whom Hitler declared unsuitable for procreation were put to work, and when they became unable to work, they were put to death. Eugenics, in its practical application, led to the ultimate outcomes of genocide and forced sterilization, regardless of how politely the concept was discussed in theory inside a lecture hall.

The concentration camps started the end of the study of the field of eugenics as it had existed before the war, and thus they helped smooth away the academic rift between the eugenicists and geneticists. The more passive field of genetics took over as the way to study the inheritance of traits. The new culture was to study inherited traits as neutral rather than applying judgments of the value of the person based on those traits, rather than using traits to decide who deserved to exist or die. At least, that is the aspirational perspective regarding inherited features to which the pure science hopes to bring society. One day.

The refugees of the UCL lab had not known about the concentration camps during the war. However, none of them seemed surprised when they learned the news. They had all fled their home countries because they knew that when prejudiced, forceful leaders were able to engage followers who did not question or regulate the leaders' behavior, the leaders never decided to regulate themselves. Foresight into that truth was what led the UCL lab members to leave, even when their family members refused.

Nussy Kalmus's brother and his wife had come to visit her family in England in June 1939; Hans and Nussy tried to convince them to stay. The duo brushed off the Kalmuses, insisting to themselves and others through a powerful cloud of denial that there would be no war. Three months later the war broke out. Nussy's brother was drafted to be an officer in the Yugoslav Army, and he spent five years as a POW in a prison camp. His wife was murdered in a concentration camp. Hans and Nussy Kalmus did not know until after the war that Hans's parents had been taken to Auschwitz, where they were both murdered. The last time Hans spoke to his father had been on his father's seventieth birthday.

For the rest of their lives, Hans and Nussy Kalmus continued to be the archetype of a married couple so loving, so genuinely devoted to each other that their relationship was perhaps more suitable for the movies than reality. The athletic and lanky Hans never expressed a desire to return to Czechoslovakia despite his earlier sorrow about leaving the mountains behind; instead he formed the new habit of walking with his children in the forests of England rather than along the rivers of continental Europe. Eventually, he did the same with his grandchildren. After the war Nussy gave birth to their daughter, Elsa, who finalized the family's quintet along with Peter and George. All three of the children rose to wonderful levels of scientific success and made contributions to the world. Peter and George became physicists, and thanks in part to encouragement from Nussy, who had always wanted a career but been stymied by the era, Elsa became a physician. Hans Kalmus never expressed bitterness toward Germans in either Germany or Czechoslovakia. But in what was perhaps an inadvertent metaphor for the way his country had treated his family, adult Peter reported that during a visit to his former hometown, the sausage snack that he had been obsessed with as a child left a bad taste in his mouth.

On October 5, 1946, along with sixteen other immigrant families, Hans, Nussy, Peter, and George Kalmus stood in front of the proper English legal authorities and swore the oath of allegiance to receive formal approval of their new status as officially naturalized citizens. Haldane had written in support of their application, and he had described Hans and his war work together. "It was not entirely safe," he mentioned, "and Dr. Kalmus was unconscious on several occasions."

On the inside of the family's naturalization application, unbeknownst to them, was a scribbled handwritten note reading: "I remember Prof. Haldane mentioning Kalmus on the BBC as a fellow scientist who experimented with him in pressure tests for the Admiralty in connection with submarine warfare. This work was of an extremely dangerous nature. Proceed." Based on the timeline of the family's approval process compared to that of others, the implication that their application got fast-tracked is plausible.

Hans Kalmus worked as a lecturer and researcher at UCL, and shortly after the war, he wrote a textbook about genetics that became so widely used and popular that one time, after Kalmus had gotten atypically lost on a ramble in the woods, the owner of the first house he stumbled upon recognized him immediately from his author photo. His granddaughter fondly remembered their last walk together before his death. She spoke with love and admiration of her grandfather, and of that final stroll through a field of purple flowers with the generous, determined man who used to do experiments on her when she was a child, because he was researching the genetics of color blindness and she was a handy subject.

———

Jim and Joan Rendel also lived out their lives in a cloud of genetics and science, although Jim would always struggle with his collapsed lung. The lanky scientist with the talent for breeding laboratory ducks for mathematical study would respond with shame and embarrassment when others had to lift his luggage for him. He needed the help for fear he would have a relapse of the wound he had incurred through participation in a heroic and noble war effort, but that worthwhile cause was not externally evident to others. It was an invisible wound. He would transfer his research focus from genetics of fowl to the inheritance patterns of larger forms of livestock, and eventually the duo would move to Australia, where he ran the genetics unit for the Commonwealth Scientific and Industrial Research Organisation (CSIRO). His name is still emblazoned on a laboratory in memory of how he encouraged employees to follow up with creative scientific ideas that danced across their brain waves, and thus inspired innovation all around him.

During the height of the Blitz, Joan Rendel had written from her new-mother isolation in the attic of the Piggery that "I still keep thinking of Mrs. [Virginia] Woolf—especially when the sirens go," and "I want to live through this, so that I can write about it, or learn from it at least. But learn isn't the right word somehow." However, by the end of the war, the ingenious and sensitive "girl poet" started to show glimmers of schizophrenia, which would sadly only increase in that era with no effective medical treatment. Jim loved her and cared for her in every way he could, which included seeking help when her disease became too much for him to handle alone. Jim and Joan were lucky enough to have three children together, and Jim another three in his second marriage, followed by ten grandchildren. The current family members are a tight-knit group bonded by love and memories of Jim and Joan.

Explosives expert Horace Cameron Wright would carry on his blast experiments after the war, and he would carry them on with his typical extremity. The scientist put together roughly a half dozen trials using a variety of setups, configurations, and charge types in order to puzzle out the ongoing enigma that was human tolerance for underwater blasts. For most of the experiments, he recruited other people, but he was always one of the test subjects in the water with them. In his most dramatic experimental series, he looped together ten thousand pounds of TNT in an array on the seafloor, and tested to see how far away such an explosion could cause a dramatic impact for "tourists" on a beach. The answer was two miles. He was the "tourist." He crawled out onto the sand after the vicious pop, unwounded but impacted. His experiments remain the largest and most useful body of human data in the history of underwater blasts.

Cameron Wright's lack of a medical degree haunted him for life, but his career speaks to how he did not need the degree to become wildly accomplished. After the war he continued to take joy in horses, racing, and his car-show-worthy vintage Alfa Romeo. His former colleagues at the Royal Naval Physiological Laboratory spoke about him with love, respect, and just enough of a sense of intimidation to make him seem fun to work with. At one point, the brusque and efficient physiologist decided to slice a small tumor off his own left hand, and he did so with skill.

Wright left behind no direct family members, so after his death, his estate was distributed among distant relatives via the real-life version of the overused plotline of bad fictions wherein government representatives show up at random on those relatives' doorsteps with briefcases and the news of surprise inheritances. The unrealistic yet also clichéd surprise seems like one he would have enjoyed. Horace Cameron "Cam" Wright died on April Fools' Day of 1979, which his friends took as a grand last joke of his and which made it into his obituary.

Also in his obituary is one haunting sentence, a hint that Cam Wright might never have married because of a tragedy, yet more details about his life story that may never be known. As Wright approached his final breaths, his eyes dimming beneath shaggy white eyebrows, a dear friend of his said that he knew Horace Cameron Wright was seeing one person in particular amid the coming shadows—Elizabeth. No record of who Elizabeth was, or how she had influenced his life, has ever been found.

———

Edwin Martin Case and his brothers continued, long after the war, to live lives of "extraordinary sense" rather than plebeian common sense. Robert Alfred Martin Case ended up at the Royal Naval Physiological Laboratory studying diving problems, with a view to improving submarine escape. He is perhaps best known for going to the Antarctic, where, while using an oxygen rebreather and a giant knife, he excised a sample of whale blubber from a living whale. His goal was to study the way the blubber absorbed gas compared to the way tissues of the human body did so, and his project was the first step in a field of research that remains an active area of work in diving. The whale was not consulted. Robert seemed to have been successful, despite being pulled from the freezing water unconscious after his rebreather sprang a leak. Brother Ralph Case became a physician, then eventually a lecturer. Perhaps most important, none of the three Case siblings seemed to have experienced any ill effects from their childhood self-experiments with chloroform and ice baths.

The X-craft that Martin Case had helped develop were used by crews to fight the war in the Pacific after VE Day. The shipyards of England built a new, air-conditioned model called the XE. On one particularly notable mission among the many exploits of the XEs, rebreather divers swam out underwater from the wet and dry to remove a section of the enemy's telegraph cable to render it useless. They were never caught, and the Japanese lost substantial communications in the Pacific as a result. The XE crews were ready and waiting to scout the beaches in Japan for a D-Day-esque landing that would precede a drive to Tokyo, but instead, atomic bombs were dropped on Hiroshima and Nagasaki, and on September 2, 1945, the war finally ended.

———

Joan Elizabeth Jermyn had "dropped Communism like 'a hot cake'" in 1939 after the Hitler–Stalin Pact, disillusioned by the fact that Stalin would sign a pact with such an obvious loon, but by then she was already secure in her job in the UCL lab. She was a born scientist, and as the UCL lab's scientific work at Siebe Gorman came to its close in 1944, she decided she wanted to be a physician. Nariman Bamji planned to return to India someday, and Elizabeth Jermyn did not think her physiology degree would be useful there when she went with him, so she chose to return to school to study medicine. She had to get special permission from the government to apply to medical school because the Ministry of Labour said that women over the age of twenty were already skilled in clerical ways that could contribute to the war and were therefore needed, and so she would not be permitted to quit her job with Haldane.

However, Jermyn was not one to be deterred. She plotted out the medical school class schedule and Haldane's normal work schedule, and she molded her medical-course load around his always nocturnal research patterns. She planned a calendar of 132-hour workweeks. The Ministry of Labour relented in the face of her extreme persistence, but on the condition that she did not quit her job at the lab. Haldane wrote her a glowing letter of recommendation, effusing that she was an intelligent person and a good worker who knew what it felt like "to be poisoned by a gas." Coming from Haldane, who had refused to write a letter of recommendation for another employee because they had made a mathematical error once—exactly once—the words were lofty praise.

Dr. Joan Elizabeth Jermyn graduated from the London School of Medicine and qualified as a physician in 1948; then she got married to Bamji the next year. The job post that Nari had been offered in India turned out to be an empty promise, so they stayed in England. Bouncing baby boy Andrew followed the year after that, and most preestablished medical practices refused to employ now Dr. Bamji because she was

Joan Elizabeth Bamji, née Jermyn, and her son, Andrew.

married and a mother, so she started her own. Nari Bamji eventually also retrained as a physician and joined her practice, where the two worked together until retirement.

The unstoppable woman spent the rest of her life moving at the speed of light, working, treating patients, caring for and tutoring her son, and eventually caring just as much for her grandchildren. Her spinal fracture caused her pain and problems until the day she died, but she never let it slow her down or get between her and running her innumerable house calls. She never disclosed the purpose of her classified wartime work to her son, Andrew, who would also become a medical doctor, but she spoke of it to him in generalities with pride. Joan Elizabeth Jermyn died in December 2010 at the age of ninety-four.

Ken Donald outlived most of the UCL team and he worked directly for the government, which meant that in 1992 he was able to get clearance to publish a textbook called *Oxygen and the Diver* without there being anyone left to protest the claims within. The textbook is still, as of this writing, considered the seminal text about diving and oxygen because of the data-based risk curves for oxygen toxicity that it contains, and it is a mandatory presence on any bookshelf dedicated to diving or hyperbaric medicine. The risk curves printed inside are plagiarized without alteration from Spurway and Haldane's appendix to Donald's original report to the government. The statistical discussion and explanations of where the curves came from are also plagiarized—verbatim—from Helen Spurway without naming her or giving her credit.

Helen Spurway remains one of the most intriguing, controversial, and arguably wrongly forgotten figures in the history of diving. She has been omitted or deleted from most text records, so much of her life has been cloaked in a shroud of mystery. It was difficult to find many people who knew her, and impossible to find one living person who spoke positively about her. However, she was a female scientist working in a time period when that was a rarity in the profession, and she was a person who was both vocal about her goals and irrepressible in making her presence known in every room. Perhaps the best evidence of that was her socially clumsy announcement on the first day of class at UCL that she intended to marry the professor, but ultimately, she was right. She and J. B. S. Haldane married in December of 1945, then took a trip to Paris to lecture at the Sorbonne; that trip seems to have passed for their version of a honeymoon. Joan Rendel provided the only known record of another woman discussing her thoughts about Helen Spurway, and she wrote highly of her, saying, "She is an extraordinary creature, but we all like her very much."

Otherwise, the first adjective nearly every source uses to describe Helen Spurway is some synonym of "shrill," which is an insult disproportionately—and usually undeservedly—applied to women, but in Spurway's case, it seems to have been true. Even J. B. S. Haldane, who loved and respected her, used to tell her that her voice sounded "like a circular saw." However, at one point she privately confessed that she had suffered profound hearing loss. She was mortified about the sound of her own voice. The feature that caused her to receive such mockery and dismissal was the result of a hearing disability.

After Haldane and Spurway married, the two continued to conduct scientific experiments together at UCL while becoming infamous for their raucous boozing at a pub called the Marlborough Arms, which was roughly across the street from the campus. Haldane expressed in letters to colleagues his high hopes that they would be able to publish their studies on diving, DCS, and oxygen toxicity, but for decades after the end of the war, those results remained cloaked in layers of red stamps screaming classification levels. Haldane and Spurway wrote a draft textbook about the work that never received permission to see the light of day.

A year after the end of the war, J. B. S. Haldane applied to register as legally disabled. His lower back, at the site of the vicious seizure-induced fracture, never stopped hurting. He always had to sit with a small pillow or cushion to ease the pain. He was granted the disability status.

In 1956, while leaving a pub after drinking 3.5 pints of bitter beer with her dinner, Helen Spurway, described by some who knew her as an alcoholic by that point, decided to stomp on the tail of a police dog seated on the sidewalk. After being arrested and charged, she protested in court that since she had been wearing rubber-soled shoes, her action was no worse than "a caress to the dog." However, she also announced to the magistrate that she had done it to "let the police know they are hated and despised." Unsurprisingly, she was found guilty, and she was fined. But she refused to pay the fine, so she was taken to a jail cell.

UCL fired her, Haldane resigned in protest, and eight months later, in July 1957, she and Haldane boarded a flight for Calcutta, India. Haldane declared to the press before leaving that they were in search of "a free country." The tail-stomping incident was part of the stated reason for their departure, although depending on when he was asked, he gave various other reasons, too, like the continued presence of American troops in England. On the plane, they carried with them several jars of live fish for the sake of science and future study of their genetics.

By the time of J. B. S. Haldane's death in 1964, he had renounced Communism. His earlier belief in the economic model had been founded in the idealistic and romantic hope that if all human beings were treated as equal, they could eradicate poverty, discrimination, and suffering. However, it was through genetics and a man named Lysenko that Haldane finally realized he had been sucked in by the Soviet propaganda machine.

Soviet pseudobiologist Trofim Lysenko was a square-jawed steel barrier against outside opinions. He firmly believed that the "gene" as an item did not exist, and he spoke vociferously against the concept. Instead, he concocted his own theories of inheritance with little to no supporting data—but with the support of the Soviet government. Those who disagreed with him were executed. His theories guided agriculture in both the USSR and China. The resulting famines caused the deaths of tens of millions of people.

Haldane died on December 1, 1964, of colorectal cancer. Spurway continued to work in genetics in India and replaced Haldane as editor of a genetics journal until she died in 1978 at the age of sixty-two. Ever the animal handler, she died of tetanus, which she contracted from a bite from her pet jaguar.

Over the course of the war, the members of the UCL lab completed 611 experiments on themselves during a total of 284 grueling days inside the chambers. After seeing how they enabled the X-craft, the LCOCU, and the diving work of the P-parties in the harbors, the Americans got to work on midget submarines and oxygen rebreathers in earnest as well, and by the 1950s they had their own rebreather models strapped to the chests of their intrepid amphibious humans. Midget submarines and oxygen rebreathers have been a fixture of military diving since, and at least within the Allied world, all dive teams still use the original guidelines for oxygen toxicity traceable to these studies during World War II.

After the official end of the war in Europe, a new employee of the French military named Jacques Cousteau asked for an introduction to whoever in England could teach him the physiological principles behind oxygen toxicity and the high-oxygen breathing gases used in the beach landings of his own country of France. Cousteau had spent the war period tinkering, playing with other ways to let human beings breathe underwater, alongside his partner in invention, Émile Gagnan. After Cousteau agreed that the information he learned would be for French military use only, not public release, William Shelford of the Admiralty Experimental Diving Unit introduced Cousteau to the Haldane lab. After the meeting, Jacques Cousteau wrote that it was an "honor" to meet the lab personnel, and Cousteau asked Haldane to pass on his thanks to Haldane's scientific team for teaching Cousteau about underwater physiology.

It is a truth well established that all generals are experts at fighting the last war. However, in the underwater world, the scientific advancements brought about during World War II would dictate the face of future warfare, keeping undersea experts shockingly relevant in the decades to come. Providing human beings with the ability to survive underwater created an entire third sphere of combat to add onto the previous realms of land and air. When people were no longer limited to what they could achieve in the length of time they could hold their breath, the ocean all of a sudden was a place to hide, and a good one at that.

Meanwhile, the world fractured again after the end of the war had made it seem almost healed. The Allied countries and the Soviet Union had stayed uneasy bedfellows for the sake of defeating the Axis, but once their mutual enemy was vanquished, their different political ideologies caused an irreparable fissure that turned into the Cold War.

Perhaps nowhere is that fissure more evident than inside the files of MI5. Everyone, according to them, with any sort of Communist contact was a spy. Vehement expectorations proclaiming suspicions of espionage, often with little supporting evidence, decorate the surveillance files of many British citizens of the time period, including the MI5 file of J. B. S. Haldane. At worst, he was accused of sending government secrets in the form of his underwater test results to the Soviet Union. It's an accusation that has gotten him labeled a traitor by some. However, he had express written permission from the Admiralty to do so, and he got that permission before he seems to have sent anything. When compiled, the agencies' documents read more like, as usual, they were not talking to one another, rather than that Haldane had done anything deliberately illegal. It's an incident that shows how sensitive the political wound had become.

The Cold War was also the perfect environment for divers and submarines to thrive in. The operatives for the two sides could swim along silently in the aptly frigid ocean and monitor one another from afar with minimal open signs of aggression.

In one of the most fascinating and secretive acts of the Cold War, United States military submarines prowled the frigid ocean between Russia and Alaska. They dredged the ocean floor, searching for a telegraph cable that might have carried uncoded transmissions. The Russians had built upon the cryptographic success of the German Enigma machine, and their radar transmissions were coded to an unparalleled degree. But telegraph cables were old technology; they had graced the seafloor since the 1800s. So the logic was that most governments would have been less likely to use encryption on a physical connection. In the

1970s, American submarine crews found exactly such a Russian cable. Because of the advanced state of knowledge about how to process the air inside their submarine, they were able to stay submerged for long enough to creep into the hostile region undetected and stay for long enough to search an immense expanse of blank ocean for the thread of the cable without needing to surface.

Divers were able to lock out of the submarines using techniques similar to the wet and dry compartments of the X-craft. Tethered to the submarines for their gas supply, they attached a listening device to the Soviet cable. Returning every month or so, they were able to retrieve and change out recorded tapes filled with uncoded messages and listen to highly classified Soviet government plans.

At those depths, even air has a high enough fraction of oxygen to throw a diver into fits of seizures. The work, known as Operation Ivy Bells, took place during the 1970s, but because of the research done during World War II, the dive planners knew they would have to create special gas blends with lower percentages of oxygen to keep the divers alive outside the subs. Modern submarine actions suggest that those types of deep-sea espionage operations remain ongoing today.

Rebreathers have also stayed and progressed. The oxygen-filled Draeger Lar V rebreather is the official, open, not-even-secret tool of special operations divers who crawl along the shallows and across the beaches of the world. It uses pure oxygen in a design that is sturdier than the Siebe Gorman machines of WWII, but that is otherwise virtually unchanged.

With the advent of computers, rebreathers, too, joined the digital age. Now the mixed-gas principles developed for the P-parties in the harbors of Normandy drive the electronics packages of so-called mixed-gas rebreathers. These breathing machines measure the depth of the diver and inject oxygen to maintain an elevated level of the gas to minimize decompression time, while also minimizing the risk of oxygen toxicity. Specific designs with minor alterations between them abound, but rebreathers' primary use by the military remains as a tool for ordnance-disposal divers. No bubbles mean minimum physical motion to disturb bombs. Rebreathers mean divers have to carry less gear and are not encumbered by the tether of an umbilical providing them with gas.

Rebreathers of both types are key to the way SEALs, Marines, Rangers, and military personnel of similar ilk prowl the ambiguous region between the land and the sea in the name of security and freedom. Such existence underwater would not be possible without the secret wartime science of J. B. S. Haldane, Martin Case, Hans Grüneberg, Elizabeth Jermyn, Hans Kalmus, Ursula Philip, Jim Rendel, and Helen Spurway.

Their achievements were buried under classification markings until 2001. The world never knew what the chamber divers had achieved, what they had contributed to D-Day, to science, to the world as a whole. Their families never knew; their children never suspected.

Until now.

Acknowledgments

I would like to thank, first and foremost, all of the people who have taken or tried to take credit for my own scientific work. Without these men, I would not have known how to dig for this story, how to find the under-credited women and refugees within it, those people who were similarly under-titled and underpaid, and whose under-pay and under-titling were used to justify shifting their achievements into the portfolios of people against whom they had no standing to argue. In 2023, a team of historians concluded that because Rosalind Franklin continued working with the same groups and universities that unabashedly stole credit for her work on DNA, she therefore must not have minded the theft; to that I say nonsense. Certain groups of people have never had the choice to fight. Through this book I have tried to fight for some of them instead.

On a more genuine note, I also want to acknowledge my family. My mother, Denise Lance, is my original role model in standing up for myself against the forces of nonsense. I plan to dedicate an entire book to her one day in thanks for her gift of fortitude, but it needs to be the right story. My dad, Ron Lance, also deserves credit for giving me my love of history. He taught me that history is not just names and dates to be memorized in anticipation of a scantron, and instead he talked with me about stories and sat me down in front of documentaries and showed me that history is a compilation of relatable stories about real people. They are the first two people to read my writing once it has been typed, and although I don't always agree with my dad's suggestions, I always enjoy the discussions.

This book was a challenge for me for several reasons, and many people deserve thanks for helping me overcome those challenges. I started the research in 2019, then I decided to have knee surgery in January 2020 so that after the roughly six-month recovery I would be better able to walk around London and Edinburgh to complete the necessary archival research. We all know what happened next. The book was delayed. I also conduct scientific research out of a hospital, and my work life at the hospital became difficult. I included stories of the medical workers during the Blitz in recognition of my brave coworkers, many of whom still bear internal scars throughout their lungs from caring for patients in the year before vaccines. I brought you the lasagnas I could. I wish I could have done more.

I also owe huge thanks to the many people around the world who helped me scrounge together archival documents until I was able to safely travel and review the rest myself. Robert Winckworth copied documents for me from the University College London College Archives; Tina Hampson proved a marvelous and speedy help with documents from the National Archives of the United Kingdom; and Lorna Kinnaird was an amazing, colorfully haired wizardess of research on my behalf at the National Library of Scotland. David L. Dekker of the Netherlands was kindly willing to scan the pages of his copy of Siebe Gorman's 1946 niche, hard-to-find publication, which he holds in his endless library of diving history books (catalogued online at divescrap.nl).

And finally, Stephen Wordsworth of the Council for At-Risk Academics (CARA) helped me find and get copies of the immigration records of Ursula Philip. However, more important, CARA continues to not only exist but to help people like Dr. Philip escape danger. They provided me with this statement to allow me to describe their mission, and to thank them properly through doing so:

Cara (https://www.cara.ngo) was founded by academics and scientists in the UK in 1933 as a rescue mission for their colleagues who were being expelled from their posts by the Nazis, and later by fascists elsewhere in Europe. Almost 90 years later the organisation is still at work, now helping academics fleeing from persecution, violence and conflict anywhere in the world, yet still guided by the task set by their founders— *"the relief of suffering and the defence of learning and science."*

I also want to thank my agent, Laurie Abkemeier, as well as the entire literary team at Dutton. Laurie knows that I need her to tell me the truth without hedging, and because of her willingness to do so, I head into our every interaction with the excited confidence that I will leave the conversation with a better understanding of books, publishing, and writing in general, all gleaned from her magnificent brain. The topic of this book was originally going to be one chapter in a more scattered compilation about scientists experimenting on themselves, but it was Laurie who had the wisdom to tell me to focus on one story start to finish.

The team at Dutton has similarly been marvelous. My editor, Stephen Morrow, lets me know when I have wandered too far off the point, and I can always rely on him to gush with me about the tiny details of historical human beings. Grace Layer, his assistant, has proved to be an organized and kind helper, and I appreciate her reacting with understanding whenever I am at fault in messing up her clear instructions, which is often. Frank Walgren is known to me only through his name in the comments on the manuscript I submitted, but he has been my copy editor. He has fixed my typos, caught my constant errors in calculating ages, and done his best to rein in my confusing tendency to invert sentences. Any remaining mistakes are mine, not his, and I have appreciated the extraordinary number he has caught and fixed.

I also want to thank the people with specialty backgrounds who proofread text and wording for me to make sure the terminology is consistent with their experience and experience in their fields, especially those who helped me translate from British English to American English and back again. David Poole and Alex Blackford gave me an education in British military terminology, and Jane Wickenden was an extraordinary help with terminology related to British nautical terms and history. When I could not for the life of me figure out what a "trot" was, Lesley

Blogg came to the rescue by connecting me with the expertise of former submariner Colin Williams, who cleared up my confusion.

Russell Worth Parker, himself an extraordinary writer, is a veteran of the Marine Raiders and a close enough friend that he knew he could call me and tell me directly that I should "stop anthropomorphizing the damn ships." Instead, he said, give credit to the soldiers and sailors themselves. He was right, and this book is stronger for that reality check in the language I used. In part because of him, and in part because of my own family members who fought in this war, I was reminded to tell this story through individual people as much as possible, instead of zooming out. I am proud to have been able to include this many stories of real-life warriors.

Rob Crane of the website COPP Survey has also turned into a friend through this process. He began his website about COPP history as a pandemic project, and it has bloomed into an invaluable resource about the history of these specialized units. It has been fun trading documents, theories, and COPP stories with him throughout the writing, and I have appreciated his willingness to chat about obscure missions and help me find resources that I lacked.

Finally, I want to thank the family members of the scientists. Peter Kalmus, George Kalmus, Miriam Kalmus, and Elsa Roe were all generous with their time in talking to me about Hans and Nussy Kalmus, to the point that I consider them friends and I hope they do likewise. Andrew Bamji provided a wealth of information about his mother, Joan Elizabeth Bamji née Jermyn, and through the two of them I learned the value of parents' writing out their own life histories in case a random historian comes knocking later on the proverbial doors of their children. Reuben Grüneberg shared his memories of his father, Hans, and mother, Elsbeth, with me, even the painful ones, and I appreciate his openness and willingness to dive into those parts of the past. Chris Love, the niece of Martin Case, proved just as much a riot as her father and uncles must have been in life, and I am sad only because I could not include all of the rich depth of information she was willing to share with me about their personalities. Cathie Wright, the daughter of Ursula Philip, not only gave me her time, but I learned she had no idea what her mother had done during the war, and I was honored to be able to share that information back. I appreciate Cathie's willingness to share her story, and her memories of her mother, who in my opinion is one of the most fascinating people in science. Irene Stoller, niece of Ursula and daughter of Manfred Philip, was also generous with her time. I would have liked to include more about the journeys of Manfred and George but was worried about wandering too far off topic. Perhaps in another book. Historian Jane Wickenden and Wright's distant relative Paul Nowlan were my partners-in-investigation in tracking down the elusive life history of Horace Cameron Wright. Our email threads tracing the story of this under-documented but jaw-droppingly heroic man trained me to have an adrenaline spike of excitement whenever I saw either Wickenden's or Nowlan's name in my inbox, because it always meant a new discovery. Finally, I want to thank Jane Robertson, Celia Robertson, Sandy Rendel, and Polly Woods, all family members of Joan and Jim Rendel. I know that this process was difficult for them, and I am touched and flattered that they decided to talk to me. I did everything I could to treat their story with the care, love, and respect that both Jim and Joan deserved, and I hope that they are proud of the result.

It took all of these people, and many more, to create this book. I may have organized the words onto the pages, but the story in truth belongs to them.

Notes

A note about the notes: References are cited in short form here and should be cross-referenced with the full publication details provided in the bibliography.

PROLOGUE: "A Piece of Cake"

1   **simple cup of stew:** The breakfast served on the approach to the raid is mentioned in Porteous, Interview, 1987.

1   **Captain Patrick Anthony Porteous:** All details of Captain Porteous's personal story are taken from Porteous, Interview, 1987.

1   **small, flat-bottomed boats:** Davies, Interview, 1989.

1   **slight breeze from the stern:** Hughes-Hallett, "Commander's Narrative," 1942, Enclosure 1, 6.

1   **sloping, pebbly shore:** C. C. Mann, "Operation Instructions," 1942, appendix B, 2.

1   **"Yellow," "Red," "White," "Green," and "Blue":** Reese, "Dieppe Raid," 1942.

2   **lighthouse was the primary means:** Hughes-Hallett, "Commander's Narrative," 1942, Enclosure 3, 3.

2   **Orange Beach party was divided:** Reese, "Dieppe Raid," 1942, 1.

2   **They used handheld tubes:** Davies, Interview, 1989; Porteous, Interview, 1987; Fussell, Interview, 1991.

2   **moonless sky:** The moon set at 12:18 a.m. and did not rise again until the afternoon. Phases calculated using timeanddate.com.

3   **No more than twenty:** Davies, Interview, 1989.

3   **A medic volunteered:** Davies, Interview, 1989; Porteous, Interview, 1987.

3   **"regardless of anything":** Porteous, Interview, 1987.

3   **Gunboat 315 looked an awful lot:** Hughes-Hallett, "Commander's Narrative," 1942, Enclosure 1, 3.

3   **punctual smoke had begun:** The timing and use of smoke screens is described at multiple points in Hughes-Hallett, "Commander's Narrative," 1942.

3   **Aerial planning photographs taken:** Mountbatten, "The Dieppe Raid," 1942.

4   **Royal Regiment landed on Blue:** C. C. Mann, Lecture Notes, 1942, 7.

4   **officers on the ships couldn't see:** Referenced multiple times in Hughes-Hallett, "Commander's Narrative," 1942.

4 **The pilots in the air:** Rockets existed at the time of Dieppe, but were in early stages of development as a technology. Experiments and models had focused on firing them from ground or shipboard locations. Dieppe sparked new meetings and the technological development of more advanced rocket types in preparation for the next beach landing, per Crow, "The Rocket as a Weapon," 1948.

4 **Not a single soldier made it:** Hughes-Hallett, "Commander's Narrative," 1942, Enclosure 3, 4.

4 **casualty rate of 97 percent:** O'Keefe, *One Day in August,* 2020, 6.

4 **thousand of the Allied landing troops:** Adjutant General Office, "'Jubilee,'" 1944.

4 **one of the most horrifying:** O'Keefe, *One Day in August,* 2020.

4 **inexperienced Canadian troops:** The troops sent for the raid were inexperienced per comments indicating that they had no prior experience in Hughes-Hallett, "Commander's Narrative," 1942.

4 **were convinced that tanks alone:** Reese, "Dieppe Raid," 1942.

4 **Everyone would jump out:** The battle plans for the landings at Red and White beaches at Dieppe are mentioned in multiple places. The first landing wave on Red and White included the 14th Canadian Army Tank Battalion per Reese, "Dieppe Raid," 1942, which also says the LCTs were hit hard as they were off-loading the tanks. Also referenced in C. C. Mann, "Lessons Learnt," 1942.

5 **"a piece of cake":** Granatstein, *The Weight of Command,* 2016, 140.

5 **Laurens Pals of the 2nd Canadian:** Details of Laurens Pals's story are taken from his oral history recording with Pals, Interview, 1980.

5 **tanks had been waterproofed:** "All the tanks had been prepared for wading up to five or six feet deep but this depended on a special fabric which of course was not bullet proof," in C. C. Mann, Lecture Notes, 1942, 9.

6 **process to unload the tanks:** Reese, "Dieppe Raid," 1942.

6 **combat engineers carried explosives:** C. C. Mann, Lecture Notes, 1942, 9.

6 **"The road blocks were never destroyed . . .":** C. C. Mann, "Answers," 1942, 8.

6 **Churchills had never been used:** Henry, "Calgary Tanks," 1995.

6 **waterfront was made of pebbles:** In British parlance, the beach was shingle. Referenced in all after-action reports, e.g., Reese, "Dieppe Raid," 1942, on multiple pages; Hughes-Hallett, "Commander's Narrative," 1942, Enclosure 3, 7; and C. C. Mann, "Answers," 1942, 14.

6 **Every single tank was destroyed:** C. C. Mann, "Answers," 1942, 14.

6 **these, too, sank feebly:** C. C. Mann, "Report on Various Equipment," 1942.

7 **There were others also charged:** Laurens Pals provides his narrative in his oral history (Pals, Interview, 1980), but the most complete reconstruction of the espionage and code-breaking missions at Dieppe is by historian David O'Keefe in *One Day in August,* 2020.

7 **But the official explanation for:** C. C. Mann, Lecture Notes, 1942, 2. The explanation for the lack of bombing is so ridiculous, I need to quote it: "(a) High level bombing with very large bombs at night is not accurate enough to expect to destroy prepared positions such as existed at Dieppe. (b) Day bombing would have had to be done in advance of the assault, and for such bombing to be sufficiently accurate would require a special form of prolonged training which was not possible in preparing for this operation." Point (c) said that it would be hard to navigate the tanks through the fires started by bombings.

7 **intact, beautiful residences:** Hughes-Hallett, "Commander's Narrative," 1942, Enclosure 3, 7.

7 **bloodstained waves:** Pals, Interview, 1980.

7   **continued to send wave after wave:** The sending of wave after wave is documented in multiple places in these after-action reports: Hughes-Hallett, "Commander's Narrative," 1942, and Roberts, "Comd 1 Cdn," 1942.

7   **new model of Enigma:** the four-rotor Enigma. Erskine and Weierud, "Naval Enigma," 2010.

7   **At 1100, after hours:** Reese, "Dieppe Raid," 1942.

7   **convoy offshore never sent:** Hughes-Hallett, "Commander's Narrative," 1942, Enclosure 3, 8. The group of sailors who made the effort was on an LCA that was swamped.

7   **Under pressure from:** Churchill's attempts to rewrite his relationship to the Dieppe Raid are documented in detail in O'Keefe, *One Day in August*, 2020, 298–300. Interestingly, most of the veterans of the raid who left behind oral history recordings described Churchill as being willing to let them die in the raid to prove his point, e.g., Nissen, Interview, 1989, and Porteous, Interview, 1987.

8   **Churchill had already been:** Aspinall-Oglander, *Gallipoli*, 2013.

8   **fled his home country:** Zuehlke, *Tragedy at Dieppe*, 2012, 367.

8   **Allied news media lied:** Howe, "The World Today," 1942.

8   **"Lessons Learnt":** C. C. Mann, "Lessons Learnt," 1942.

8   **"As soon as it is known . . .":** Mann, "Lessons Learnt," 1942, 10.

8   **had encountered underwater obstacles:** Hussey, COHQ Docket, October 26, 1942.

8   **"Naval reconnaissance methods . . .":** C. C. Mann, "Lessons Learnt," 1942, 10.

9   **only knowledge of the composition:** Mountbatten, "The Dieppe Raid," 1942. The scouting photograph used is in the insert between pages 20 and 21.

CHAPTER 1: Professor Haldane and Dr. Spurway's Problem

11   **A few days later in London:** This experiment took place on August 24, 1942, and provided data points number 239 and 240 on experimental day 107. Unless otherwise noted, the descriptions of this test and all other tests come from multiple sources, specifically lab notebooks, reports, and academic publications that were aligned and compiled into narratives using the test date, experiment type, and experimental subject name. Each test narrative will cite all the sources used to build the narrative for that test. The sources for this test were J. B. S. Haldane, Papers, Part 1, and UCL Lab, Miscellany of Papers, ca. 1940–44, with entries for the date of August 24, 1942.

11   **a mere four feet:** Descriptions of this and all other hyperbaric chambers are taken from a combination of the descriptions in UCL Lab, Miscellany of Papers, ca. 1940–44, as well as from photographs for *Life* magazine that were never published.

11   **Six foot one:** Haldane's height is taken from J. B. S. Haldane, Application, 1934. The rest of his description is taken from photographs.

11   **The gas roared around them:** Descriptions of the experience of being inside a hyperbaric chamber are taken from the test notes as cited, but have occasionally been augmented by the author's personal experience working inside hyperbaric chambers at the Duke University Center for Hyperbaric Medicine & Environmental Physiology and at the University of Southern California, Catalina Hyperbaric Chamber.

11   **Twenty-seven-year-old Spurway:** Helen Spurway was born on June 12, 1915. General Register Office, "England & Wales, Civil Registration Birth," 1915.

11   **a PhD geneticist:** Gordon et al., "An Analysis," 1939.

11   **was as lanky:** Descriptions of Spurway are taken from photographs.

12   **triple digits Fahrenheit:** The temperature reached is taken from how hot comparably sized or larger chambers at Duke get when pressurized at the same rate described in the test documents.

13 **dancing purple lights:** Spurway described having dazzle in the documentation for this experiment, but the fact that she saw her dazzle as the color purple is stated in UCL Lab, "Diving Notes," Part 2, July 1942–May 1943, and J. B. S. Haldane, "Effects of High Pressure Oxygen," ca. 1942, 5.

13 **a thirteen-year-old boy peered:** Boycott et al., "Prevention of Compressed-Air Illness," 1908, 436. Page 436 is specific to J. B. S. Haldane's age, but the rest of the document contains the description of the chamber on board the ship and the test being conducted. The description of the chamber is on page 429.

14 **born in 1892:** General Register Office, "England & Wales, Civil Registration," 1892.

14 **homes have turrets:** L. K. Haldane, *Friends and Kindred*, 1961, the photo insert between pages 16 and 17; also, N. Mitchison, *Among You*, 2000, 12, describes turrets.

14 **Stately portraits of ancestors:** Photo inserts of portraits in L. K. Haldane, *Friends and Kindred*, 1961.

14 **"Jack" in his youth:** L. K. Haldane, *Friends and Kindred*, 1961, 215.

14 **bathtub full of tadpoles:** L. K. Haldane, *Friends and Kindred*, 1961, 141.

14 **pedigree back to the 1200s:** Pirie, "Haldane," 1966, 219.

14 **She was a brilliant young woman:** The description is from image inserts in L. K. Haldane, *Friends and Kindred*, 1961.

14 **occasional cigarette:** L. K. Haldane, *Friends and Kindred*, 1961, 134.

14 **stumbled across John Scott:** L. K. Haldane, *Friends and Kindred*, 1961, 105.

14 **lanky young man:** The description is from available images of J. S. Haldane when he was young.

15 **He couldn't convince her:** L. K. Haldane, *Friends and Kindred*, 1961, 105.

15 **It stormed . . . being 'given away'":** L. K. Haldane, *Friends and Kindred*, 1961, 152.

15 **both firm agnostics:** N. Mitchison, *Small Talk*, 1988, 12.

15 **gave himself a hernia:** R. W. Clark, *JBS*, 1969, 13.

15 **Naomi was born:** Naomi Haldane was born on November 1, 1897. N. Mitchison, *Among You*, 2000, 16.

15 **along with Kathleen's mother:** L. K. Haldane, *Friends and Kindred*, 1961, 203–4.

15 **"comfortable and ugly":** R. W. Clark, *JBS*, 1969, 18.

15 **bathtubs be made of lead:** N. Mitchison, *Small Talk*, 1988, 15.

15 **refused to allow:** N. Mitchison, *Small Talk*, 1988, 15.

16 **scarab beetles:** N. Mitchison, *Small Talk*, 1988, 15.

16 **painted Chinese bowls:** N. Mitchison, *Small Talk*, 18.

16 **encrusted in a mosslike coating:** R. W. Clark, *JBS*, 1969, 19.

16 **room-length wooden table:** The description of the office and laboratory configuration are in N. Mitchison, *Small Talk*, 1988, 123; the step down is in Goodman, *Suffer and Survive*, 2008, 215.

16 **"the coffin.":** Acott, "A Brief History," 1999.

16 **family cat slept curled:** N. Mitchison, *Small Talk*, 1988, 18; L. K. Haldane, *Friends and Kindred*, 1961, 174.

16 **dollhouse was decorated:** L. K. Haldane, *Friends and Kindred*, 1961, 174.

16 **physicist Niels Bohr:** N. Mitchison, *Small Talk*, 1988, 18.

16 **Carved into the entryway:** R. W. Clark, *JBS*, 1969, 19.

16 **chased globules:** N. Mitchison, *Small Talk*, 1988, 23.

16 **effects on their voices:** N. Mitchison, *All Change Here*, 1988, 26.

16 **dancing Viennese waltzes:** Dronamraju, *Popularizing Science*, 2017, 17.

16 **"Nou.":** This nickname is evident in his letters home to Naomi during WWI, e.g., J. B. S. Haldane, Letter to Nou, 1917.

16 **She addressed him:** N. Mitchison, *Small Talk*, 1988, 29.

16 **"As children we were:** N. Mitchison, "Beginnings," 1968, 302.

16 **Naomi's job:** Norton, *Stars Beneath the Seas*, 1999, 108.

16 **By age three:** R. W. Clark, *JBS*, 1969, 14.

17 **by age four he . . . London Underground:** R. W. Clark, *JBS*, 1969, 14.

17 **"uffer":** N. Mitchison, *All Change Here*, 1988, 9; R. W. Clark, *JBS*, 1969, 22.

17 **exploring coal mines:** Norton, *Stars Beneath the Seas*, 1999, 105.

17 **known among the miners:** N. Mitchison, *All Change Here*, 1988, 26.

17 **John Scott Haldane's idea:** J. B. S. Haldane, "Scientific Work," 1960; Norton, *Stars Beneath the Seas*, 1999, 106.

17 **"other human beings who . . .":** R. W. Clark, *JBS*, 1969, 15.

17 **physicians joked openly:** e.g., Le Poer Trench, "Congress on Tuberculosis," 1901.

17 **designed a canary cage:** Goodman, *Suffer and Survive*, 2008, 33.

17 **send home periodic telegrams:** R. W. Clark, *JBS*, 1969, 24; N. Mitchison, *Small Talk*, 1988, 49.

18 **sealed up his home gas chambers:** Haldane records the carbon monoxide levels of exposure of a small number of his experiments in J. S. Haldane, *Respiration*, 1922, 238–42. "Severe" carbon monoxide poisoning is defined including Haldane's reported levels in Olson and Smollin, "Carbon Monoxide," 2008.

18 **Naomi wasn't invited:** N. Mitchison, *Small Talk*, 1988, 24.

18 **heart-shaped jaw:** N. Mitchison, *Small Talk*, 1988, the photograph used as cover art.

18 **born a disappointment:** N. Mitchison, *Small Talk*, 1988.

18 **most biographers omit:** e.g., R. W. Clark, *JBS*, 1969; Dronamraju, *Popularizing Science*, 2017; Goodman, *Suffer and Survive*, 2008.

18 **"My Son":** L. K. Haldane, *Friends and Kindred*, 1961.

18 **"Certain avenues of understanding . . .":** N. Mitchison, *Small Talk*, 1988, 24.

18 **"You come in here . . .":** L. K. Haldane, *Friends and Kindred*, 1961, 217.

19 **diary kept by Naomi:** This diary is held in the private collection of Dr. Agnes Arnold-Forster, Naomi Mitchison's great-granddaughter; she provided images of the referenced entries.

19 **When Jack was eight years:** R. W. Clark, *JBS*, 1969, 28.

19 **Naomi had begun keeping:** N. Mitchison, *All Change Here*, 1988, 61–63.

19 **front lawn of their home:** R. W. Clark, *JBS*, 1969, 29.

19 **"the darling silky mice":** N. Mitchison, *Small Talk*, 1988, 13.

19 **"would like to Mendel them.":** J. B. S. Haldane, Letter to Nou.

20 **Haldanes were on this ship:** Boycott et al., "Prevention of Compressed-Air Illness," 1908.

CHAPTER 2: Brooklyn Bridge of Death

21 **On February 28, 1872:** Except where otherwise cited, Joseph Brown's story is from A. H. Smith, *High Atmospheric Pressure*, 1873, 45.

21 **work-issued, thigh-high black:** The boots worn by the Brooklyn Bridge caisson workers are described in *Brooklyn Daily Eagle*, "The Bridge," 1872.

21 **Each caisson used:** 31 × 52 meters is equal to 102 × 170 feet. Green, *A Complete History*, 1883, 95.

21 **divided into six:** Green, *A Complete History*, 1883, 15. The internal wooden walls ran parallel to the short side of the caisson, and each one formed a trench in which to move wheelbarrows.

22 **"sand hogs":** The name of sand hogs was applied years after the building of the Brooklyn Bridge, but it is commonly used retroactively to describe all caisson workers, e.g., *New-York Tribune*, "Caisson's Depths," 1897.

23   **peeled off their shirts:** McCullough, *Great Bridge*, 1972, 279.

23   **The air temperature:** A. H. Smith, *High Atmospheric Pressure*, 1873, 16, describes the environment inside the caissons.

23   **roughly sixty-person:** *Brooklyn Union*, "Life in the Caisson," 1872.

23   **end of their third hour:** A. H. Smith, *High Atmospheric Pressure*, 1873, 45.

23   **A particularly brutal winter:** Green, *A Complete History*, 1883, 94, shows the timeline of the organization of the construction, which started in January 1867; the newspapers of Brooklyn were replete with coverage of the ice and demands for a bridge, e.g., *Brooklyn Union*, "The River," 1867.

23   **broad, active tidal basin:** Hilgard, "On Tidal Waves," 1875.

24   **need to be the tallest:** Claim of the supports being the tallest construction is in McCullough, *Great Bridge*, 1972, 11. The towers were 271 feet 6 inches (82.8 meters) above water at the high-level mark. Green, *A Complete History*, 1883, 95.

24   **Joseph Brown's caisson:** Brown was in the Manhattan caisson based on the dates of construction, as the Brooklyn side was finished by the time of his accident, per Green, *A Complete History*, 1883, 94.

24   **The lock was a sturdy chamber:** The description of the lock, the upper window, and the spiral staircase out of the caisson are taken from A. H. Smith, *High Atmospheric Pressure*, 1873, 13–14.

24   **hot coffee:** There was company-brewed coffee available upon exit, per Butler, "Caisson Disease," 2004, and *Brooklyn Union*, "Life in the Caisson," 1872.

24   **First, a crew member turned on:** The use of steam coils on the New York side is in A. H. Smith, *High Atmospheric Pressure*, 1873, 16.

24   **cracked open the brass valve:** McCullough, *Great Bridge*, 1972, 192.

25   **At first he thought . . . "like the thrust of a knife":** Joseph Brown's medical case information is taken from A. H. Smith, *High Atmospheric Pressure*, 1873, case XII.

25   **"drachm" of ergot:** Use of ergot was documented by bridge physician Andrew Smith. A. H. Smith, *High Atmospheric Pressure*, 1873, 40, 46.

25   **the medical cases seen:** The medical case reports—including those for John Roland, Hugh Rourke, John Myers, Patrick McKay, James Hefferner, Thomas Kirby, and Bruno Wieland—are from A. H. Smith, *High Atmospheric Pressure*, 1873.

26   **stretched beyond sixty feet:** *New-York Tribune*, "Another Death," 1872, cites the pressure as 32 psi, which is 72 feet, per physician Andrew Smith, who was interviewed about the death.

26   **"feeling oppressed by . . .":** *New-York Tribune*, "Death in the Bridge," 1872.

26   **"congestion of the brain":** *Brooklyn Times-Union*, "Death in the Caisson," 1872.

26   **sand hogs began to strike:** *Buffalo Morning Express*, "Strikes," 1872.

26   **They got their wage increase:** *Buffalo Commercial*, "The News," 1872.

26   **"Reardon, England, 38." . . . "Began work . . .":** A. H. Smith, *High Atmospheric Pressure*, 1873, 49.

27   **Mr. Daniel Reardon:** Mr. Daniel Reardon's death certificate from 1872 is in the public record and states his manner of death.

27   **"was taken with very severe . . .":** A. H. Smith, *High Atmospheric Pressure*, 1873, 49.

27   **"The Caisson Again":** *Brooklyn Union*, "The Caisson Again," 1872.

27   **lasted only three weeks:** McCullough, *Great Bridge*, 1972, 203.

27   **"the Bridge of Death":** *Brooklyn Daily Eagle*, "Frightful Accident," 1872.

CHAPTER 3: Devil's Bubbles

28   **"completely bent up":** *Brooklyn Union*, "Another Victim," 1872.

28   **name originally started as a joke:** McCullough, *Great Bridge*, 1972, 182.

28   **The devil of DCS:** The mechanisms of decompression sickness are explained in Bur-
     bakk et al., *Bennett & Elliott's Physiology*, 2003.
29   **Andrew Smith had deduced:** A. H. Smith, *High Atmospheric Pressure*, 1873, 34.
30   **The doctors knew that the risk:** Butler, "Caisson Disease," 2004.
30   **tall, strong, self-described "fat" boy:** Case and Haldane, Report, 1940.
30   **older boys physically beat him:** R. W. Clark, *JBS*, 1969, 21.
30   **They chanted mocking rhymes:** R. W. Clark, *JBS*, 1969, 17.
30   **Jack missed seeing him:** R. W. Clark, *JBS*, 1969, 21.
30   **"and attacked his tormentor . . .":** R. W. Clark, *JBS*, 1969, 16.
31   **"he just cried and said . . .":** L. K. Haldane, *Friends and Kindred*, 1961, 222.
31   **records have spun and whipped:** For example, Kathleen Haldane wrote in her
     memoir that on her honeymoon with John Scott Haldane, she tried to get to know
     him better during the days by taking dictation from him for a book he was working
     on; L. K. Haldane, *Friends and Kindred*, 1961, 154–55. Some sources have warped this
     story, saying instead that on their wedding night, Haldane forced her to stay locked
     in a bedroom with him, taking dictation instead of engaging in the usual wedding-
     night activities. Norton, *Stars Beneath the Seas*, 1999, 102, is one source to repeat this
     rumor, but far from the only one.
31   **more joyful, than any drug:** J. B. S. Haldane wrote, "I have tried morphine, heroin,
     and bhang and ganja (hemp prepared for eating and smoking). The alterations of my
     consciousness due to these drugs were trivial compared with those produced in the
     course of my work." J. B. S. Haldane, *Science and Life*, 1968, 203.
32   **travel at a steady, constant pace:** Strauss and Yount, "Decompression Sickness,"
     1977.
32   **installing fans in the caissons:** *Brooklyn Daily Eagle*, "Rapid Transit Tunnel," 1901.
32   **John Scott's idea:** Boycott et al., "Prevention of Compressed-Air Illness," 1908.
32   **goats went first:** Boycott et al., "Prevention of Compressed-Air Illness," 1908.
32   **Around dinnertime:** The stories of these dive trials, unless otherwise cited, are from
     Boycott et al., "Prevention of Compressed-Air Illness," 1908.
33   **surprisingly intact medieval castle:** Called Rothesay Castle, it still stands as of
     2022.
33   **The cliffs plummet:** Clyde Cruising Club, *Firth of Clyde*, 2020.
33   **Each metal boot:** Goodman, *Suffer and Survive*, 2008, 197.
34   **Jack and Naomi had started out:** N. Mitchison, *Small Talk*, 1988, 119.
34   **He was allowed on board:** L. K. Haldane, *Friends and Kindred*, 1961, 174.
34   **"I wish I were a boy":** Naomi quotes her own childhood diary in N. Mitchison, *Small
     Talk*, 1988, 120.
34   **he did not complain because:** Goodman, *Suffer and Survive*, 2008, 198; Boycott et
     al., "Prevention of Compressed-Air Illness," 1908, 434–36.
35   **August heat wave:** Burt, "Exceptional Hot Spell," 1992.
35   **should have spent eighty-seven minutes:** Boycott et al., "Prevention of Compressed-
     Air Illness," 1908, 374.
35   **John Scott was offered a knighthood:** Goodman, *Suffer and Survive*, 2008, 272.
35   **grand master of the Mediterranean fruit fly:** In addition to her dissertation, she
     was publishing in studies of *Drosophila subobscura*. Gordon et al., "An Analysis,"
     1939.
36   **inherent talent for sustaining:** John Maynard Smith, technically a student of J. B. S.
     Haldane, said he was in practice Helen Spurway's student instead, and he frequently
     spoke of her skill with animals. J. M. Smith, *Theory of Evolution*, 1993; J. M. Smith,
     Interview, 1997.

36   **She feuded with lab mate:** Rosenhead, "Swords into Ploughshares," 1991, 483.

36   **an extraordinary paper:** Gordon et al., "An Analysis," 1939.

36   **she had alienated many:** R. Grüneberg, Interview, 2020.

CHAPTER 4: Subsunk

37   **duty officer Commander George Fawkes peered in:** Fawkes's name is from Booth, *Thetis Down*, 2009, 42–43.

37   **Bayne finished his talk:** Except where otherwise cited, the details of Bayne's account of the day are in Bayne, "HMS/M 'Thetis'—Detailed Statement," 1939.

37   **stately, boxy government gray:** Bayne mentioned that his planning took place in the commander in chief's office, which at that time in Portsmouth would have been in the Admiralty House.

37   **At 1356 that afternoon:** Bayne, "HMS/M 'Thetis'—Detailed Statement," 1939.

37   **sub was overloaded with 103:** Nasmith Committee, Methods of Saving Life, 1939.

39   **Telegraphist William Harold Bradshaw:** Shepherd, Interview, 1990; Bradshaw, Interview, 1989.

39   **once they reached Liverpool Bay:** Shepherd, Interview, 1990.

39   **For a while the stern:** Nasmith Committee, Methods of Saving Life, 1939.

39   **Helpers on board a tugboat:** The efforts of the tugboat are in official records, but the best description is in Shepherd, Interview, 1990. Jesse Shepherd was on board the HMS *Eskimo* and aided in the efforts.

39   **"The Admiralty regrets . . .":** Booth, *Thetis Down*, 2009, front matter. I have cited Booth's book because he deserves credit for tracking this source down and confirming it was the first communication. His book is also the seminal text for the full story of the sinking.

40   **Less than four hours after:** R. H. Davis, Letter to Haldane, June 6, 1939.

40   **circling surface ships took turns:** Shepherd, Interview, 1990.

40   **four *Thetis* submariners managed:** Nasmith Committee, Methods of Saving Life, 1939.

40   **"was the day the war started.":** Bradshaw, Interview, 1989.

40   **the British had assumed:** McCartney, *British Submarines of World War I*, 2013, multiple places in the introductory chapter.

41   **JBS had signed up early:** J. B. S. Haldane, WWI Field Service Notebook, 1914–18; he was a second lieutenant by October 24, 1914, when he signed a disciplinary form as that rank. J. B. S. Haldane, Disciplinary Form, 1914.

41   **"teaching bombing . . .":** J. B. S. Haldane, Letter to Naomi Haldane, ca. 1914.

41   **"the bravest and dirtiest officer":** Norton, *Stars Beneath the Seas*, 1999, 124.

41   **"the dirt is too far . . .":** J. B. S. Haldane, Letter to Naomi Haldane, 1915.

41   **"one of the happiest . . ." . . . "a very enjoyable experience":** Norton, *Stars Beneath the Seas*, 1999, 124.

41   **twice physically torn apart:** L. K. Haldane, *Friends and Kindred*, 1961.

41   **He locked himself inside:** J. S. Haldane, *Respiration*, 1922, 241.

41   **John Scott also sealed off:** J. B. S. Haldane, *Callinicus*, 1925, 63–65, description of experiment; Cowell et al., "Chlorine," 2007, 232, quotes a primary source confirming that the anonymous "physiologist" in the Haldane description is John Scott Haldane.

41   **He first helped to identify:** J. S. Haldane, "War with Poisonous Gases," 1915.

41   **He used his contacts to have:** Norton, *Stars Beneath the Seas*, 1999, 117.

42   **Naomi and Kathleen were determined:** Norton, *Stars Beneath the Seas*, 1999, 117; Lloyd, "Witness to a Century," 2005, 38.

42   **Naomi, at age seventeen, also wrote:** J. B. S. Haldane et al., "Reduplication," 1915. Alexander Dalzell Sprunt was mortally wounded on March 10, 1915, at the Battle of Neuve Chapelle, and died a week later. His war memorial is reference no. 14590 at the Imperial War Museum.

42   **later biographers would downplay:** e.g., Dronamraju, *Popularizing Science*, 2017, 16.

42   **"if she wants it.":** R. W. Clark, *JBS*, 1969, 39.

43   **vibrations of wings humming:** Naomi Mitchison described the lab as ringed with containers of flies. N. Mitchison, *Among You*, 2000, 116.

43   **scholars of UCL were firmly divided:** Magnello, "Biometrics and Eugenics, Part I," 1999, and "Biometrics and Eugenics, Part II," 1999.

43   **higher-ups at University College London:** H. Kalmus, "50 Years of Exiles," 1984, 5–6.

44   **April 7, 1933, Hitler banned:** The Law for the Restoration of the Professional Civil Service (*Gesetz zur Wiederherstellung des Berufsbeamtentums*) excluded Jews and other political opponents of the Nazis from all civil service positions. Jewish people were also expelled from public universities.

44   **organizations like the Rockefeller Foundation:** Fosdick and Wheatley, *Rockefeller Foundation*, 1989.

44   **and the Academic Assistance Council:** Nossum, "Emigration of Mathematicians," 2012. The Academic Assistance Council was renamed after a few years to Society for the Protection of Science and Learning.

44   **By November, sixteen exiled:** H. Kalmus, "50 Years of Exiles," 1984, 6.

44   **other British universities quickly joined:** Other universities who participated, including but not limited to Oxford and Cambridge, are listed in full in Fosdick and Wheatley, *Rockefeller Foundation*, 1989.

44   **His fights with:** Edwards, "Scientific Interactions," 2017.

44   **a sufficiently clear essay:** He gave that lecture several times per his MI5 file, and his thoughts are written in essay form by hand in J. B. S. Haldane, WWI Field Service Notebook, 1914–18.

44   **he could see no genetic issue:** J. B. S. Haldane, *Heredity & Politics*, 1938.

44   **He would later jokingly declare:** H. Kalmus, "50 Years of Exiles," 1984, 10.

45   **Hans Grüneberg said that:** Unless otherwise cited, Hans Grüneberg's experiences and quotations are taken from his oral history recording at the Imperial War Museum, H. Grüneberg, Interview, 1979.

45   **high forehead, an oval face:** Descriptions are taken from surviving photographs.

45   **roughly five foot six in height:** R. Grüneberg, Interview, 2020.

46   **to become a dental surgeon:** Grüneberg-Capell, *Systeme und Methoden der Orthodontie*, 1934.

46   **"failed to trace down the cause . . .":** Miller, Entry January 28, 1937.

47   **Anna Ursula Philip went by Ursula:** Unless otherwise cited, descriptions of Ursula Philip and her family story are from images and interviews provided by her daughter, Cathie Wright.

47   **welcomed Ursula into the world:** Landesarchiv Berlin, *Personenstandsregister*, 1908.

48   **on August 27, 1933:** Philip, General Information, 1934.

48   **She spoke English, French, and Italian:** Philip, General Information, 1934.

48   **onto the calendar on July 20, 1933:** Vogt, the entry for Ursula Philip, *Wissenschaftlerinnen*, 2008. (Translation into English was provided by Frauke Tillmans, who is the best.) Philip's dissertation is Philip, *"Drosophila melanogaster,"* 1934.

48   **Rumors about Ursula's flight:** Vogt, "Ursula Philip," 1942.

48   **immediately admired and liked her:** Philip, General Information, 1934.

49   **"invaluable" . . . "of the greatest value":** Philip, General Information, 27, 1934.

49 **pay Philip and Grüneberg the same:** Both started out making 187 GBP at the time of hiring, but Haldane gave Grüneberg a raise to 250 GBP in 1934 when he found out that Hans had gotten married. H. Grüneberg, Interview, 1979; Miller, Entry January 28, 1937. Philip did not get her raise to 250 GBP until 1939, when she asked for help getting her brother out of Germany (cited later in the text), so there was a period of difference, but it seems based on marital status rather than sex.

49 **the 1930s was characterized:** Turnbull, "Attitude of Government," 1973.

49 **In one particular march:** Now referred to as the Battle of Cable Street.

50 **Communist party's supporters were joined:** J. B. S. Haldane was an ardent supporter of the party at this time, but not yet a formal member. MI5, Report, February 14, 1938.

50 **"the remarkable advances of scientific theory . . .":** *Daily Worker*, "100 Scientists," 1939.

50 **Britain's elite spy agency MI5:** J. B. S. Haldane, 1924–53.

50 **he volunteered to fight overseas:** Stories of Haldane during the Spanish Civil War are legion, but by far the best account is from the perspective of Copeman, Interview, 1980, which is the source of the next sentence, too.

50 **Royal Navy trainees:** Sheppard, Interview, 1993.

51 **retrieval personnel used breathing apparatuses:** Williams, Interview, 1990.

51 **Those removing the bodies:** Sheppard, Interview, 1993.

51 **The four who got out:** Royal Courts of Justice, Testimony, 1939.

51 **first stepped onto a submarine:** Dronamraju, *Popularizing Science*, 2017, 13.

52 **Haldane wrote to his contacts:** R. H. Davis, Letter to Haldane, July 7, 1939.

52 **Siebe Gorman had a magnificent:** Davis, Davis, and Davis, *Record of War*, 1946, 6–7.

52 **Haldane chose the largest gas-tight:** All details of the following experiments, including quotations, are from W. Alexander et al., "After-Effects of Exposure," 1939.

53 **we have not evolved to survive inside submarines:** Don't be silly. Of course there's no citation for this one. I am, however, actively working to repair this mistake of evolution.

54 **He had five major shrapnel:** *Daily Worker*, "Haldane and Four Spain Brigadiers," 1939.

55 **Haldane took the stand:** Royal Courts of Justice, Testimony, 1939, 22–29.

55 **At least one of the dead:** Royal Courts of Justice, Testimony, 1939, 15; E. R. Hughes, "Post-Mortem," 1939.

56 **In the newspapers, Haldane:** *Times* (London), "Last Hours," 1939.

56 **By the end of that same July:** Dunbar-Nasmith, Letter to Haldane, 1939.

56 **"facilities are unreservedly . . .":** R. H. Davis, Letter to Haldane, June 6, 1939.

56 **MI5 quietly added:** *Daily Worker*, "Haldane and Four Spain Brigadiers," 1939.

56 **Haldane's secretary was one:** Harvey, Letter to Haldane, 1947.

57 **One Rockefeller representative:** Miller, Diary June–December 1934, entry October 22.

CHAPTER 5: Love, Cooking, and Communism

58 **Clutching the flimsy pages:** Hans Kalmus's perspective of his own and his family's immigration story is taken from his memoir, H. Kalmus, *Odyssey of a Scientist*, 1991, as well as from interviews with his sons, G. Kalmus, Interview, 2020a; P. Kalmus, Interview, 2020b.

58 **lanky, athletic frame:** As described by his sons, G. Kalmus, Interview, 2020a; P. Kalmus, Interview, 2020b.

58 **Hans Kalmus had learned:** P. Kalmus, Interview, 2020b.

59 **"lick his ass.":** G. Kalmus, Interview, 2020a.

59 **Hans Kalmus had fallen in love:** M. Kalmus, Interview, 2021.

59 **even though the brash:** M. Kalmus, Interview, 2021; P. Kalmus, Interview, 2020b.

61 **wintry, snow-covered:** The weather and snow level are not mentioned in Hans Kalmus's account, but given the time of the journey (late December) and the average snowfall of that region, it is reasonable to assume that those conditions prevailed.

63 **Ursula Philip's two younger brothers:** Manfred Ernest Philip, born March 26, 1911. Naturalization Records, Manfred Ernest Philip, 1942. George (Georg) Orkin Philip, born December 21, 1924. Public Record, George O. Philip.

63 **Ursula began pleading:** J. B. S. Haldane, Letter to the UCL Provost, March 25, 1939.

63 **Rockefeller Foundation stepped forward:** J. B. S. Haldane, Letter to "To Whom It May Concern," June 15, 1939.

63 **bought George his life:** Passenger List, Georg Philip, 1940.

63 **Her parents began to make:** Passenger List, Dagobert Philip, 1941.

63 **"Bomb-o":** R. W. Clark, *JBS*, 1969, 37.

63 **His wife, Charlotte, would write:** C. Haldane, *Truth Will Out*, 1949, 188.

64 **"taking every possible opportunity . . .":** Miller, Diary May–December 1936, entry July 9–10.

64 **"a Jewess":** Miller, Diary May–December 1936, entry July 9–10.

64 **by then JBS had been rumored:** The dissolution of J. B. S. Haldane's marriage to Charlotte Haldane is typically reported to have occurred in 1939 in practice, although they did not get divorced until 1945. Sarkar, Centenary Reassessment, 1992.

64 **Sister Naomi and her husband:** J. Calder, *The Burning Glass*, 2019.

64 **some reports, JBS and Charlotte:** Copeman, Interview, 1980.

64 **his writings advocate:** J. B. S. Haldane, "The Dark Religions," 1968, 148–53.

64 **She expanded her zoological repertoire:** Spurway, "Newt Larvae," 1943.

64 **She explored every body:** She even wrote to the occupants of houses with ponds that she thought might be suitable for newts, asking for permission to come explore. Spurway, Letter to "The Occupier," ca. 1940.

64 **Joan Elizabeth "Betty" Jermyn:** Uncited personal details and imagery about Elizabeth Jermyn are provided by interviews and photographs from her son, Andrew Bamji.

65 **in the days following her arrival:** Pike, "Rainfall Records," 2004.

65 **passing the entrance exams:** Jebb, Letter of Acceptance, 1934.

65 **"an extremely nice girl" . . . "above the average":** (signature illegible), headmistress, Benenden School, Letter of recommendation, 1934.

65 **she would row:** J. E. Bamji, Letter #1 to Cramer, 2007.

65 **learn chemistry, physiology:** "Bedford College," 1937.

66 **Elizabeth Jermyn and her friends:** Specifically, on the corner of Oxford Street and Tottenham Court Road. J. E. Bamji, Letter #2 to Cramer, ca. 2007.

66 **"no doubt she tolerated . . .":** J. E. Bamji, Letter #1 to Cramer, 2007.

66 **Jermyn's grades at Bedford:** "Bedford College," 1937.

66 **a librarian job:** Jermyn, Letter to Jebb, 1937.

66 **she was offered the job:** J. B. S. Haldane, Letter to Mr. Tanner, 1938.

66 **He had picked a scientist secretary:** Haldane left records of his requirements for his secretaries during other hiring periods, most abundantly after Elizabeth Jermyn quit in 1945. He wrote that the position required not only a degree in zoology with a background in mathematics, but an honors degree. He described the job of his

secretary as "most of the work would require quite high technical qualifications," J. B. S. Haldane, Letter to Frank Jackson, 1945. Other secretaries, specifically a Miss Suley, are described as "doing research in her spare time" in Haldane's lab. Secretary to Haldane (unnamed), Letter to HM Inspector, 1947. His letters to the Communist party further outline his expectations that his secretary should be able to help with scientific papers and articles, e.g., J. B. S. Haldane, Letter to Harry Pollitt, 1945. He rejected one woman, Ruth Freedman, with a high-placing biology degree from Oxford as insufficient for the position. MI5, Telephone Transcript, 1945.

67   **Jermyn started her job in 1939:** J. B. S. Haldane, Letter to Mr. Tanner, 1938, is where Haldane asked that she be hired. Her exact start date is not known, but the first document with her as an employee is dated August 15, 1939. R. H. Davis, Letter to Jermyn, 1939.

67   **neighbor Hans Kalmus:** Hans Kalmus arrived in England on January 12, 1939. UK Immigration, Registration Card, 1939.

67   **They formed a subcommittee:** Dunbar-Nasmith, Letter to Haldane, 1939.

67   **The subcommittee started with:** Rainsford and Frederick, Letter to the Secretary, 1939.

67   **The previous March . . . a similar discussion:** R. H. Davis, Letter to Commander Cumming, 1939.

67   **"carbon dioxide poisoning, its effects . . .":** Rainsford and Frederick, Letter to the Secretary, 1939.

67   **"I expect by this time . . .":** W. S. Jameson, Letter to Haldane, 1939.

68   **"help by drawing up . . .":** Buckingham, Letter to Haldane, 1939.

68   **"was not the time . . .":** Grüneberg's quotations in this section are from H. Grüneberg, Interview, 1979.

68   **Haldane began to concoct plans:** J. B. S. Haldane, Letter to J. Buckingham, 1939.

69   **lines of volunteers:** *Times* (London), "Women Volunteers," 1939.

69   **"the building of world order":** *Times* (London), "Why Britain Goes to War," 1939.

69   **"breaking the Nazi yoke of tyranny":** *Times* (London), "Breaking the Nazi Yoke," 1939.

69   **"Britain's fight to save the world":** *Times* (London), "Britain's Fight," 1939.

69   **Hans Kalmus made his children:** G. Kalmus, Interview, 2020a.

69   **Elizabeth Jermyn was horrified:** J. E. Bamji, Letter #1 to Cramer, 2007.

69   **Hans Grüneberg felt a wave . . . "be an absolute catastrophe":** H. Grüneberg, Interview, 1979.

69   **already had one son, Reuben:** General Register Office, "England and Wales Birth Registration," 1936.

69   **visibly pregnant with their second:** General Register Office, "England and Wales Birth Registration," 1940.

69   **They had permanent permission:** H. Grüneberg, Interview, 1979.

70   **British papers lampooned:** *Times* (London), "We Stand with Britain," 1939.

70   **"by a thoroughly responsible person":** J. B. S. Haldane, Letter to Hore-Belisha, 1939.

71   **"Divers using self-contained dresses . . .":** All quotes by Haldane in this section are from J. B. S. Haldane, Letter to Hore-Belisha, 1939.

CHAPTER 6: Drunk and Doing Math

73   University College London got prompt: Tanner, Letter to Haldane, 1939.

73   "dispose of": J. B. S. Haldane, Letter to Grüneberg, 1940.

73   piles of dead turtles: H. Kalmus, *Odyssey of a Scientist*, 1991, 56.

73   "the college should be absolutely . . .": Tanner, Letter to Haldane, 1939.

73 **they shut down the electricity ... "full-height wooden barrier":** Miller, Diary June–December 1939, entry November 6.

73 **Guards patrolled against wily:** H. Kalmus, *Odyssey of a Scientist*, 1991, 56.

74 **accidentally left the heat on:** Miller, Diary June–December 1939, entry November 6.

74 **a record-breaking winter:** *Nature*, "Severe Winter," 1940.

74 **"succeeded in retaining a foothold":** Miller, Diary June–December 1939, entry November 6.

74 **"Say the Ministry of Food":** J. B. S. Haldane, Letter to Michael Fry, 1940.

74 **He nailed to his door a poster:** H. Kalmus, "50 Years of Exiles," 1984, 56.

74 **Over the winter Hans Kalmus:** H. Kalmus, *Odyssey of a Scientist*, 1991, 56.

75 **"most anxious to obtain ... whether [Haldane] could arrange ...":** R. C. Frederick, Letter to Haldane, January 9, 1940.

75 **UCL's physiological laboratory was not only closed:** Haldane, Letter to Frederick, January 10, 1940.

75 **had received the same:** R. H. Davis, Letter to Haldane, January 11, 1940.

75 **Haldane made the arrangements:** Haldane, Letter to Davis, January 11, 1940.

75 **Edwin Martin Case:** Unless otherwise cited, personal details about Martin Case come from an interview with his niece, who also provided family paperwork and documentation of anecdotes. Love, Interview, 2020.

75 **first decade of the 1900s:** Edwin Martin Case was born on February 26, 1905. General Register Office, "England and Wales Death Registration," 1978.

78 **One day, Case got cited:** The story about Case, the safety officer, and Haldane also appears in R. W. Clark's seminal biography of J. B. S. Haldane. However, Clark says Case was driving a car. The Case family provided additional, contemporary writings of the story that clarify that Case was riding a bicycle. The citation type could theoretically be applied to either type of vehicle, but it was rarely used for bicycle riders, so it is plausible Clark made that assumption or misunderstood. The rest of the story is not in dispute.

78 **he wrote to Haldane ... "slight danger":** J. B. S. Haldane, Letter to Case, January 17, 1940.

78 **He followed that letter up with ... "I consider that ...":** J. B. S. Haldane, Letter to the Undersecretary, 1940.

79 **"My Dear Boss," Case wrote:** Case, Letter to Haldane, 19, 1940.

79 **small children would simply assume:** J. Robertson, Interview, 2020. She said she made this assumption as a child.

80 **Naomi had married:** General Register Office, "England and Wales Marriage Registration," 1916; N. Mitchison, *Among You*, 2000, 16.

80 **personal acquaintance Virginia Woolf:** Southworth, *Leonard and Virginia Woolf*, 2012.

80 **"The girl poet is my discovery":** Woolf, Letter to Walpole, 1975.

80 **Joan was a totally unsophisticated:** Unless otherwise cited, personal details and descriptions of Joan and Jim Rendel are from an interview with their daughter (J. Robertson, Interview, 2020). Joan and Jim's granddaughter, Celia Robertson, also wrote a memoir about her grandmother's life (C. Robertson, *Who Was Sophie?*, 2008). Because Celia had done research for that book, family members had many personal details ready and available when we talked, and her book provides beautiful and in-depth details about Jim and Joan.

81 **"hanging on the back of the door":** C. Robertson, *Who Was Sophie?*, 2008, 71.

81 **he had already worked with Haldane ... "an extra toe":** J. B. S. Haldane, Letter to Pease, 1938.

82   "where one can go on working . . .": Rendel, Letter to Haldane, 1939.

82   "Eggatron": Franklin et al., "James Meadows Rendel," 2004.

82   within a few months: J. B. S. Haldane, Letter to Adrian, 1940.

82   Ursula Philip had offered: J. B. S. Haldane, Letter to "Sir," 1939.

82   Hans Grüneberg had been asked: Weiner, Letter to Haldane, 1939.

82   government issued ration books: *Times* (London), "Rationing in December," 1939.

83   By April 11, 1940, the scientists: April 11 was the first experiment on record, and is described in Case, Lab Notebook, 1940.

83   On Friday, April 19, 1940, Martin Case: The descriptions of the ball game and the arithmetic calculations to assess manual and mental dexterity, respectively, are in Case and Haldane, "Human Physiology," 1941. The rest of the experiment is described in Case, Lab Notebook, 1940, in the entry for that date.

83   the fourth day of testing: The first three tests were on April 11, 17, and 18, 1940.

84   The first thing on their checklist: The introduction of the letter from Rainsford provides the timeline for this request as February 7, 1940, when Haldane wrote to Rainsford, "Today you asked me to investigate the effects of nitrogen narcosis of breathing air under high pressure." J. B. S. Haldane, Letter to Rainsford, February 7, 1940. Haldane then additionally clarifies that this was their first goal in a letter to Case: "If . . . you succeed in clearing up the question whether nitrogen makes people rather drunk at moderate pressure and the like, it seems probable that you would get a job on another problem of the same character." J. B. S. Haldane, Letter to Case, February 7, 1940 The experiments started after Haldane finished a lecture series he was scheduled to give. J. B. S. Haldane, Letter to Frederick, February 16, 1940.

84   "I should be glad to carry . . .": J. B. S. Haldane, Letter to Rainsford, February 7, 1940.

85   "Slight feeling akin . . .": Quoted in the above-cited records for the test on Friday, April 19, 1940.

86   "felt somewhat mystical": Case and Haldane, "Human Physiology," 1941, 236.

86   "I believe [the toothache] is . . .": J. B. S. Haldane, Letter to Frederick, April 22, 1940.

86   After many more tests: More tests for narcosis were done on April 22, 23, 24, 26, 29, and 30.

86   their conclusion was the same: J. B. S. Haldane, Letter to Frederick, May 7, 1940.

87   "it is quite imperative . . .": Case and Haldane, "Human Physiology," 1941, 236.

87   Sir Humphry Davy: Davy, *Collected Works*, 1839.

87   Around 1767, surgeon John Hunter: Cabot, *General Considerations*, 1918, 241.

87   epidemiologist Joseph Goldberger: Kraut, *Goldberger's War*, 2021.

88   In 1939, the Royal Institution: *Times* (London), "Humphry Davy's Discoveries," 1938.

88   the latest in a long list of doctors: The doctor who died was Julian Castelian (*Times* [London], "Doctor's Death," 1939), but there were many earlier such articles, e.g., *Times* (London), "An Anaesthetist's Experiment," 1927.

88   another "scientist" was put: Vincent Blumhardt Nesfield. *Times* (London), "Charges Against Doctor," 1932.

89   Dozens in England volunteered: Mellanby, MRC Form of Application, 1942.

89   One man named Horace Cameron Wright: See the chapter titled "Lung Maps."

89   Quakers signed up for: Keys et al., *Biology of Human Starvation*, 1950.

89   Haldane had loudly offered: His public statements about air raid shelters were many, several of which are in J. B. S. Haldane, Multiple Letters and Articles, 1938.

89   on day eight, it was Jim Rendel's turn: Experimental day eight was on April 26, 1940, and Jim Rendel's experiment was number fourteen. See the entry for that date in Case, Lab Notebook, 1940, for all quotes and details.

89   **Tubes of air run:** Ear problems during pressure changes are described in Burbakk et al., *Bennett & Elliott's Physiology*, 2003, 227–56.

CHAPTER 7: Everyday Poisonous Gases

92   **Helen Spurway's first test:** Spurway's first test was on April 30, 1940. Except for the two details otherwise cited, information about her test is from the notebook entry on that date in Case, Lab Notebook, 1940.

92   **Spurway could not do math silently:** Case and Haldane, "Human Physiology," 1941.

92   **Fruit flies brought to the same depths:** They brought in fruit flies during the experiment on May 5, 1940 (Case, Lab Notebook, 1940), and also discussed them in Case and Haldane, "Human Physiology," 1941.

93   **"For 300 years the rulers . . .":** *Times* (London), "Hitler's Proclamation," 1940.

93   **atypically balmy:** Historical weather data from the Oxford weather station (the closest station open at that time) shows a maximum temperature of 18.6 C and a minimum temperature of 7.8 C for that May. That was the highest maximum temperature in May by almost a full degree for at least the previous ten years. UK Meterological Office, "Historic Station Data."

93   **Martin Case and Jim Rendel prepared:** May 27, 1940, was experimental day twenty-five, with experiments forty-six and forty-seven. The details are in Case, Lab Notebook, 1940, at the entry for that date.

93   **The English military had delayed:** Deferring Case's military service has already been cited. Deferring Rendel's was the subject of a month of argument between Haldane and the draft board, starting on May 9, 1940, with the board finally agreeing in writing to delay his draft indefinitely on June 15. Patterson, Letter from the Medical Director General, June 15, 1940.

93   **"Should he be called . . .":** J. B. S. Haldane, Letter to the Manager, 1940, to the draft board. Haldane also wrote to the project sponsor, Seymour Grome Rainsford, who agreed to help. Rainsford, Letter to Haldane, 1940.

94   **registration number 7080:** J. B. S. Haldane, Letter to the Manager, 1940.

95   **seven months pregnant:** Jane Rendel was born on July 22, 1940. *Times* (London), "Births," 1940a.

95   **"intended to save the lives of sailors":** J. B. S. Haldane, Letter to Winfrey, May 15, 1940.

95   **"a young and vigorous man":** J. B. S. Haldane, Letter to the Manager, 1940.

95   **But the test had placed:** The full story of Rendel's medical struggles as a result of this experiment is chronicled in Multiple, Letters Regarding James Rendel's Lungs, 1940; UKNA, "Dr. Rendell's Claim," 1940–44.

95   **Jim Rendel thought the new pang:** J. B. S. Haldane, Letter to Frederick, June 18, 1940.

96   **He asked Rendel to remove:** Morland, "Report on J. M. Rendel," 1942.

96   **Eventually, if its growth had continued:** This type of event is called a tension pneumothorax. It and other types of pulmonary barotrauma are discussed in Burbakk et al., *Bennett & Elliott's Physiology*, 2003, 557–73.

97   **Haldane continued to give public:** In 1938 he spoke to twelve thousand people in Trafalgar Square, then led a march of people to the German embassy to protest. MI5, Cross-Reference P.F. 45762, 1938. Over the course of the war, he gave many speeches that are documented in his MI5 file.

98   **One MI5 informant . . . was the Royal Navy:** A. G. Anderson, Letter from Mrs. A. G. Anderson, 1940.

98   **Beneath a ragged scar, one:** This experiment falls slightly out of sequence, in order to keep Jim Rendel's experiences together in the text. Jim Rendel's lung injury

occurred on May 27, 1940. He went to the doctor on June 12, 1940, and this experiment took place on May 31, 1940. It is described in the entry for that day in Case, Lab Notebook, 1940.

98   **The wound was throbbing from:** The experiment that made his wound throb took place on May 29, 1940; it was experiment fifty, experimental day twenty-seven. It is described in the entry for that date in Case, Lab Notebook, 1940.

98   **The Salvus device had originally:** R. H. Davis, Davis, and Davis, *Record of War*, 1946, 27; Quick and Shelford, "History of CCUBAs," 1970.

102   **the term du jour for the phenomenon:** Burbakk et al., *Bennett & Elliott's Physiology*, 2003, 358–418.

102   **Oxygen Pete:** The references to Oxygen Pete from the reports of this time period are many, but my favorite is in Waldron and Gleeson, *The Frogmen*, 1955, 50. The description is from Sydney Woollcott, a British policeman turned wartime Charioteer who participated in the experiments as a subject. He wrote down the only two lines anyone could remember of a poem someone had scribbled on a chalkboard at Siebe Gorman near the experimental tanks:

> *For down at a depth of seventy feet*
> *Lives a guy by the name of Oxygen Pete.*

103   **our bodies manipulate it:** The scientist who figured out the key chemical reactions by which our bodies achieve this manipulation was named Sir Hans Adolf Krebs, and the Krebs cycle bears his name. Krebs was a German Jewish refugee who before WWII fled to the UK, where he was given asylum. His escape was funded by the Rockefeller Foundation. He was later awarded the Nobel Prize.

103   **undesirable reactions of oxygen:** Liochev, "Reactive Oxygen Species," 2013.

103   **Certain substances—like tea:** Arts et al., "Catechin Contents, 1," 2000, and "Catechin Contents, 2," 2000.

104   **108 kilograms:** His weight is reported as seventeen stone, which is 108 kg, or 238 pounds.

104   **eighty-one hundred milliliters of blood:** Blood volume was calculated using the online tool provided by Medscape at https://reference.medscape.com/calculator/648/estimated-blood-volume.

104   **an extra 4,023,000,000,000,000,000,000:** By breathing 100 percent oxygen on the surface, Haldane's oxygen partial pressure (ppO_2) would be 760 mm Hg. ScyMed provides a handy calculator for dissolved arterial oxygen at http://www.scymed.com/en/smnxpr/prgdb276.htm, and the arterial content when breathing room air is 22 mL/dL. The arterial content when breathing 100 percent O_2 is 24.1 mL/dL, for an increase in dissolved oxygen of 2.1 mL/dL (21 mL/L) when going from room air to 100 percent O_2. Haldane's 8,100 mL of blood = 8.1 L; therefore he gets a calculable 170.1 mL of additional arterial oxygen content when breathing 100 percent O_2 compared to when breathing room air. For some reason dissolved-oxygen-concentration calculations are done as a function of oxygen volume, which doesn't make sense to me, but I'm not in charge, so from there we can use the ideal gas equation ($PV = nRT$) to calculate that 170.1 mL of oxygen under those conditions is 0.00668 mols, or 4,022,815,568,200,770,000,000 molecules. Peter Kalmus, ever the diligent scientist, correctly and with humor pointed out that that number contains too many significant figures and said that if I had been his student, he would have marked me down for it, so I rounded, because Peter is the best and he was right.

104   **bottle and a third of Merlot:** Merlot catechins weigh 6.049 mg/mM (6,049 mg/M), as calculated using the volumetric density of 44.1 mg/L (Arts et al., "Catechin Con-

tents, 2," 2000) divided by a mass density of 7.29 mM CE/L (Katalinić et al., "Antioxidant Effectiveness," 2004). The 0.00668 mols in the bloodstream times 6,049 mg/M equals that 40.4 mg catechins are needed to counteract 100 percent O_2 at 1 atmosphere. With a volumetric density of 44.1 mg/L, that equals 0.92 L of Merlot per breath, because the dissolved arterial oxygen level gets replenished with each breath inhaled.

104 **192 liters of Earl Grey tea:** Earl Grey tea contains 139.1 mg/L total catechins (Arts et al., "Catechin Contents, 2," 2000). Mass density numbers specific to the catechins in Earl Grey tea could not be found, so a mean molecular weight for average catechins was used of 4,000 g/M (Ünlü et al., "Polymerization of Catechin," 2017). Alternatively, he could have eaten 44 kg of dark chocolate or 2,407 kg of blueberries per breath.

104 **a gas of even 75 percent:** At ninety feet of pressure, Haldane was at 3.7 atmospheres of pressure (90/33+1). By breathing a gas of 75 percent, he was at a pO_2 of 2.8 atmospheres (3.7*0.75). The additional dissolved arterial oxygen levels at pO_2 of 3 atmospheres of oxygen were measured as 57 mL/L (Leach et al., "Hyperbaric Oxygen Therapy," 1998). From there, the math is the same as above.

105 **rats and mice start to die:** Demchenko et al., "Similar but Not the Same," 2007.

105 **percentages need to be throttled:** Lin et al., "Dual Effects," 2022.

105 **At depths like three hundred feet:** The National Fire Protection Association (NFPA) has extensive standards relevant to this section that describe flammability.

105 **The Admiralty had started using:** Damant, "Deep Sea Diving," 1937.

105 **The earliest oxygen seizures:** Bean and Rottschafer, "Reflexogenic and Central Structures," 1939.

106 **"I think we have now enough . . .":** J. B. S. Haldane, Letter to Rainsford, June 3, 1940.

106 **"underestimated the effect of oxygen":** J. B. S. Haldane, Letter to Frederick, April 13, 1940.

106 **A week after Haldane's losing battle:** This experiment by Martin Case took place on June 7, 1940, and is described in the entry for that date in Case, Lab Notebook, 1940.

106 **US Navy physiologists had theorized:** Behnke and Wilmon, "USS *Squalus* Medical Aspects," 1939. Haldane clipped and saved this newspaper article, proving he knew about the use.

CHAPTER 8: The Blitz

107 **Jim Rendel still in the hospital:** Rendel's hospitalization lasted at least through the window of June 18 to 29, 1940, per J. B. S. Haldane, Letter to Frederick, June 18, 1940, and J. B. S. Haldane, Letter to Winfrey, June 29, 1940.

107 **Haldane and Case climbed:** This test occurred on June 27, 1940 (experiment sixty, experimental day thirty-seven) and is described on that date in Case, Lab Notebook, 1940.

107 **Helen Spurway tried adding:** Spurway's test was the next day, on June 28. Case, Lab Notebook, 1940.

107 **Haldane told the Admiralty:** Haldane's message is not preserved, but the Admiralty responded on July 6, 1940, that because of the UCL experiments, they were going to limit submariners to 0.7 percent carbon dioxide. Rainsford, Letter to Haldane, July 6, 1940.

108 **"I think it is an open . . .":** J. B. S. Haldane, Letter to J. Buckingham, 1939.

108 **"I think that experiments of . . .":** J. B. S. Haldane, Letter to Frederick, June 18, 1940.

108   **"Perhaps you will be good enough . . .":** J. B. S. Haldane, Letter to Winfrey, June 29, 1940.

108   **Case was receiving a modest:** J. B. S. Haldane, Letter to Walsham Brothers, June 11, 1940.

108   **newspapers advertised war bonds:** *Times* (London), "Subscriptions," 1940.

109   **Kalmus family received:** G. Kalmus, Interview, 2020a; P. Kalmus, Interview, 2020b.

109   **people nonetheless treated the idea:** A comprehensive and diverse collection of opinions, observations, and comments from people with a variety of attitudes can be found in Mass Observation, Observations Gathered August 1940, 1938–45.

109   **Woven in among the clouds:** My favorite resource for the Battle of Britain is Malcolm Brown's 2015 book *Spitfire Summer*, which uses primary-source accounts to tell what the events were like.

110   **"'That Village' Raided Again":** *Times* (London), "'That Village' Raided Again," 1940.

110   **Some adhered meticulously:** Diarist 5094, Diary, entry August 8, 1940. None of the scientists left behind a detailed record of what the bombings were like for them. Luckily, however, plenty of the people around them did. These diaries, maintained and preserved by a group called Mass Observation, provide a window into what it was like to try to survive, let alone achieve science, as the city around the research group began to burn.

110   **"There has been a good deal of talk . . .":** Diarist 5368, Diary, entry August 12, 1940.

111   **"89 Raiders Down . . .":** *Times* (London), "89 Raiders Down," 1940.

112   **"It's amazing what a capacity . . .":** Diarist 5368, Diary, entry August 15, 1940.

112   **parents brought back children:** e.g., Welch, Interview, 2000.

112   **campaigned loudly and publicly:** Haldane's public campaign for better shelters that were accessible to all are well-documented in his MI5 surveillance file. MI5, J. B. S. Haldane, 1924–53, for example on dates February 17, 1937; December 2, 1937; May 30, 1938; and October 3, 1939.

112   **thick, billowing clouds:** Observer K.B., Air Raid Alarm, 1940.

112   **Less than thirteen kilometers north:** Diarist 5368, Diary, entry August 15, 1940.

113   **When the sirens first yelled:** Observer (unnamed), Air Raids, 1940.

113   **At least one man on . . . ". . . bound to happen somewhere else":** Observer R., Air Raid Warning in Battersea, 1940.

113   **One woman reportedly remarked:** Diarist 5368, Diary, entry August 20, 1940.

114   **"I heard more than one . . .":** Diarist 5368, Diary, entry August 16, 1940.

114   **"so quiet that I couldn't find anything . . .":** Diarist 5368, Diary, entry August 20, 1940.

114   **On August 23, stray bombs:** *Times* (London), "Little Damage," 1940.

114   **German bombers with what they claimed:** Levine, *Strategic Bombing*, 1992, 26.

114   **"the personnel are fully insured . . .":** R. C. Frederick, Letter to Haldane, June 17, 1940.

114   **But after honest disclosure:** J. B. S. Haldane, Letter to Walsham Brothers, June 11, 1940, and Letter to Rainsford, July 1, 1940.

115   **"payments of compensation . . .":** Griffiths, Letter to the Secretary, 1940.

115   **The government also removed:** Duncan, Letter to the Secretary, 1940.

115   ***Get started, please hurry:*** Requests emphasizing new urgency are in Multiple, Letters Regarding James Rendel's Lungs, 1940.

115   **Young Peter and George Kalmus began:** G. Kalmus, Interview, 2020a; P. Kalmus, Interview, 2020b.

115   **Mr. M. Paine wrote:** Diarist 5166, Diary, entry August 27, 1940.

115   **named Mariel Bennett described:** Diarist 5250, Diary, entry August 28, 1940.

115 **"Blimey they didn't half . . .":** Observer H.P., Air Raid (Night of 28.8. 40), 1940.

115 **"close enough to shake the windows,":** Diarist 5166, Diary, entry August 30, 1940.

115 **Mariel Bennett wrote that she was:** Diarist 5250, Diary, entry August 30, 1940.

116 **Londoners dragged mattresses down:** Diarist 5250, Diary, entry August 25, 1940; Diarist 5438, 1940, entry September 24, 1940.

116 **they learned to identify:** Diarist 5166, Diary, entry August 31, 1940; Diarist 5150, Diary, entry August 30, 1940.

116 **The papers quietly announced:** *Times* (London), "London Air Raid Casualties," 1940.

116 **Some curious people:** Observer S.S., Maida Vale Air Raid, September 6, 1940a; Observer H.P., Air Raid (Night of 30.8. 40), 1940.

116 **"From the start people seem" . . . "loss of sleep":** O. Smith, "Some Blitz Reactions," 1940.

117 **"if it's coming . . .":** Observer C.F., Indirect Report Dated September 25, 1940.

117 **"The worst area to suffer was the East End" . . . improvised refugee camp:** Observer L.E., Evacuees at Oxford, 1940, 1.

117 **couple walking their dog . . . north of London was on fire:** Diarist 5166, Diary, entry September 9, 1940, referring to the Hampstead Heath fires.

118 **Rescue workers labored to pull:** Observer S.S., Maida Vale Air Raid, September 9, 1940b.

118 **On Tuesday, an unexploded bomb:** Diarist 5438, Diary, entry September 10, 1940.

118 **medical student named D. E. Marmion:** Diarist 5142, Diary, entry September 10, 1940.

118 **Almost exactly halfway:** Observer T.H., Untitled Investigator Report, 1940.

118 **Peter and George Kalmus:** G. Kalmus, Interview, 2020a; P. Kalmus, Interview, 2020b.

118 **"a great adventure":** Welch, Interview, 2000.

118 **Mrs. L. C. Tavener wrote:** Diarist 5438, Diary, entry September 18, 1940.

119 **Shrapnel hunting escalated:** Almost everyone who was a child during the Blitz seems to have reported shrapnel hunting, which I find kind of charming. Examples include Welch, Interview, 2000; Frankland, Interview, 1999.

119 **"It was a house . . .":** Hills, Interview, 1998.

121 **187 Westminster Bridge Road:** R. C. Frederick, Researches by Haldane, 1940.

121 **"a poorer area" . . . "lovely":** Welch, Interview, 2000.

121 **Lilian Emmeline Welch was a resident:** Lilian Welch's story and quotes are from her oral history at the Imperial War Museum, Welch, Interview, 2000.

122 **medical student Arthur Walker:** His story and quotes are from his oral history, Walker, Interview, 1998.

124 **they had welcomed screaming:** General Register Office, "England and Wales Birth Registration," 1940.

124 **meant they were exempted:** Home Office, "Exemption (Elspeth Grüneberg)," 1939, and "Exemption (Hans Grüneberg)," 1939.

124 **his own research funding:** Grüneberg's story and quotes are from his oral history recording, H. Grüneberg, Interview, 1979.

124 **(VAS) of the Royal Navy:** The tank tests took place September 21 and 22, 1940, as well as on October 2 to 4, 1940. The experiment write-ups, including all quotes and the fact that the tests were requested are in J. B. S. Haldane and Case, Untitled report labeled "SECRET," 1940.

126 **"honeycombed with air raid shelters":** Hills, Interview, 1998.

126 **teen Doris Eileen Connolly:** Connolly, Interview, 1999.

126 **Bombs fell inside the submarine base:** An interactive map of the bombing sites in Portsmouth, with the dates associated with each bomb, is available online at the Portsmouth Bomb Map, 2020. The Portsmouth Central Library also has indexed lists

of where bombs dropped in Portsmouth in X108A/3/2/12/12. The effects of the Blitz on Portsmouth are described alongside photographs in Easthope, *Smitten City*, 2010.

126   **middle of one of:** Norton, *Stars Beneath the Seas*, 1999, 138–39.

127   **"of a secret character.":** J. B. S. Haldane, Letter to Frederick, October 8, 1940.

127   **Horton had been suggesting:** Zetterling and Tamelander, *Tirpitz*, 2009, 61.

127   **The government raised more:** Diarist 5438, Diary, entry September 22, 1940.

127   **Londoners who traveled:** Diarist 5280, Diary, 1940, entry September 20, 1940.

128   **"long and uninterrupted":** Diarist 5438, Diary, entry September 22, 1940.

128   **By the height of the Blitz:** Fremlin, "Air Raid Shelters," 1939, 3.

129   **Women abandoned traditionally:** Observer L.E., Tube Shelter Observations, 1940; multiple places in MASTER, Mass Obs Tube Report, 1940.

129   **underground reminded her of:** Observer H.P., Air Raid Shelterers, 1940.

129   **Belgian woman was denied:** Observer Vaugan, Untitled Report, 1940, 3.

129   **"cruel warble" of the sirens:** Diarist 5166, Diary, entry September 9, 1940.

129   **"It's wonderful down here . . .":** Observer D.H., Stations/Tube Shelters, 1940, 4.

129   **"I feel lovely and safe . . .":** Observer C.F., Indirect Report Dated October 22, 1940.

129   **"I feel completely safe there . . .":** Fremlin, "Air Raid Shelters," 1939, 11.

129   **"You can't hear that plane . . .":** Fremlin, "Air Raid Shelters," 1939, 11.

129   **"Bit hot and dirty . . .":** Fremlin, "Air Raid Shelters," 1939, 11.

129   **"It may be the attitude . . .":** Fremlin, "Air Raid Shelters," 1939, 19.

130   **Eventually, the government issued:** *Times* (London), "More and Better Shelters," 1940.

130   **Bombs penetrated the road:** The Balham tube station was bombed on October 14, 1940.

130   **what we know today about:** For a comprehensive history of ordnance disposal during the Blitz, and the wartime efforts to learn how to dismantle bombs as they fell, please see Leatherwood, *Nine from Aberdeen*, 2012.

130   **during the daytime, people:** Diarist 5250, Diary, entry August 24, 1940.

130   **Fred Astaire and Ginger Rogers:** A Ginger Rogers and Fred Astaire movie was playing at the Coronet Cinema on October 17 during an air raid near the Holland Park station, per Observer H.P., Air Raid Warning, 1940.

130   **"had both of her legs blown off":** Diarist 5257, Diary, 1940, entry October 9.

130   **soldiers drove through:** Diarist 5245, Diary, entry October 16, 1940.

131   **research group had tried to do . . . "Air raids have hampered . . .":** J. B. S. Haldane, Letter to Frederick, October 8, 1940.

131   **through contacts with collaborators:** J. B. S. Haldane, Letter to Grüneberg, 1940; H. Kalmus, *Odyssey of a Scientist*, 1991, 58.

131   **Haldane rented a home:** Known as the Crossways, the house was at 25 Salisbury Avenue in Harpenden. Salvesen. Letter to Haldane, 1940.

131   **On July 17, 1940:** C. Robertson, *Who Was Sophie?*, 2008, 97; *Times* (London), "Births: Rendel," 1940b.

131   **He was not allowed:** J. Robertson, Interview, 2020.

131   **Ursula Philip decided to plead:** Marie, "The Importance of Place," 2004, 52–53.

132   **"the Piggery":** N. Mitchison, *Among You*, 2000, 116; P. Kalmus, Interview, 2020b; H. Kalmus, *Odyssey of a Scientist*, 1991.

CHAPTER 9: The Piggery

133   **Joan and Jim Rendel's bedroom:** Details about their bedroom and the rental house are taken from Frederick et al., "Inventory," 1940.

133   **Joan was even less flattering:** Rendel, Letter to Mitchison, December 5, 1940.

133   **right lung had ruptured:** Rendel, Letter to Mitchison, October 25, 1940, for lung status and quote.

133   **Rendel did what he could:** J. Robertson, Interview, 2020. Haldane left behind letters in which he applied to funding agencies for Jim Rendel's salary following his lung injury. Haldane laid on the patriotic guilt thickly to get enough funding to help the Rendels stay afloat. He wrote to the Sir Halley Stewart Trust, "If however you can see a way to subsidizing him on a monthly basis during this [health-recovery] period, I shall be very grateful to you. You will realise that in doing so you are helping a man who, besides doing genetical work, has risked his life and injured his health in doing unpaid work designed to save the lives of members of the Navy." J. B. S. Haldane, Letter to Winfrey, June 20, 1940.

134   **the adult "girl poet" . . . "the atmosphere . . .":** Rendel, Letter to Mitchison, December 5, 1940.

134   **"London conversation and London letters.":** Rendel, Letter to Mitchison, October 8, 1940.

134   **hinted of bombs:** Frederick et al., "Inventory," 1940.

134   **The flies' food:** C. Wright, Interview, 2020.

134   **common but sexist expectation:** J. B. S. Haldane, Letter to "Sir," 1944.

135   **she and Nussy Kalmus suddenly found:** H. Kalmus, *Odyssey of a Scientist*, 1991, 59.

135   **On one occasion, she magically:** P. Kalmus, Interview, 2020b.

135   **Haldane always sat in a studded:** G. Kalmus, Interview, 2020a.

135   **He rather belligerently declined:** J. B. S. Haldane, Letter to Salvesen, 1941; Salvesen, Letter to Haldane, 1941.

135   **Haldane demanded that a guest:** A. Bamji, Interview, 2020. The guest was Nariman Bamji, but he has not yet been introduced at this point in the text.

136   **Rendel bonded when:** J. Robertson, Interview, 2020.

136   **Helen Spurway kept:** H. Kalmus, *Odyssey of a Scientist*, 1991, 60.

136   **George, who slept:** H. Kalmus, *Odyssey of a Scientist*, 1991, 59–60.

136   **"not to bring extra beds.":** J. B. S. Haldane, Letter to Salvesen, 1940.

136   **Scientist-secretary Elizabeth Jermyn joined:** J. E. Bamji, Letter #1 to Cramer, 2007.

136   **Ursula Philip settled into her own:** Ursula Philip had a slightly different timeline for her evacuation, but the details are not sufficiently relevant to the story for inclusion in the main text. Following the shutdown of UCL in the fall of 1940, her actions are unclear for a few months until she rejoins the lab in Rothamsted. What is known for certain is that she participated in testing with the group on-site at Siebe Gorman until their London factory was bombed in May 1941, and she was settled into the same remote lab with them at Rothamsted on November 8, 1941, where she worked in room number seventy-one. Jermyn, Letter to Barnicot, 1941.

136   **Haldane's sister, Naomi Mitchison, shipped:** Rendel, Letter to Mitchison, January 13, 1941.

136   **J. B. S. Haldane and Martin Case traveled:** This testing took place in March 1941 and is documented in that date range in entries in UCL Lab, Miscellany of Papers, ca. 1940–44; J. B. S. Haldane and Case, Compiled Report, 1941.

137   **"I don't know quite what is happening . . .":** Rendel, Letter to Mitchison, March 31, 1941.

137   **"one takes oneself away with . . .":** Rendel, Letter to Mitchison, March 31, 1941.

137   **Joan was terrified of failing:** She stated her tremendous desire not to fail everyone in Rendel, Letter to Mitchison, March 31, 1941.

138   **get a special dispensation:** Details of multiple incidents of damage to his car as well as his repeated petrol requests are in Car Records, 1940–52. Those files also contain

records that in 1947 he had to have the cylinders rebored in the car he drove during the war, which is a classic sign of not changing the oil enough.

138 **On experimental day fifty:** This experiment took place on March 14, 1941, and is documented in detail in J. B. S. Haldane and Case, Compiled Report, 1941, and UCL Lab, Miscellany of Papers, ca. 1940–44.

140 **Less than two weeks later:** Jermyn's test took place on March 31, 1941, and was experiment eighty-six on experimental day fifty-two. UCL Lab, Miscellany of Papers, ca. 1940–44.

140 **Ursula Philip went through the test, too:** Philip's first dive was on April 15, 1941. UCL Lab, Miscellany of Papers, ca. 1940–44.

141 **back would hurt:** (signature illegible) Manager, Letter to Haldane, 1946.

141 **still they repeated:** They repeated the same test on too many dates to list. The records of their experiments, with dates, are in UCL Lab, Miscellany of Papers, ca. 1940–44, "Diving Notes," Part 2, July 1942–May 1943.

141 **"too bloody afraid . . .":** MI5, Telephone Check, 1942.

141 **Research work had kept:** Home Office, "Exemption (Hans Grüneberg)," 1939, and "Exemption (Ursula Philip)," 1939.

142 **organized killing systems:** Preston, *Spanish Civil War*, 2007, 103.

142 **Just over forty years old:** His description is compiled from MI5, Re Hans Georg Kahle, 1941, and Kahle, Application for Certificate, 1939.

142 **Kahle's most lasting claim:** Kvam, "Ernest Hemingway and Hans Kahle," 1983.

142 **a high-ranking member:** MI5, Cross-Reference of Form P.F. 47192, April 23, 1940.

143 **more phone taps for Communism:** MI5, Memo, 1939–42.

143 **down unarmed citizens "wholesale":** MI5, Copy of P.F. 48682, 1940.

143 **Kahle was picked up:** Jenkins, Letter to Divisional Detective Inspector, 1940.

143 **apartment ransacked for Communist:** Wallich, Interview, 1979, is an oral history description of the same roundup that Kahle was caught in, by a speaker who was a professor at Cambridge. Confirmation that Kahle's arrest was the same general sweep-up of "B"-graded aliens is in Bagot, Letter, April 29, 1941.

143 **certificate number 708001:** Jenkins, Letter to Divisional Detective Inspector, 1940.

143 **Kahle's friend J. B. S. Haldane:** Confirmation of their close relationship through the International Brigade is in multiple points in both their MI5 files, e.g., MI5, Letter to Deputy Assistant Coordinator, February 4, 1942.

143 **"The work would be more valuable . . .":** Haldane's records indicate that he wrote to an MP named Fletcher. The only MP in office at that time with that name was Reginald Thomas Herbert Fletcher, of the Labour and Liberal parties, which means that essentially Haldane was joking that they could throw one of Fletcher's political opponents in the chamber as a substitute if they couldn't get Kahle back. J. B. S. Haldane, Letter to Fletcher, July 25, 1940.

143 **"If Mr. Kahle is not used . . .":** J. B. S. Haldane, Letter to Frederick, August 1, 1940.

143 **"I may add that I do not know . . .":** J. B. S. Haldane, Letter to Frederick, May 17, 1940.

143 **"Of all the subjects he . . .":** J. B. S. Haldane, Letter to Frederick, May 17, 1940.

144 **"with Kahle interned our position . . .":** J. B. S. Haldane, Letter to Frederick, June 18, 1940.

144 **the government relented:** MI5, Cross-Reference of Form P.F. 47192, July 18, 1940.

144 **"If the experiments are successful . . .":** Bagot, Letter, May 21, 1941.

144 **Without further objection, Hans:** MI5, Letter to Home Office, 1940.

144 **Hemingway had written:** Kvam, "Ernest Hemingway and Hans Kahle," 1983.

144 **"on his return to the United Kingdom . . .":** MI5, Letter Reference Number K. 14233/3, 1940.

144 **The assessment ranking that indicated:** Before he was interned, he was considered a "B" refugee, whereas after he returned he was considered a "C"-grade refugee, which is a lower-risk status. Chief Inspector, Letter to Aliens Registration Office, 1944.

144 **"this man is rigidly precluded . . .":** L. L. Smith, Letter on Behalf of the Medical Director-General, 1941.

144 **Haldane promised:** J. B. S. Haldane, Letter to Frederick, March 14, 1941.

144 **"He [Kahle] has no access to any documents . . .":** [signature illegible] for ADNI, Letter to MI5, 1941.

144 **Haldane pulled the same trick:** MacLeod, Letter to Negrín, 1940.

144 **participated in exactly one:** This experiment was on April 23, 1940, described in Case, Lab Notebook, 1940.

144 **participated in three more:** The dates were February 24, 1941; March 11, 1941; and November 24, 1943—all described at those dates in UCL Lab, Miscellany of Papers, ca. 1940–44.

145 **Haldane's "rabbits.":** Use of the term "rabbits" by Haldane to refer to his volunteers is in J. B. S. Haldane, Letter to Dr. Morland, 1941.

145 **get a dental bridge replaced:** Kahle, Letter to Haldane, 1941.

145 **the British government paid:** Haldane made the request in J. B. S. Haldane, Undated Notes and Bills, 1940, and confirmation it was paid is on the last page of UKNA, "Investigations," 1943.

145 **yet another lung surgery:** His surgery date was January 28, 1941, per J. B. S. Haldane, Letter to Winfrey, 1941.

145 **"The whole tenor of Kahle's career . . .":** Hanley, Note on Kahle, 1949.

145 **"precise information on this question . . ." . . . "to the risk of personal injury":** Admiralty, "Methods of Saving Life," 1939–41.

146 **government-employed researchers:** Pratt, "Organization of Naval Physiological Research," 1943.

146 **three urgent goals:** The goals are outlined in Admiralty, "Methods of Saving Life," 1939–41. Information about the work starting as submarine escape but then being redirected toward "underwater methods of attack involving shallow-water swimmers" is also in F. P. Ellis, "Personnel Research," 1950, 5.

146 **worried about disabled submarines:** The Physiological Sub-Committee on Saving Life from Sunken Submarines decided to dissolve by April to reduce the tasking of people needed for more urgent war work. But it was not stated which people or which work. UKNA, "Investigations," 1943.

146 **"Since January 1940 Professor Haldane . . .":** R. C. Frederick, Letter to Divisional Petroleum Officer, 1941.

146 **"anxious":** R. C. Frederick, Letter to Haldane, January 9, 1940.

146 **"fear":** R. C. Frederick, Letter to Divisional Petroleum Office, 1941.

146 **"unsafe":** R. C. Frederick, Letter to Divisional Petroleum Officer 1941.

147 **Siebe Gorman was hit on May 17:** J. B. S. Haldane, Letter to Frederick, May 21, 1941.

CHAPTER 10: The Littlest Pig

148 **Gino Birindelli crawled:** Unless otherwise cited, the details of Birindelli's account are taken from his oral history recording at the Imperial War Museum, Birindelli, Interview, 1994.

148 **full-body rubber suit:** *Maiale* riders' suits were made of rubber. Extended Defence Officer Gibraltar, "Report on A/S Trials," 6.

150 **jet-black new moon:** Phases of the moon in Alexandria found using holidays-info.com.

150   **British sailors stationed there:** The experience of what it was like to be a British sailor stationed in Alexandria is taken from the oral histories Keen, Interview, 2006, and Innes, Interview, 2001.

150   **had intel they might:** Holloway, Interview, 2005.

150   **decided, against all prudence:** The best, most complete historical account of the *Maiale* attack is in O'Hara and Cernuschi, "Frogmen Against a Fleet," 2015. The historical details of the attack, unless otherwise cited, are from this excellent paper.

150   **nickname slapped on:** Kagan and Hyslop, *Secret History*, 2016, 259.

151   **seven-meter-long craft:** Dimensions and details of the nomenclature of the *Maiale* are taken from Flag Officer Gibraltar, Midget Submarines, ca. 1945.

151   **known to his friends as Gigi:** Birindelli, Interview, 1994.

151   **Bianchi's position as copilot:** Birindelli, Interview, 1994.

151   **They squeezed processed food:** Holloway, Interview, 2005.

151   **"rather worrisome":** O'Hara and Cernuschi, "Frogmen Against a Fleet," 2015.

152   **waves against their heads:** The bioluminescence caused by the heads of *Maiale* riders was documented and described by British divers during their "testing" (in quotation marks because the report reads more like the divers were joyriding) of a captured *Maiale* vessel. Extended Defence Officer Gibraltar, "Report on A/S Trials," 1944.

153   **Midshipman Adrian George Holloway:** Holloway, Interview, 2005. Holloway told the story from the perspective of the HMS *Valiant*.

153   **as large as three:** Holloway, Interview, 2005.

153   **whoosh erupted . . . opposite walls:** Bigg-Wither, Interview, 1991.

153   **sailors on other ships . . . "listing quite badly":** Roxby, Interview, 1992.

153   **Italians had destroyed the core:** O'Hara and Cernuschi, "Frogmen Against a Fleet," 2015.

154   **"the most perfect attack.":** Holloway, Interview, 2005.

154   **"Bang went the Mediterranean Fleet . . .":** Dunn, Interview, 2000.

154   **"Please report what is being done . . .":** Churchill, *The Hinge of Fate*, 2014, 750. A reprint of a January 18, 1942, letter from the prime minister to General Hastings Ismay, for the Chiefs of Staff Committee.

154   **"the need for an immediate underwater . . .":** Shelford wrote his own story in Warren and Benson, *Above Us the Waves*, 1953, 15.

154   **On a cold, sunny afternoon . . . "The Admiralty, [Horton] explained . . .":** Shelford, *Subsunk*, 1960, 158.

154   **The Royal Navy had Gino Birindelli's:** Admiralty, "Miniature Torpedoes," 1943.

CHAPTER 11: Lung Maps

156   **Horace Cameron Wright donned:** The experiment described took place on May 13, 1942, at the Spithead Range, off the southern edge of Portsmouth. The details are taken from several references that have been blended together to make the full narrative. Except where noted, the information is from H. C. Wright et al., "Effects of Underwater Explosions," 1950.

156   **depth of three feet:** H. C. Wright, "Subjective Effects," 1947, trials dated May 1942.

157   **using similar small charges:** The effects of the bombs, their impact on the operators of small submersibles, and their use to prevent underwater operators from entering harbors were discussed in a lecture by H. C. Wright, "Review of Underwater Blast Research," 1957, 2.

157   **"Cam" Wright:** Horace Cameron Wright's preference of the nickname "Cam" is in Amoroso, "Memorial," 1979.

157 **prominent cleft chin:** Description written from a photo in Harris, "He Risks Life," 1950.

157 **He was known for being:** Amoroso, "Memorial," 1979.

157 **tendency to hoist heavy objects:** Bennett, Interview, 2021.

157 **specifically requested to work:** Wright requested scientific and physiology duty in January 1941 (Dansy, Letter to Wright, January 19, 1941). By March 1941 he received notice that he would be assigned to be a physicist for the Admiralty, per (signature illegible) for Director of Scientific Research, Letter to Wright, 1941. He joined the staff in the Admiralty Service of HMS *Vernon*, Portsmouth, in May 1941, working directly for A. B. Wood, the superintendent of scientists of the Admiralty Scientific Research and Experiment Department. Wood, Testimonial, 1944.

157 **when granted leadership over:** He was the leader of the research with the diving experiments, R. Cameron, Testimonial, 1964.

157 **had insisted on being:** Authority, "Central Chancery," 1950.

158 **a shock wave is mighty enough:** For the physics of shock waves and explosions in water, please see Cole, *Underwater Explosions*, 1948. All of the research in that text-book dates back to experiments from WWII, as the scientists were trying to figure the problems out midwar.

158 **saw the explosion before they felt:** This description of underwater explosion was not recorded by Wright, but instead is taken from author's personal experience with underwater bombs.

158 **like a punch or a stab:** Gordon-Taylor, "Injuries of the Large Intestine," 1943.

159 **was at least their fifth:** The first four explosions took place on February 19, February 24, and two on February 25, 1942, and are cited in H. C. Wright et al., "Effects of Underwater Explosions," 1950.

159 **One and a quarter pound of TNT:** Details of these experiments are in H. C. Wright, "Investigations by RNPL," 1947, and Wright's participation in this experiment is mentioned by name in UKNA, RNPRC Minutes 1943.

159 **sat astride wooden logs:** H. C. Wright et al., "Effects of Underwater Explosions," 1950.

160 **"six operational Chariots . . .", "the early Spring . . .":** A. V. Alexander, "Chariots, X-Craft, Etc.," 1942.

160 **"The DSEA afforded . . .":** H. C. Wright et al., "Effects of Underwater Explosions," 1950, appendix I.

161 **often get in the water:** Wright wrote: "As the writer is responsible for controlling the severity of the conditions and safety of the divers the policy has always been for him to experience the underwater sensation before and during the trial. This has proved to be a wise and profitable procedure and one that will be adhered to in future experiments of this kind." H. C. Wright, "Underwater Blast Trials," 1949.

161 **"undergoing instruction in diving":** S.5, Confidential Message, 1942.

161 **diving attendant reported:** Warren and Benson, *Above Us the Waves*, 1953, 24.

161 **"From Admiralty. Deeply regret . . .":** Hobbs, Government Telegram, 1942.

161 **On May 13, 1942, UCL experimental day:** Details of this experiment are from UCL Lab, "Diving Notes," Part 1, May–June 1942, unless otherwise cited.

162 **restarted testing earlier that month:** The lab conducted two tests on February 3, 1942 (UCL Lab, "Diving Notes," Part 3, June 1943–December 1944), and then restarted testing in earnest on May 4, 1942 (UCL Lab, Miscellany of Papers, ca. 1940–44).

162 **saw golden yellow:** UCL Lab, "Diving Notes," Part 3, June 1943–December 1944, the experiment dated November 26, 1943.

162 **Spurway hallucinated in purple:** UCL Lab, "Diving Notes," Part 2, July 1942–May 1943, the experiment dated July 20, 1942.

162 **flashed him spots of red:** UCL Lab, "Diving Notes," Part 3, June 1943–December 1944, the experiment dated June 10, 1943.

163 **"great fun":** Harris, "He Risks Life," 1950.

163 **"Loved by many . . .":** Amoroso, "Memorial," 1979.

163 **Wright was born:** General Register Office, "Horace Cameron Wright," 1902.

163 **father was a lithographer:** General Register Office, "Class: RG13," 1901.

163 **mother died in 1932:** General Register Office, "England and Wales, National Index," 1932.

163 **father died in 1948:** General Register Office, "GRO Reference," 1948.

163 **Brangwyn was a close friend:** Brangwyn, Christmas Card, 1954. The Wrights also lived with Brangwyn and his wife for a while. General Register Office, "London, England, Electoral Registers," 1918.

163 **mother, Laura, had been described:** General Register Office, "Marriage Registration of Edward," 1900.

163 **live as the artist son:** General Register Office, "Marriage Registration of Edward," 1900.

163 **nineteen-year-old Cam was living at home:** General Register Office, "Census Returns of England," 1921.

163 **by 1935, he had earned:** A. J. D. Cameron and Wright, "Short Wave Therapy," 1935.

163 **researching radiation treatments:** Wakely, Letter to the Royal College, 1977.

163 **Wright researched carcinoma:** H. C. Wright, "Concerning Dr. H. C. Wright," ca. 1972.

164 **He wrote to other universities:** Matthey, Letter to Wright, 1938.

164 **Wright omitted the part:** Wright wrote his translation on the bottom of the original letter, cited above.

164 **"I have explained to him . . .":** R. V. Jones, Letter to Wright, 1941.

164 **"This appointment should enable . . .":** (unsigned) from Air Ministry, Letter to Wright, 1941. Wright reported for Admiralty work on March 12, 1941, under the supervision of Dr. R. V. Jones. Dansey, Letter to Wright, March 6, 1941.

165 **"It is a surprising fact that . . .":** H. C. Wright, "Injurious Effects," ca. 1950.

165 **"many of the survivors reported . . .":** H. C. Wright, "Review of Underwater Blast Research," 1957.

165 **The medical director general convened:** H. C. Wright, "Review of Underwater Blast," 1957.

165 **"His striving to make discoveries . . .":** Amoroso, "Memorial," 1979.

165 **For a small fee:** Needham, Letter to "Sir," 1942.

165 **thrilled to mail Wright:** Needham, Diploma from Intercollegiate University, 1942, and Letter to Wright, 1942.

165 **Needham, however, was a bishop:** Kersey, "Members of the San Luigi," 2012.

165 **On February 2, 1942, Wright received:** Needham, Letter to Wright, 1942.

166 **"world-famous scientist":** *The Springfield Herald* (IL), "Feeling Old?," 1956.

166 **roughly a decade wandering:** DeSeblo has advertisements for his tours starting at least in 1949, *The Eugene Guard* (OR), "Feeling Old?," 1949.

166 **one major run-in:** FDA, "Notices," 1974.

166 **murky forward plodding:** The details of this experiment, unless otherwise cited, are from H. C. Wright et al., "Effects of Underwater Explosions," 1950. The date was October 8, 1942.

166   **six feet and one half inch:** Cowan et al., "Respiratory and Other Responses," 1947, 10.

166   **A thick, black, durable:** Description of the Sladen suit, aka the shallow water dress, is from Bain, "Description of 'Sladen' Dress," 1943.

167   **Reports from other days:** H. C. Wright, "Underwater Blast Trials," 1943.

168   **worked out a new:** H. C. Wright, "Injurious Effects," ca. 1950; H. C. Wright et al., "Effects of Underwater Explosions," 1950.

168   **After nearly every one:** H. C. Wright et al., "Effects of Underwater Explosions," 1950.

169   **American divers stood on the bottom:** F. P. Ellis, "Present Position," 1944, 4.

169   **they thought the surface of the water "shredded":** Draeger et al., "Blast Injury," 1946; the British blast scientists suspected this number was incorrect from the start, per F. P. Ellis, "Present Position," 1944, 7.

169   **Five hundred psi is roughly:** Lance et al., "Human Injury Criteria," 2015.

169   **guidelines that the risk:** Lance and Bass, "Underwater Blast Injury," 2015.

170   **blast records make it seem:** A lot of these studies have been republished in English in Hornberger, "Decompression Sickness," 1950.

CHAPTER 12: Nerves

171   **Elizabeth Jermyn was puttering:** Jermyn's test took place on June 1, 1942, and the details, unless otherwise cited, are from the entries on that date in UCL Lab, Miscellany of Papers, ca. 1940–44, and UCL Lab, "Diving Notes," Part 1, May–June 1942. Her test was experiment 170 on experimental day seventy-seven. On that day, they ran five experiments, which I mention because it is an insanely high number. I get exhausted after running two such experiments in one day.

171   **four meters tall:** Height of twelve feet, given in Donald, *Oxygen and the Diver*, 1992.

171   **sturdy curved rail:** Images of the tank, including the ladder and piping, are in both Donald, *Oxygen and the Diver*, 1992, and R. H. Davis, Davis, and Davis, *Record of War*, 1946, on multiple pages.

172   **system of pulleys:** Donald, *Oxygen and the Diver*, 1992, 35.

172   **rubber fishing waders:** Donald, *Oxygen and the Diver*, 1992, 32.

172   **stretched his arms out:** The diameter of that chamber was six feet. Donald, *Oxygen and the Diver*, 1992, 32, and Waldron and Gleeson, *The Frogmen*, 1955, 46.

172   **"like one of Mr. Churchill's famous hats.":** The quote is from Sydney Woolcott, as he described the testing in Waldron and Gleeson, *The Frogmen*, 1955, 46.

173   **Joan Elizabeth Jermyn was young:** A. Bamji, Interview, 2020.

173   **She left a written record:** J. E. Bamji, Letter #1 to Cramer, 2007.

173   **"if [Haldane] wanted you . . .":** A. Bamji, Interview, 2020.

173   **Jermyn and Spurway's being actively:** R. H. Davis, Letter re Jermyn, 1942, and Letter re Spurway, 1942.

174   **field hockey and lacrosse:** A. Bamji, Interview, 2020.

174   **patrolling the grounds:** J. E. Bamji, Letter #1 to Cramer, 2007.

174   **Jermyn's sister Molly learned:** J. E. Bamji, Letter #2 to Cramer, ca. 2007.

174   **He was a botanist:** A. Bamji, Interview, 2020.

174   **chief fire officer:** H. Kalmus, *Odyssey of a Scientist*, 1991, 60.

174   **worked on compost formulations:** The information about Nariman Bamji comes from long letters about her own life that Elizabeth Jermyn (by then Elizabeth Bamji) wrote to a friend named Cramer, who was a local butcher. The letters were intended as a sort of autobiography between the two. J. E. Bamji, Letter #1 to Cramer, 2007, and Letter #2 to Cramer, ca. 2007.

175 **somewhere around that time:** No record exists of the exact timing of their engagement. However, their son, Andrew Bamji, approximates their being engaged at around this time based on his memories of their stories about their wedding and life planning.

176 **"pretty incompetent.":** J. B. S. Haldane, Letter to "Dear Sir," 1943.

176 **After about two hours had passed:** Case's experiment on the same day (June 1, 1942) is in the entries for that date in UCL Lab, Miscellany of Papers, ca. 1940–44, and UCL Lab, "Diving Notes," Part 1, May–June 1942.

176 **Her fifth thoracic vertebral:** E. B. Jones, Letter to Haldane, 1942.

177 **only one half inch:** The exact amount of compression needed to cause a burst fracture (described here colloquially as "exploding" the bone) varies from person to person. Values in the literature span between a quarter inch of compression (Bozic et al., "Three-Dimensional Finite," 1994: 3,400 N fracture load multiplied by 0.5 MN/m stiffness of the intervertebral discs allowing compression) to an inch and a quarter (Yoganandan et al., "Strength and Kinematic Response," 1991), but in the interest of brevity, I've provided a median value. A half inch is the general rule of thumb used by biomechanists as an estimate.

177 **The human spine is so flexible:** The extreme flexibility of the cervical spine is discussed in detail in Huelke and Nusholtz, "Cervical Spine Biomechanics," 1966.

177 **Haldane asked the insurance company . . . "and still holds some":** J. B. S. Haldane, Letter to "Dear Sir," 1943.

178 **Less than a month earlier:** The test dated May 5, 1942, is described in UCL Lab, Miscellany of Papers, ca. 1940–44, and UCL Lab, "Diving Notes," Part 1, May–June 1942.

178 **Chief Petty Officer and engine cleaner:** Brownrigg's full name and occupation are from his service records, Admiralty, Lieutenant P. C. A. Brownrigg, June 8, 1942.

178 **looking "rather blue":** Rowkins's test took place on May 11, 1942, and is described in UCL Lab, "Diving Notes," Part 1, May–June 1942, and Donald, "Variation in the Oxygen Tolerance," 1944, 7.

178 **Almost everyone who knew:** Donald, "Continued Report," 1942; Shelford, Letter to Professor Haldane, 1942.

178 **The Admiralty analyzed his breathing gas:** Haldane's measurements and calculations show the breathing gas had 0.1 percent CO_2. At a pressure of seventy feet, that provides a CO_2 partial pressure of 3 percent. It's not an ideal level, but it's well below the level required to cause those types of symptoms. The calculations are immediately following the notes for the experiment, as cited above.

179 **symptoms of oxygen toxicity:** Burbakk et al., *Bennett & Elliott's Physiology*, 2003, 358–418.

179 **principle that has dictated:** This point needs no citation for those in the diving field, but strict guidelines about oxygen toxicity are found in the *US Navy Diving Manual* as well as the diving educational literature for every diving certification program.

179 **limit of two hours:** Donald, "Variation in the Oxygen Tolerance," 1944, and Donald, *Oxygen and the Diver*, 1992, 23.

179 **guidelines had gone largely untested:** Donald, *Oxygen and the Diver*, 1992, 192, says: "With regard to the safety times and depths when breathing oxygen under water, it was tacitly assumed that these could be inferred from the oxygen tolerance determined in the 'dry' hyperbaric chambers."

179 **"I have also heard at last . . .":** J. B. S. Haldane, Letter to Commander Sladen, 1942, and Donald, *Oxygen and the Diver*, 1992, 40.

180 **had written a letter . . . "In conjunction with [Lucan]":** Sladen, Letter to Davis, 1942.

180 **Renouf invited ... "at what depth" and "for how long":** Blackett, Letter to Haldane Labeled "Secret," 1942.

181 **Haldane, Spurway, and Case all climbed:** The day of the meeting was July 8, 1942, and the experimental notes are on that date in UCL Lab, Miscellany of Papers, ca. 1940–44, and J. B. S. Haldane, Papers, Part 1.

181 **demand for their science:** Brickhill, *The Dam Busters*, 1951, chapter 4, page 2.

181 **On June 18, 1942, Surgeon Lieutenant:** The record of this dive is on June 18 to 19, 1942, in UCL Lab, "Diving Notes," Part 1, May–June 1942.

181 **Donald was tall and trim:** Description of Ken Donald is taken from the photos of him in Donald, *Oxygen and the Diver*, 1992.

181 **He later claimed ... he "took over":** Donald, *Oxygen and the Diver*, 1992, 193. Kenneth Donald's book claims that he arrived at Siebe Gorman in May 1942, but all contemporary records indicate he went there for the first time at the end of June 1942.

181 **"There I was, a Surgeon Lieutenant ...":** Donald, *Oxygen and the Diver*, 1992, 194.

181 **he did have a commanding officer:** The documentation that Shelford was the one really in charge is abundant, e.g., UPS, UPS Report 1, 1943; Shelford, Letter to Haldane, 1942; R. H. Davis, Letter to Haldane, 1943.

182 **"expert technical help ...":** Donald, *Oxygen and the Diver*, 1992, 193.

182 **"A further piece of extreme ...":** Donald, *Oxygen and the Diver*, 1992, 194.

182 **Martin Case ran the chamber:** Case's notebooks show his chamber operations for the end of June 1942. UCL Lab, "Diving Notes," Part 1, May–June 1942, and the same citations show Haldane's handwriting for the gas analyses.

182 **acted as their dive tenders:** Hans Kalmus was tender for PO Derrick on September 8, 1943. UCL Lab, Notebook, September 1942–June 1944.

182 **Donald's first trial on himself:** June 22, 1942. UCL Lab, "Diving Notes," Part 1, May–June 1942.

183 **helped the first twelve:** June 18–19, 1942; UCL Lab, "Diving Notes," Part 1, May–June 1942.

183 **"These divers were operational personnel ... minimal":** Donald, *Oxygen and the Diver*, 1992, 36.

184 **"On occasions, the next diver stepped ...":** Donald, *Oxygen and the Diver*, 1992, 36.

184 **dealt with decompression sickness:** Tests resulting in mild symptoms of decompression sickness occurred on April 22, 23, and 26, 1940, as well as on May 27, 1940; June 18, 1940; and many other dates not listed. Case, Lab Notebook, 1940.

184 **Decompression sickness, however, has another:** Burbakk et al., *Bennett & Elliott's Physiology*, 2003, 530–650.

185 **diver known as Miss C:** Liow et al., "MRI Findings," 2014.

185 **twenty-nine-year-old diving instructor:** F. M. Davis, "Element of Doubt," 2003.

186 **A fifty-year-old man diving:** I. M. Calder et al., "Spinal Cord Degeneration," 1989.

187 **On July 7, 1942, Martin Case:** The information for experiment 205 is compiled from the entries on that date in J. B. S. Haldane, Papers, Part 1, and UCL Lab, "Diving Notes," Part 2, July 1942–May 1943.

187 **"clammy death":** M. White, *Australian Submarines*, 2015, 1152.

187 **Each of Case's shoes was fitted:** The renaming of the Sladen suit to SWD and details of its construction are taken from Bain, "Description of 'Sladen' Dress," 1943.

187 **bottle that had been liberated:** Use of a Luftwaffe bottle is in UCL Lab, "Diving Notes," Part 2, July 1942–May 1943, in the entry for the experiment dated July 7, 1942.

187 **aluminum trophies:** Material of aluminum. Waldron and Gleeson, *The Frogmen*, 1955, 43.

187 **After wrapping up his final few:** Case joined the RNVR in July 1942. J. B. S. Haldane and Spurway, "Contributions to the Physiology," ca. 1943, 1.

188 **Helen Spurway took . . . "during a fit.":** J. B. S. Haldane, Letter to Case, 1942.

188 **Jermyn climbed into the chambers:** Tests were on June 29 and July 1, 1942. UCL Lab, Miscellany of Papers, ca. 1940–44.

188 **On the advice of her doctor:** E. B. Jones, Letter to Haldane, 1942.

188 **Haldane's desk-drawer stash:** J. B. S. Haldane, Letter to Jermyn, 1942.

CHAPTER 13: The Inevitable Solution

Note: Facts of the Dieppe raid that were cited in the prologue are not cited a second time here.

189 **total of 3,367 Allies:** Contemporary records state 3,363 casualties (Adjutant General Office, "'Jubilee,'" 1944), but modern records cite 3,367, which indicates an additional four died or were counted in the following years. Government of Canada, "Dieppe Raid," 1942.

189 **one every thirty-five seconds:** Gettysburg numbers: 7,058 dead total, battle starting 8:30 a.m., July 1, through 7:00 p.m., July 3. This equates to a death every thirty seconds. Danto, "Gettysburg," 1987.

189 **previously cleared by minesweepers:** Porteous, Interview, 1987.

190 **Cyril Joseph Thornberry bobbed:** Thornberry's story and quotes are taken from his oral history recording at the Imperial War Museum. Thornberry, Interview, 1999.

190 **George Cook woke up thirsty:** Cook's story is from his oral history recording at the Imperial War Museum. Cook, Interview, 1987.

192 **"And what did they achieve?":** Porteous, Interview, 1987.

192 **Clogstoun-Willmott also buzzed into action . . . "Party Koodoo-Inhuman.":** A. Hughes, "History of COPP," 1946, pinpoints this as "end of August 1942."

192 **Courtney and his wife, Doris, had:** Roger Courtney and Dorris Courtney (née Butt) were married in the spring quarter (April to June) of 1938. General Register Office, "Marriage Registration of Dorris," 1938.

192 **spent their honeymoon paddling:** Owen, *Commando*, 2012, xxiii.

193 **In Egypt, in the spring of 1941:** James Sherwood left behind an oral history of this incident (Sherwood, Interview, 1987). In it he said the ship was the *Medway*, but that is a depot ship, not a submarine. The correct ship was the *Triumph*. Clogstoun-Willmott, War Diary, 1943, 4.

193 **"pitch dark night":** Sherwood, Interview, 1987.

193 **tiny civilian canoe:** Clogstoun-Willmott, War Diary, 1943, 3.

193 **roughly a mile to paddle:** Sherwood, Interview, 1987.

193 **Once close enough, they:** A. Hughes, "History of COPP," 1946, 3.

193 **infrared beacon to guide them:** Clogstoun-Willmott, War Diary, 1943, 4.

193 **"the true value of such . . .":** Clogstoun-Willmott, War Diary, 1943, 1.

193 **By mid-September 1942, less than . . . submarine's periscope:** Clogstoun-Willmott, War Diary, 1943, 8.

194 **first troops were surprised:** Clogstoun-Willmott, C.O.P.P. Training Book, 1943, 3.

194 **Party Koodoo-Inhuman had mapped:** Clogstoun-Willmott, War Diary, 1943, 8.

194 **"Beach reconnaissance should become . . .":** A. Hughes, "History of COPP," 1946, 1.

194 **units would be named COPP:** Clogstoun-Willmott, C.O.P.P. Training Book, 1943, 1.

195 **COPP teams recruited professors:** One of them was Professor John Desmond Bernal, a known Communist. Hastings, C.O.P.P. Report 58, 1944.

195 **methods of getting ashore:** "As no really winter operations had previously been done it was essential to evolve some sort of protective swimsuit. After hasty trials a

somewhat inefficient and uncomfortable but possible one was put into production at Siebe Gormans Ltd." Clogstoun-Willmott, War Diary, 1943, 12.

195 **On October 26, 1942, at 12:14 in the afternoon:** Wilson, Title, 1942, 4.

195 **On board were ten men:** The names of the Norwegian personnel involved in Operation Title were Leif Larsen (quartermaster), PO Bjørnøy (engineer), R. Strand (wireless operator), and J. Kalve (assistant engineer and deckhand). Chariot crew no. 1: Sublieutenant Brewster and Able Seaman Brown. Chariot crew no. 2: Sergeant Donald Craig and Able Seaman Robert Evans. Dressers: Able Seaman Tebb and Able Seaman Causter. Larsen, "Report by Quartermaster," 1942, 1.

195 *Arthur* **had a trapdoor:** Wilson, Operation "Title," 1942, 3.

195 **covered by bags of peat:** Admiralty, Instructions, 1942.

196 **would pose as fishermen:** Wilson, Operation "Title," 1942, 4.

196 **waterproof maps printed on silk:** London SOE HQ, Title, 1943, 2.

196 **Tied beneath the trawler:** Strand, "Title Report," 1942, 2.

196 **with their forged papers:** MPB, Operation Title, 1942.

196 **slipped past a German guard:** Strand, "Title Report," 1942, 3; Larsen, "Report by Quartermaster," 1942, 8.

196 **weather sent the crew bucking:** Larsen, "Report by Quartermaster," 1942, 9.

196 **Evans donned his diving suit:** Strand, "Title Report," 1942, 4.

196 **"The towing wires were still . . .":** Larsen, "Report by Quartermaster," 1942, 10.

196 **For the third time:** The engine stopped twice en route. Strand, "Title Report," 1942, 1; then the last time, Larsen, "Report by Quartermaster," 1942, 2.

196 **scuttled the fake *Arthur* poorly:** SOE Agent in Trondheim, "Translation of Report," 1943, 4.

196 **splitting into two parties:** SOE Agent 8627, Title, 1942, 2.

196 **Nine of the ten crew:** PO Bjørnøy lost several toes to frostbite and needed a blood transfusion. Wilson, Telegram to Øen, 1943.

196 **but they starved:** Strand, "Title Report," 1942, 6.

196 **thigh-deep Nordic snow:** The snow depth was three to four feet. Strand, "Title Report," 1942, 6.

197 **frigid river crossings:** Brewster, Brown, and Bjørnøy fell in a river before getting to Kjløhaugene. Strand, "Title Report," 1942, 7.

197 **Charioteer Robert Evans was shot:** C. Mann, *British Policy and Strategy*, 2012, 26.

197 **German harbor guard who:** Naval Commandant Moller. SOE Agent in Trondheim, Title, 1942.

197 **Operation Title also brought to the forefront:** SNL, Title, 1942.

197 **The trek, should it have proved:** SOE Agent 8628, Title, 1942, 2.

197 **The same day as the departure:** National Archives Prague, Entry for Elsa Kalmusovà, 1942, and Entry for Arnošt Kalmus, 1942.

197 **took turns in the chamber:** The experiments were on October 30, 1942, and the test details are on that date in UCL Lab, Miscellany of Papers, ca. 1940–44, and Notebook, September 1942–June 1944.

198 **The new Siebe Gorman factory:** Information about the layout of the Siebe Gorman factory, employees, and the contents of its buildings are from the book they printed about their own activities during the war, R. H. Davis, Davis, and Davis, *Record of War*, 1946.

198 **"a couple of the lads soon . . .":** The quote is from Sydney Woolcott's report of his own activities, published in Waldron and Gleeson, *The Frogmen*, 1955, 45.

198 **Quickly after the Admiralty's decision:** "Prior to the Party's return [to the UK in December 1942] it had been decided that beach reconnaissance should become an

integral part of future amphibious operations, and at a meeting held in December the Chiefs of Staff ruled that there were to be ten units formed, equipped and trained to be able to undertake this work." A. Hughes, "History of COPP," 1946, 2. Before that mission, they had improvised "a somewhat inefficient and uncomfortable but possible" cold-water suit. Clogstoun-Willmott, War Diary, 1943, 12. The new COPP teams placed a massive supply order including new Siebe Gorman suits on February 11, 1943. Chief of Combined Operations, Enclosure to SR 380/43, 1943, line item 516.

199 **new suits would be custom fitted:** Measurements of each COPP war fighter were provided with equipment orders, e.g., Commander R.N. for Chief of Combined Operations, Letter to D. of P., 1943.

199 **eleven cargo pockets:** Pockets counted from photos in Combined Operations Headquarters, Bulletin T/18, 1945.

199 **workers glued coils:** Director of Victualling, Enclosure, 1943.

199 **"inefficient and uncomfortable but possible":** Clogstoun-Willmott, War Diary, 1943, 12.

199 **swimmers chose to pee:** Galwey, "Geoff's Opus," 1992, 107–8.

199 **Allied troops plucked:** Swim fins were developed by the rubber company Pirelli, per Shelford, *Subsunk*, 1960, 47. They were picked up off *Maiale* operators and used by Lionel "Buster" Crabb and his diving partner, Sydney Knowles. Knowles, *Diver in the Dark*, 2009, 21–22.

200 **morning of January 3, 1943:** Polacchini, Message, 1943.

200 **One of the host submarines was:** The other two submarines were HMS *311* and HMS *Trooper*. Captain S.10, Message 1111A, 1943.

200 **been caught and blown up:** This is the Admiralty's conclusion after an explosion was heard coincident with the disappearance of the submarine. Admiralty, Letter to Mrs. M. Sargent, 1944.

200 **toward the Palermo harbor:** Phillips, SM 04349/820, 1945.

200 **sank only one ship:** The cruiser *Regolo*. S.10, Message 1938A, 1943.

200 **fifteen died, and two:** Captain S.10, Message 1643A, 1943. Their names were Lieutenant Charles Ernest Bonnell, RCNVR; Lieutenant Guy Stretton-Smith, RNVR; Lieutenant Richard Thomas Goodwin Greenland, RNVR (taken POW); Lieutenant Harold Faulkner Cook, RNVR; Lieutenant Kenneth Stewart Kerr, Royal Scots Regiment; Sublieutenant Jack Sargent, RNVR; Sublieutenant George Gilbert Goss, RNVR; Sublieutenant Rodney George Dove, RNVR (taken POW); John M. Miln, PO; Berties G. S. Rickwood, Ldg Sea; James H. Freel, A/LBG Sea; Alexander N. Ferrier, Ldg Sig; Harold V. Worthy, AB; Bernard Trevethan, AB; Walter R. Simpson, AB; Paul Mapplebeck, AB; Robert Anderson Sea RNR; Ronald W. B. Pridham, Sto.

200 **started on the idea seven years:** O'Hara and Cernuschi, "Frogmen Against a Fleet," 2015.

200 **Haldane swallowed ten milligrams:** The experiment dated December 15, 1942. J. B. S. Haldane, Papers, Part 1 and Part 2; UCL Lab, Miscellany of Papers, ca. 1940–44, and Notebook, September 1942–June 1944.

200 **Benzedrine was similar:** Bett, "Benzedrine Sulphate," 1946.

201 **trend to keep military pilots:** Pugh, "RAF and Benzedrine," 2018.

201 **discovery a few decades:** It was discovered in 1910. Barger and Dale, "Chemical Structure," 1910; Bett, "Benzedrine Sulphate," 1946.

201 **German Army was consuming it:** Reinold and Hoberman, "Myth of the Nazi Steroid," 2014.

201 **Norwegian wireless operator:** Strand, "Title Report," 1942.

201 **induce a paranoid psychosis:** McKetin, "Methamphetamine Psychosis," 2018.

201　the Royal Air Force flipped: Pugh, "RAF and Benzedrine," 2018.

202　Ken Donald thought the Benzedrine: J. B. S. Haldane, "Contributions," ca. 1942, 4.

202　COPP operators would be coming: Clogstoun-Willmott, C.O.P.P. Training Book, 1943, 2.

202　"to assist in carrying out . . .": DM, Notes on M.E. C.O.P.P.'s, 1943.

202　"had had to leave the UK . . .": Clogstoun-Willmott, Letter Z/448, 1943, 1.

202　without compasses . . . "featureless": Clogstoun-Willmott, Letter Z/448, 1943, 1.

202　Lieutenants P. R. G. "Bob" Smith and: Admiralty, "Comparison Between COPP and Chariot," (undated), 6, for distance and time.

203　An MI5 phone tap eavesdropped: MI5, Timeline of Entries, 1942. There is no evidence that he was actually unconscious for a full month, and the experimental schedule does not show any unexplained monthlong gaps.

203　"JBS Haldane has aged . . .": Miller, Entry April 23, 1942.

203　Max Horton wrote a memo: Horton, Letter to Alexander, 1942.

203　Alexander scribbled a handwritten: A. V. Alexander, Letter to Churchill, 1942.

203　The Aliens War Service Department granted: Aliens War Service Department, Letter to Haldane re Hans Kalmus, 1943b, and Letter to Haldane re Ursula Philip, 1943a.

203　MI5 granted official permission: Sams, Letter for Anderson, 1943.

203　Haldane took a dose: The test referenced is March 30, 1943, with information on that date in UCL Lab, Miscellany of Papers, ca. 1940–44, and Notebook, September 1942–June 1944.

204　The lab group had to move out: J. B. S. Haldane, Letter to Bywaters, 1942.

204　The cracked sink was likely: Rodwell & Sons, Estimate, 1942; Salvesen, "Schedule of Dilapidations," 1942.

204　whose lungs kept collapsing: J. B. S. Haldane, Letter to Professor Frederick, January 26, 1943; Morland, "Report on J. M. Rendel," 1942.

204　took a job with the Ministry: Jermyn, Letter to Tanner, 1943.

204　"everyone concerned says . . .": J. B. S. Haldane, Letter to Rendel, 1943.

204　"Our underwater work is going on . . .": J. B. S. Haldane, Letter to Rendel, 1943.

205　The same group was going: Both of these experiments took place on June 9, 1943. The notes for both are on that date in UCL Lab, "Diving Notes," Part 3, June 1943–December 1944, and Notebook, September 1942–June 1944. The second experiment of the day was data point number 433, with the amyl nitrite.

205　Amyl nitrite was first described: Fye, "T. Lauder Brunton," 1986.

205　one of the first-guessed causes: Goedert et al., "Amyl Nitrate," 1982.

205　handed around at chemistry lectures: Fye, "T. Lauder Brunton," 1986.

206　"a further urgent demand" . . . "were to have any degree of success": A. Hughes, "History of COPP," 1946, 4.

206　got his two teams . . . ". . . even in prototype.": Clogstoun-Willmott, Progress Report No. 12, 1943.

207　They brought some newfangled device: A. Hughes, "History of COPP," 1946, 5.

207　They also brought Chariots: Clogstoun-Willmott, Letter Z/448, 1943, 4.

207　"the Chariot method was slow": S.10, Report Subject C.O.P.P. Reconnaissance, January 8, 1943, 2.

207　swimming on their backs: The method of swimming on the back is shown in Combined Operations Headquarters, Bulletin T/18, 1945, 8.

207　hour and a half of oxygen: Admiralty, "Comparison Between COPP and Chariot," (undated), 2.

207　Before dawn on July 10, 1943: Huxen, "Operation Husky," 2017.

207　"succeeded in marking . . .": A. Hughes, "History of COPP," 1946, 5.

207   **clear and unbridled acrimony:** Admiralty, "Comparison Between COPP and Chariot," (undated).

207   **"Bad weather causes . . .":** S.10, Report Subject C.O.P.P. Reconnaissance, July 18, 1943, 2.

208   **bombs in Messina's harbor:** Details of the bomb types, quotes, and disposal activities in Messina's harbor are taken from John Bridge's oral history at the Imperial War Museum. Bridge, Interview, 1995.

209   **Bridge reported their success on a Wednesday:** Bridge's oral history says the invasion started the next day, but the invasion of mainland Italy occurred on September 3, 1943, which was a Friday. It is not clear whether he was misremembering the number of days between, or the days of the week of the events. For this text, I have assumed he remembered the day of the week correctly, but misremembered the number of days between.

209   **submitted his first report:** Donald, "Time-Pressure Curves," 1942.

210   **He wrote a second report:** Donald, "Time-Pressure Curves," 1942, 3.

210   **hatching success of duck eggs:** Franklin et al., "James Meadows Rendel," 2004.

210   **colorful fur splotches:** Tracking and recording the fur patterns of feral cats was the subject of an adorable tiny journal carried around by J. B. S. Haldane during an unrelated trip to Italy. J. B. S. Haldane, "Gatti Italiani."

210   **Donald's data points were available:** The UCL group having done all the statistics is evident by Donald's final report, dated March 1944: Donald, "Variation in the Oxygen Tolerance." The UCL group and Donald were all on the Underwater Physiology Sub-Committee, as both their names showed up in the meeting minutes, e.g., RNPRC, Minutes of the Twenty-First Meeting, 1945.

210   **on July 9, 1943, Helen Spurway:** Experimental notes, including all quotes, are on that date in UCL Lab, "Diving Notes," Part 3, June 1943–December 1944, and Notebook, September 1942–June 1944.

212   **on August 27, 1943, COPP:** All quotes from this section are from Clogstoun-Willmott, COPP "Overlord" Report, 1943.

CHAPTER 14: The Midget Subs

214   **The plan was simple:** The story of the Japanese midget submarines is taken from Sakamaki, *I Attacked Pearl Harbor*, 2017, unless otherwise cited.

214   **Ensign Kazuo Sakamaki struggled:** Sakamaki, *I Attacked Pearl Harbor*, 2017, 36.

214   **"Learn the peculiarities . . .":** Sakamaki, *I Attacked Pearl Harbor*, 2017, 33.

214   **the maps were outdated:** The maps of Pearl Harbor used by Sakamaki and his crewmate, which were recovered from the *HA-19* after the attack on Pearl Harbor, were preserved by the US National Archives and are available online. COMSUBRON 4—Japanese Midget Submarine, No. 19—Navigational Charts, 1941. The maps show anticipated but inaccurate mooring positions for the American ships that can be compared to the known positions of the ships during the attack.

215   **Unable to navigate underwater:** Sakamaki, *I Attacked Pearl Harbor*, 2017, 33.

215   **"the air inside the submarine was . . ." . . . ". . . hatch wide open":** Sakamaki, *I Attacked Pearl Harbor*, 2017, 40–42.

216   ***HA-19* was found intact:** COMSUBRON 4—Japanese Midget Submarine, No. 19—Navigational Charts, 1941.

217   **small, stealthy submarines:** My favorite reference for the history of submarine development is Swinfield, *Sea Devils*, 2014, which has these midget sub stories and more.

217   **killed its entire crew:** Have you heard of my other book, *In the Waves*? It's got explosions!

218   "As the Italians first developed...": Director of Plans, Letter PLS/PBA, 1943.

218   Germany, Japan, and Italy built: Prenatt and Stille, *Axis Midget Submarines*, 2014, 30.

219   "ugly duckling": Gallagher, *X-Craft Raid*, 1971, 39.

219   The boats had been designed: Place, Interview, 1988.

219   not enough room: George Butler Honour said you could "just about stand" in the control room alone. Honour, Interview, 1987. Also referenced in Warren and Benson, *Above Us the Waves*, 1953, 45, and Mitchell, *Tip of the Spear*, 1993, 33.

220   maximum of six knots: Barry, SM 04349/990, 1943.

220   It smelled constantly: Mitchell, *Tip of the Spear*, 1993, 15.

220   ladies' hair dryer: Mitchell, *Tip of the Spear*, 1993, 78.

220   survivable but chronic: The crew trained for breathing high carbon dioxide. F. P. Ellis, "Personnel Research," 1950, 17.

220   X-craft could lug not one: Barry, SM 04349/990, 1943.

220   gray explosive granules: Mitchell, *Tip of the Spear*, 1993, 64.

220   containing six hundred: Barry, SM 04349/990, 1943.

220   be released 150 miles: Admiralty, "Naval Cypher," 1943.

221   A crack team of welders: The story of Vickers-Armstrongs is told through multiple oral histories and records compiled in Mitchell, *Tip of the Spear*, 1993.

221   "we worked hard and played hard...": Mitchell, *Tip of the Spear*, 1993, 17.

221   "long working hours and anxieties": Mitchell, *Tip of the Spear*, 1993, 10.

221   constructed using specialized engines: F. White, Reports on Miniature Torpedos, 1943, 2.

221   in practice it was silent: F. White, Reports on Miniature Torpedos, 1943.

221   sturdy and reliable: George Butler Honour called the engines "very reliable" and said they "never gave trouble." Honour, Interview, 1987.

221   engines of London buses: Gallagher, *X-Craft Raid*, 1971, 42.

221   commander in chief had to write: C in C Levant, Message 101031C/April, 1943.

222   The nearby island Gruinard: Swiderski, *Anthrax*, 2004, 173.

222   Admiralty established a secret: The site in Loch Cairnbawn was Port HHZ and it was operational with X-craft training there by March 1943. Banks, "Narrative of Exercise PBX4," 1943.

222   They called the newly: The organization and responsibility of *Varbel* and the 12th Submarine are described in Admiral (Submarines), "Employment of X-craft," 1943.

222   his propaganda machine: Mitchell, *Tip of the Spear*, 1993, 25.

223   dubbed the northern practice site: Banks, "Narrative of Exercise PBX4," 1943.

223   They placed to drift: The first X-craft exercise practicing an attack on the HMS *Bonaventure* in Port HHZ was in April 1943. Banks, "Narrative of Exercise PBX4," 1943.

223   The Wrens of HMS *Varbel*: Mitchell, *Tip of the Spear*, 1993, 22–23.

223   "badly wanted to get...": Angela Cooper's story was recorded in Mitchell, *Tip of the Spear*, 1993, 23. I also want to thank Colin Williams of the Royal Navy and QinetiQ for translating some of the Royal Navy submarine lingo for me, and special thanks to Lesley Blogg for finding Williams to help me.

223   "the use of X-craft": Clogstoun-Willmott, Progress Report No. 29, 1943.

224   "By COPP methods...": Mountbatten, Letter Ref. CR 4204/43, 1943, 2.

224   provided a margin: Calculations are recorded in Mountbatten, Letter Ref. CR 4204/43, 1943.

224   March 23, 1944, would be: Barry, Letter to the Secretary, 1943, 2.

225   "cold-blooded and unpleasant duty": Clogstoun-Willmott, Letter Ref. No. 26/1, 1943.

225 **"applied for a reversion . . .":** Clogstoun-Willmott, Letter Ref. No. 26/1, 1943, 1.

225 **"COPP 1 will become operational . . .":** Hastings, C.O.P.P. Report 33, 1943.

225 **X-craft training in November:** Hastings, C.O.P.P. Report 33, 1943.

226 **August 21, 1943, the idea:** Clogstoun-Willmott, Progress Report No. 29, 1943.

226 **On the twenty-seventh, Clogstoun-Willmott:** Willmott, COPP "Overlord" Report, 1943.

226 **On August 29, the thought:** Clogstoun-Willmott, Progress Report No. 30, 1943.

226 **Two days later, J. B. S. Haldane:** The test dated August 31, 1943, with notes on that date in UCL Lab, "Diving Notes," Part 3, June 1943–December 1944, and Notebook, September 1942–June 1944.

226 **the next day, Port HHZ:** Mitchell, *Tip of the Spear*, 1993, 67, describes the closure of the Loch and its associated roads by early September. The date of September 1, 1943, as the first day of the towing trials is in Grehan and Mace, *Capital Ships*, 2014, 61.

226 **COPP team 5 guided troops:** Stanbury, Report on Pilotage, 1943, 1.

226 **same day, Helen Spurway:** The tests dated September 1, 1943, were 511 and 512 on experimental day 240. Their records are on that date in UCL Lab, "Diving Notes," Part 3, June 1943–December 1944, and Notebook, September 1942–June 1944.

227 **would slowly release oxygen:** Ian Edward Fraser. Fraser, Interview, 1987, describes oxygen cylinders bleeding oxygen into the X-craft environment to keep it breathable.

227 **each minisub carried twelve:** Mitchell, *Tip of the Spear*, 1993, 126.

227 **"marvelous jobs":** Waldron and Gleeson, *The Frogmen*, 1955, 43.

227 **To remove the carbon dioxide:** Fraser, Interview, 1987, describes the boxes, which are also referenced in Hudspeth, Letter to Captain, 1944, 6.

228 **crewmembers opened new boxes . . . Siebe Gorman packed the boxes:** Hudspeth, Letter to Captain, 1944, 6.

228 **"special and hazardous duty":** *Times* (London), "Richard Kendall," 2006, and "Personal Tribute," 1945.

228 **"hazardous underwater work":** Honour, Interview, 1987.

228 **Hugh Michael Irwin:** Irwin's experience in the escape tower and his quotes are taken from his oral history at the Imperial War Museum. Irwin, Interview, 1987.

230 **An ascent from three feet:** Kumar et al., "Pulmonary Barotrauma," 1973; Lindblom and Tosterud, "Pulmonary Barotrauma," 2021.

231 **This was no small task; submarine escape training remains:** Edgar et al., "Case Series," 2021.

232 **Irwin and the others were sent:** Irwin was sent for full COPP training the week ending January 23, 1944, so his departure to Scotland is slightly out of the timeline here. Hastings, C.O.P.P. Report 51, 1944. He was working with X-craft there by March. P. W. Clark, Progress Report No. 1, 1944, 6.

232 **"No nation had ever towed . . .":** Broome, Interview, 1993.

232 **All six would go on the mission:** Warren and Benson, *Above Us the Waves*, 1953, 123.

232 **"W. and D.":** Warren and Benson, *Above Us the Waves*, 1953, 45.

232 **The divers had spent plenty:** Waldron and Gleeson, *The Frogmen*, 1955, 83.

233 **there had been oxygen casualties:** Waldron and Gleeson, *The Frogmen*, 1955, 28–29; Warren and Benson, *Above Us the Waves*, 1953, 122.

233 **under strict quarantine:** UCL Lab, Notebook, September 1942–June 1944, in notes for September 17, 1943.

233 **eight continuous days:** Warren and Benson, *Above Us the Waves*, 1953, 122.

233 **sprung from quarantine:** Those experiments took place on August 31, 1943; they are described in UCL Lab, "Diving Notes," Part 3, June 1943–December 1944, and Notebook, September 1942–June 1944, on that date.

233 **Chamber number three:** Spurway, "Personnel and Equipment," ca. 1940–44.

234 **X-9's tow rope broke:** W. Jameson, *Submariners VC*, 2004, 176.

234 **X-8's rope snapped:** Warren and Benson, *Above Us the Waves*, 1953, 123–25.

234 **"became impaled . . .":** Warren and Benson, *Above Us the Waves*, 1953, 127.

234 **Submarines do not work as well:** No, of course there's no citation for this sentence! Get back to reading. It's about to get dramatic.

234 **Later analysis of German documents:** Warren and Benson, *Above Us the Waves*, 1953, 144.

234 **"mountainous" . . . "not going to come back.":** Broome, Interview, 1993.

235 **X-6's periscope flooded:** Konstam, *Tirpitz in Norway*, 2019, 48.

235 **"So now we've had it . . .":** Warren and Benson, *Above Us the Waves*, 1953, 130.

235 **In the growing light:** Donald Cameron was the captain of the *X-6* and he told the story of this attack directly, including quotes, in Waldron and Gleeson, *The Frogmen*, 1955, 87.

235 **at 7:15 a.m.:** Warren and Benson, *Above Us the Waves*, 1953, 131.

236 **The X-7 got stuck sooner:** Godfrey Place commanded the *X-7*, and he told his story and provided the relevant quotes in Warren and Benson, *Above Us the Waves*, 1953, 129.

236 **"there was a tremendous explosion":** Warren and Benson, *Above Us the Waves*, 1953, 133.

236 **Sublieutenant Bob Aitken:** Aitken told his own story in Warren and Benson, *Above Us the Waves*, 1953, 134–35.

238 **Aitken, Godfrey Place, and the surviving crew:** Warren and Benson, *Above Us the Waves*, 1953, 135; Broome, Interview, 1993.

CHAPTER 15: Airs of Overlord

239 **Randy Lovelace's charm:** Unless otherwise cited, this story about Randy Lovelace is taken from his official report on the test, Lovelace, "High Altitude Flight Testing," 1944. Some of the visual details describing the appearance of Earth below a plane are filled in from the author's personal skydiving experiences.

239 **his mid-thirties:** William Randolph Lovelace II was born on December 30, 1907.

240 **Fighter pilots can get decompression:** Burbakk et al., *Bennett & Elliott's Physiology*, 2003, 461.

241 **Paralysis is rare but possible:** Gribble, "Comparison of the 'High-Altitude,'" 1960.

241 **previous wars had been limited:** Ferris and Engel, "Clinical Nature," 1951, 4–5.

241 **Lovelace and the others:** Seminal work in oxygen prebreathing was done during this era by Boothby and his colleagues. Boothby and Benson, "High Altitude," 1940, and Boothby and Lovelace, "Necessity for the Use," 1938.

241 **the military awarded him:** Department of Defense Office of Public Information, "Press Release," 1955.

241 **two teammate countries traded:** e.g., Yarbrough, Letter to Haldane, 1943.

242 **German physician Theodor Benzinger:** Hornberger, "Decompression Sickness," 1950, 354.

242 **Roughly six feet tall:** Benzinger, "Explosive Decompression," 1950, 397–99.

243 **"As this problem [of explosive decompression] . . .":** Benzinger, "Explosive Decompression," 1950, 395.

243 **"The American authors probably . . .":** Benzinger, "Explosive Decompression," 1950, 399.

243 **Above about thirteen thousand feet . . . less than fifteen seconds:** Benzinger, "Explosive Decompression," 1950, 401.

244 **Hans Kalmus also climbed into:** Test date August 27, 1943, experiment 507 on experimental day 238. The notes on his eye chart are in Helen Spurway's handwriting. J. B. S. Haldane, Papers, Part 1; UCL Lab, "Diving Notes," Part 3, June 1943–December 1944, Miscellany of Papers, ca. 1940–44, and Notebook, September 1942–June 1944.

244 **Kalmus and his family:** The lab notes specify that Kalmus had returned from a holiday on August 27, 1943. UCL Lab, Notebook, September 1942–June 1944. The rest of the information and photos of the holiday were provided by Peter Kalmus. P. Kalmus, Interview, 2020b.

244 **She had been hosting friends:** N. Mitchison, *Among You*, 2000, 120.

244 **even found a camera:** The photographs are still held by the Kalmus family, and the portrait of Nussy Kalmus in the small boat on page 60 of this book is from that trip.

245 **Six days later, Kalmus repeated:** Test date September 2, 1943. Kalmus was the subject for two experiments, 515 and 516, on experimental day 241. Notes are on that date in J. B. S. Haldane, Papers, Part 1; UCL Lab, Miscellany of Papers, ca. 1940–44, and "Diving Notes," Part 3, June 1943–December 1944, and Notebook, September 1942–June 1944.

245 **same test a few weeks later:** The test took place on November 18, 1943, and was experiment 563 on experimental day 263. Notes are on that date in UCL Lab, Miscellany of Papers, ca. 1940–44, and Notebook, September 1942–June 1944.

245 **different Germans did similar work:** The Germans' research on oxygen toxicity is evident through their parallel, independent development and use of oxygen rebreathers. MacIntosh and Paton, "Notes on the 'Tauchretter,'" 1950.

246 **"the only complete unit . . .":** Chief of Staff (X), "Note of a Meeting," 1943.

246 **The COPP training personnel would need:** Clogstoun-Willmott, Progress Report No. 32, 1943.

246 **"In connection with beach . . .":** Barry, Letter to the Secretary, 1943.

246 **thirty operational *Maiale*:** Fotali, Naval Cypher XD, 1943.

246 **"These midget submarines were . . .":** A. Hughes, "History of COPP," 1946, 7.

246 **Meanwhile, elsewhere in Norfolk House:** (signature illegible) for Chief of Staff, Underwater Obstacles, 1943.

247 **evidence the Germans had tried:** Admiralty, "Minutes of the Meeting," May 5, 1943, 5.

247 **Some personnel had tried to make:** Hallett, "Underwater Obstacles," 1942.

247 **"There is no user demand . . .":** Lieutenant Colonel R.E., "Some Notes," 1943.

247 **"The evolution of a technique . . .":** Hussey, COHQ Docket, October 26, 1942, point VII.

247 **throwing depth charges at them:** Hussey, COHQ Docket, September 23, 1942, point I.

247 **the biggest hurdle would turn out:** The argument over who would be responsible for the underwater obstacles is preserved in a lengthy series of documents in the National Archives of the United Kingdom, in file DEFE 2/963. They start on or before Hallett, "Underwater Obstacles," 1942, on June 17, 1942, and run through at least October 19, 1942, per Hussey, COHQ Docket, October 19, 1942, point V.

248 **Boom Commando Party:** Hussey, COHQ Docket, October 19, 1942, point I.

248 **Keith May Briggs:** Briggs's story and quotes are from his oral history recording, preserved at the Imperial War Museum. Briggs, Interview, 1996.

248 **Party would eventually be given:** Combined Operations Headquarters, "Minutes of Meeting," 1944.

248 **A series of reconstructed replicas:** R. M. Ellis, "Boom Commando Party Proposed," 1942, 3; Combined Operations Headquarters, "Minutes of Meeting," 1944, 3.

248   **loosely constructed Appledore:** Briggs, Interview, 1996.

249   **Firing bundles of detonating:** Admiralty, "Minutes of the Meeting," May 5, 1943, 5.

249   **Dragging long, thick tubes:** Admiralty, "Minutes of the Meeting," May 5, 1943, appendix D.

249   **Mines thrown downward:** Lieutenant Colonel R.E., "Some Notes," 1943.

249   **"close contact between . . .":** Admiralty, "Minutes of the Meeting," May 5, 1943, appendix D.

249   **camp had a pump to fill . . . ". . . photographs of Normandy":** Briggs, Interview, 1996.

249   **Across the ocean, the Americans:** RNPRC, "Memorandum on Human Blast Trials," 1948, references American blast research conducted at Fort Pierce. The UDT/SEAL Museum in Fort Pierce, Florida, reports lingering replica obstacles in their water for over fifty years after the end of the war.

249   **lengthy catalogues of techniques:** SHAEF, SHAEF/315GX/INT, 1944.

249   **ten-foot-tall:** The dimensions of Element C, along with the directions for its destruction, are in SHAEF, SHAEF/315GX/INT, 1944, appendix C.

251   **"Placing charges accurately . . .":** Hussey, COHQ Docket, September 23, 1942, point IV.

251   **By November 18, 1943, COPP 1:** COPP 1 left to join the 12th Submarine, who were in charge of the X-craft, on October 31, 1943. Lyne, COPP Depot Report 39, November 3, 1943.

252   **At 1:26 p.m. that same day:** This was experiment 563 on experimental day 263. Notes on that date in UCL Lab, Miscellany of Papers, ca. 1940–44, and Notebook, September 1942–June 1944.

252   **"in the absence . . .":** Brown, "Minutes of a Special Meeting," 1943.

253   **"Having had personal . . .":** Director of Scientific Research, "Proposal to Dispense," 1943.

253   **"The individual experimental . . ." . . . "working of the Admiralty Experimental Diving Unit":** Brown, "Minutes of a Special Meeting," 1943.

253   **years later, he would live:** A. Bamji, Interview, 2020.

254   **They decided to tell Haldane:** R. H. Davis, Letter to Bussell, 1943.

254   **"It is almost certain that . . .":** Assistant Controller (R&D), Letter to Darke, 1943.

254   **wanted to be able:** Director of Scientific Research, "Proposal to Dispense," 1943.

254   **UCL was starting:** UCL was paying Haldane's salary during the testing (R. H. Davis, Letter to Haldane, January 23, 1940). The government was asked repeatedly, from multiple sources, to start paying for UCL (Pye, Letter Adressed "My Dear Haldane," 1943). A representative with an illegible signature declared on October 16 that they would have to terminate Haldane's services if they were asked to pay ([signature illegible], Letter Addressed "Dear Pye," 1943).

254   **had until January 31, 1944:** Barnes, Letter to Haldane, December 21, 1943b.

254   **"I am, of course, very . . .":** J. B. S. Haldane, Letter to Barnes, 1943.

254   **traded results and analyses:** Yarbrough, Letter to Haldane, 1943; F. P. Ellis, "Personnel Research," 1950, 6.

255   **A month after the decision:** Brown, Letter Addressed "Dear Haldane," 1943.

CHAPTER 16: The Operations

256   **"from the point . . .":** Clogstoun-Willmott, Enclosure No. 1, 1944, 1.

256   **Nigel Clogstoun-Willmott hunkered down:** Except where otherwise cited, the facts of Operation KJH are from Clogstoun-Willmott, Enclosure No. 1, 1944.

256   **edge of turning twenty-four:** Scott-Bowden was born on February 21, 1920.

257   **Ogden-Smith had been born:** Ogden-Smith was born on April 19, 1918.

257 wearing rain gear: Clogstoun-Willmott, Enclosure No. 4, 1944, 7.

257 robust khaki designed: The need for a light color because of camouflage on beaches is in Director of Victualling, Enclosure, 1943; photos of the resultant khaki suit are in Clogstoun-Willmott, C.O.P.P. Training Book, 1943.

257 The rope soles: Director of Victualling, Enclosure, 1943.

257 suits even had pockets: Suits and equipment described in Clogstoun-Willmott, "Recon Report 'Postage Able,'" 1.

257 rest of the crew on board: Clogstoun-Willmott, Enclosure No. 2, 1944, 2.

257 everyone was seasick: Clogstoun-Willmott, Enclosure No. 2, 1944, 3.

257 spent too much time: Clogstoun-Willmott, Enclosure No. 1, 1944, 1.

257 practicing inside X-craft: Scott-Bowden, Letter Addressed "Dear Colonel Dove," 1943.

257 place a rush order: Siebe Gorman & Co. Ltd., Letter to Chief, 1943.

257 "the high death rate of suits": P. W. Clark, Progress Report No. 1, 1944, 2.

257 newly constructed X-20 had received: P. W. Clark, Progress Report No. 1, 1944, 3.

258 COPP 1 had been: P. W. Clark, Progress Report No. 1, 1944, 4.

258 Clogstoun-Willmott had barely recovered: P. W. Clark, Progress Report No. 1, 1944, 3.

258 "It had not, at this time . . .": P. W. Clark, Progress Report No 1, 1944, 3.

258 It was too high-risk: P. W. Clark, Progress Report No. 1, 1944, 3.

258 "policy with regard . . .": C in C Portsmouth, Operation "K.J.H.," 1943, 3.

258 By 2200 hours Friday: This paragraph is from Clogstoun-Willmott, Enclosure No. 1, 1944, 1–2. The operational reports do not specifically state that Clogstoun-Willmott was in charge of navigation for Operation KJH. However, he was the expert navigator of the crew, and he conducted the navigation for their other missions, such as Operation Postage Able. It, therefore, seems a reasonable conclusion that he was the one doing the navigating, which is supported by the level of detail he gives to navigation in his operational report.

258 three distinctive flashes: Little, "Operation 'Postage Able' Appendix I," 1944, 2.

258 The sight of the beam came: Clogstoun-Willmott, Enclosure No. 1, 1944, 1.

258 An hour before midnight . . . beach town of La Rivière: Clogstoun-Willmott, Enclosure No. 1, 1944, 2.

259 Scott-Bowden thought he felt: Willmott, Enclosure No. 4, 1944, 4.

259 By the stroke of midnight: Clogstoun-Willmott, Enclosure No. 4, 1944, 7.

259 The slowly spinning beacon . . . "A gap in the wire . . .": Clogstoun-Willmott, Enclosure No. 4, 1944, 6.

260 sliver of light broke: Clogstoun-Willmott, Enclosure No. 4, 1944, 7.

260 "a belt of clean sand . . .": Clogstoun-Willmott, Enclosure No. 4, 1944, 4.

260 Ogden-Smith huddled on Scott-Bowden's . . . "the visibility was very bad": Clogstoun-Willmott, Enclosure No. 4, 1944, 8.

260 "worn smooth and flat . . .": Clogstoun-Willmott, Enclosure No. 4, 1944, 5.

260 two scouts had been briefed: Admiralty, "Minutes of a Meeting," December 31, 1943, 1.

260 "We both observed . . ." . . . ". . . out by the flooding tide": Clogstoun-Willmott, Enclosure No. 4, 1944, 8.

261 "The important object was . . .": Admiralty, "Minutes of a Meeting," December 31, 1943, 1.

261 "On these operations depends . . .": Ramsay, Memo "Operations 'K.J.H.,'" 1944.

261 crew of Operation KJH returned: (signature illegible) Major, Letter C.M.P. to S.S. Group, 1944.

261 **moon would once again:** Clogstoun-Willmott, COPP Postage Able Report, 1944, Part II, Section II, 1.

262 **UCL lab conducted eleven:** Experiments were on January 7, 1944 (3 tests); January 11 (2 tests); January 12 (3 tests); and January 14 (3 tests). The dates are listed in UCL Lab, Miscellany of Papers, ca. 1940–44, with expanded information in UCL Lab, "Diving Notes," Part 3, June 1943–December 1944, and Notebook, September 1942–June 1944.

262 **"at present there are no . . .":** (signature illegible) for Chief of Staff, Underwater Obstacles, 1943.

262 **divers of the Boom Commando:** Combined Operations Headquarters, "Minutes of Meeting," 1944.

262 **On January 14, 1944, Haldane started:** The experiment on January 14, 1944, is cited above: UCL Lab, Miscellany of Papers, ca. 1940–44., and "Diving Notes," Part 3, June 1943–December 1944.

262 **HM Trawler *Darthema* towed:** Clogstoun-Willmott, COPP Postage Able Report, 1944, Part II, Section I, 1.

262 **"Carry out Operation Postage Able":** Little, "Operation 'Postage Able,'" 1944, 3.

263 **Clogstoun-Willmott was on board again:** The crew list is in Clogstoun-Willmott, COPP Postage Able Report, 1944, Part II, Section I, 1.

263 **Nobody had much hope:** The pessimism is woven through the operational planning. The beach scouts were supposed to write reports every night in case they did not return the next. A. Hughes, "History of COPP," 1946, appendix B, 6. The layers of plans for the sinking of the X-craft are in the next paragraph of this text, and cited with each statement.

263 **The team of five's command:** Little, "Operation 'Postage Able,'" 1944, 1.

263 **take a break on night four:** Clogstoun-Willmott, COPP Postage Able Report, 1944, Part II, Section I, 7.

263 **Aircraft would be flying . . . codes snuck into radio programs:** Little, "Operation 'Postage Able,'" 1944, 3.

263 **scout "the American sector":** Clogstoun-Willmott, COPP Postage Able Report, 1944, Part I, 1.

263 **"abundant intelligence material":** P. W. Clark, Progress Report No. 1, 1944, 3.

263 **At 2158, the *X-20*:** Unless otherwise cited, all details of the operation and quotes are from Clogstoun-Willmott, COPP Postage Able Report, 1944, Part II, Section I. Page numbers have been omitted because the document is brief.

265 **"about the thickness . . ." . . . ocean flotsam:** Clogstoun-Willmott, COPP Postage Able Report, 1944, Part II, Section VII, 1.

265 **Case and Haldane had advised:** J. B. S. Haldane and Case, Untitled report labeled "SECRET," 1940, 2.

265 **never escalated to machine guns:** Clogstoun-Willmott, COPP Postage Able Report, 1944, Part II, Section VII, 1.

266 **new box of Protosorb:** Hudspeth, Letter to Captain, 1944, 6.

266 **donned their protective khaki:** Clogstoun-Willmott, COPP Postage Able Report, 1944, Part II, Section III, 1.

266 **"success depends . . .":** Clogstoun-Willmott, C.O.P.P. Training Book, 1943, C.O.P.P. 10 (Close Combat).

266 **duo reached a water:** Unless otherwise cited the details of the swimmers on the beaches, and quotes, are taken from Clogstoun-Willmott, COPP Postage Able Report, 1944, Part II, Section III.

267 **"hard well compacted sand":** Clogstoun-Willmott, COPP Postage Able Report, 1944, Part II, Section IV, 1.

267 **"slow deliberate":** Clogstoun-Willmott, COPP Postage Able Report, 1944, Part I, 2.

267 **a hemp cable that:** Clogstoun-Willmott, COPP Postage Able Report, 1944, Part II, Section I, 4.

267 **Germans and farm animals:** e.g., Clogstoun-Willmott, COPP Postage Able Report, 1944, Part II, Section IV, 3.

267 **locations of pillboxes:** Clogstoun-Willmott, COPP Postage Able Report, 1944, Part II, Section V, 1.

268 **"Withdrew. Air was getting . . ." . . . ". . . improved on the casing":** Clogstoun-Willmott, COPP Postage Able Report, 1944, Part II, Section I, 5.

268 **They sometimes drugged themselves:** Clogstoun-Willmott, COPP Postage Able Report, 1944, Part II, Section I, 8.

268 **drinking far less water:** They drank six gallons over the mission, which equates to 0.77 liters per person per day. The recommended quantity when hiking is 0.5 liters per person per hour. Hudspeth, Letter to Captain, 1944, 7.

268 **"it would appear desirable . . .":** Hudspeth, Letter to Captain, 1944, 6.

268 **"Withdrew from shore . . .":** Hudspeth, Letter to Captain, 1944, 4.

268 **"After 11 hours submerged . . .":** Hudspeth, Letter to Captain, 1944, 5.

268 **"for a first attempt" . . . ". . . showed signs of distress":** Clogstoun-Willmott, COPP Postage Able Report, 1944, Part II, Section I, 7.

269 **"one of the most . . .":** Honour, Interview, 1987.

269 **Professors, government scientists:** Professor John Desmond Bernal, a noted Communist, was involved in the soil sampling program, and even trained Scott-Bowden in taking soil samples. Hastings, C.O.P.P. Report 58, 1944.

269 **would result in medals . . . "leadership":** UKNA, "Awards," 1943.

269 **one comment remarking on:** The comparison made in the comment was against the Holland submarine of forty years earlier. Little, "Report on Operation 'Postage Able,'" 1.

269 **Ursula Philip and Helen Spurway became the last:** The final known experiments, 605 and 606, were on January 28, 1944, experimental day 281. (Note: Five experiments were recorded without known dates, and I consider them to be experiments 607 to 611.) Records are in UCL Lab, Miscellany of Papers, ca. 1940–44, and "Diving Notes," Part 3, June 1943–December 1944, and Notebook, September 1942–June 1944.

269 **no longer a priority:** "[The Lords Commissioners] have decided that, in view of the present state of the investigations in hand for the Committee, such employment is no longer necessary." Barnes, Letter to Haldane, December 21, 1943a.

269 **William O. Shelford would lead:** Shelford, Letter to the Chairman, 1944.

270 **newly forming Naval Party:** The P-parties were preparing by January 24, 1944. Donald, Letter to the Secretary, January 24, 1944.

270 **participated in at least three:** Mould got in the Siebe Gorman chambers on December 16, 1942; August 31, 1943; and September 17, 1943. UCL Lab, Notebook, September 1942–June 1944, and "Diving Notes," Part 3, June 1943–December 1944. He also served as the in-chamber attendant for others during tests with Royal Navy divers on April 28, 1943 (twice); May 3, 1943; and May 4, 1943. Those dates are not counted in his tally, as he was not the subject of the experiment. Donald, "Report Concerning the M.R.S. Suit," 1943, 2–5.

270 **Mould and his burgeoning P-parties:** F. P. Ellis, "Personnel Research," 1950, 15.

270 **"Variation in the Oxygen Tolerance . . .":** Donald, "Variation in the Oxygen Tolerance," 1944.

271 **would practice serving:** They practiced and refined their techniques for serving as beacons over several months, e.g., testing new types of transmitter beacons. Lyne,

C.O.P.P. 9 Progress Report Number 2, 1944. The biggest practice event was Exercise Fabius, which was a full-practice landing. Lyne, C.O.P.P. 9 Progress Report Number 4, 1944.

271 **team of Operation Postage Able had:** Clogstoun-Willmott, COPP Postage Able Report, 1944, Part II, Section VI, 1.

271 **At Omaha, small mistakes:** Clogstoun-Willmott, COPP Postage Able Report, 1944, Part II, Section IV, 1–9.

271 **Each planned beach location threaded:** Office of Naval Commander, "Force 'S' Orders," S.8B, 1944, appendix II, 1–2.

271 **Americans declined the service:** Galwey, "Geoff's Opus," 1992, 118; Trenowden, *Stealthily by Night*, 1997, 140; Strutton and Pearson, *Secret Invaders*, 1958, 209–10. Rob Crane of coppsurvey.uk deserves credit for compiling the sources for this point, and his website provides a more nuanced discussion of the rejection at coppsurvey.uk /June-1944/#omaha.

271 **On February 25, 1944:** P. W. Clark, Progress Report No. 1, 1944, 5.

271 **"repeatedly risked their lives . . .":** J. B. S. Haldane, *A.R.P.*, 1938, 97.

CHAPTER 17: The Longest Days

272 **on the beaches of Normandy:** Honour, Interview, 1987.

272 **Watching over the festivities from behind:** This description of Sword beach was developed through the review of many videos and photographs of Sword on D-Day, preserved and made available online by the Imperial War Museum.

273 **George Butler Honour watched:** The details of Honour's story, unless otherwise cited, are from his oral history at the Imperial War Museum. Honour, Interview, 1987.

273 **radar operator Albert Bovill:** Bovill arrived in Portsmouth on June 4, in parallel with Honour watching the Germans onshore in Normandy. Bovill, Interview, 2002.

273 **British household custom:** This custom and Gueritz's quotes and perspective are from Gueritz, Interview, 1997.

274 **most had never been under:** Of Force "S," per the official documents, "the vast majority of them under fire for the first time in their lives." Talbot, "Operation Neptune," 1944, phase II, point 104.

274 **Gueritz was twenty-four . . . beachmaster for Sword:** Gueritz, Interview, 1997.

275 **Honour and the rest of the crew:** The names of the *X-23* crew in Operation Gambit are from Honour, "Force 'S' Report," 1944.

275 **just before midnight:** Lyne, "Force 'S' Report," 1944.

275 **the crew had to turn off:** Honour, "Force 'S' Report," 1944.

275 **Honour worried about the oxygen:** Honour, Interview, 1987.

275 **had to stay submerged:** Lyne, "Force 'S' Report," 1944.

275 **They had food:** Honour, "Force 'S' Report," 1944.

276 **Lionel Geoffrey Lyne was a:** Lyne, "Force 'S' Report," 1944; C.O.P.P. 9 Progress Report Number 2, 1944; C.O.P.P. 9 Progress Report Number 4, 1944.

276 **Ken Hudspeth, the pilot:** Waldron and Gleeson, *The Frogmen*, 1955, 94.

276 **described as "difficult":** Office of Naval Commander, "Force 'S' Orders," S.18, 1944, 4.

276 **In between lay pillboxes:** Office of Naval Commander, "Force 'S' Orders," S.8B, 1944, appendix II, 1–7.

276 **the shoreline was backed:** Talbot, "Operation Neptune," 1944, phase II, point 33.

276 **Gun batteries pockmarked:** Office of Naval Commander, "Force 'S' Orders," S.7B, 1944, appendix II, 1.

277 **exact location as determined:** Talbot, "Operation Neptune," 1944, phase II, point 33.

277 **"absolutely bang on":** Honour, Interview, 1987.

277 **downward thirty-five feet:** The water depth was taken from bathymetric maps of the area, and based on the location and state of the tide.

277 **June 5, just before midnight:** Time 2315. Honour, "Force 'S' Report," 1944.

277 **crew of the *X-23* resubmerged:** Lyne, "Force 'S' Report," 1944.

277 **"See you in Paris" . . . "Before the war . . .":** Lunn, Interview, 2004.

277 **One group of the Fighting:** Talbot, "Operation Neptune," 1944, phase II, point 49.

277 **was of searchlight crews:** Bellows, Interview, 1992.

277 **Once under way, they:** The fact that the troops were not told their destination until they were under way has become a well-documented part of D-Day history. However, it is interesting that the delay in the information was such a poignant part of the experience that many mention it even decades later in their oral histories, e.g., Oakley, Interview, 1997.

277 **When the moon reached:** The timing and phase of the moon were taken from June 6, 1944, on skyandtelescope.org.

277 **first airborne troops began:** Bando, *101st Airborne*, 2007, 42.

278 **smallest of the troop-carrying:** Office of Naval Commander, "Force 'S' Orders," S.7B, 1944, 1.

278 **Someone vomited on Edward:** Gueritz, Interview, 1997.

278 **ship crews had been ordered:** Office of Naval Commander, "Force 'S' Orders," S.7A, 1944, 1.

278 **to keep radio silence:** Referred to in the parlance of the time as "wireless silence." Office of Naval Commander, "Force 'S' Orders," S.7A, 1944, 4.

278 **command ship LCH 185:** Office of Naval Commander, "Force 'S' Orders," S.1, 1944, appendix I, 2.

278 **At 0340 that morning:** Talbot, "Operation Neptune," 1944, point 93.

278 **"the whole horizon . . .":** Gueritz, Interview, 1997.

278 **citizen Michele Grimond thought:** Wheeler, *D-Day*, 1994, time point 27:50.

279 **ships of the convoy had:** Office of Naval Commander, "Force 'S' Orders," S.7B, 1944, 1.

279 **between there and the shore:** The distance from shore for the launching of the DD tanks was determined on the day of by DSOAG "One" with Brigadier Commanding the 27th Armoured Brigade. Office of Naval Commander, "Force 'S' Orders," S.7B, 1944, 2.

279 **about three dozen:** Thirty-four DD tanks were successfully launched at Sword. Talbot, "Operation Neptune," 1944, phase II, point 110.

279 **from a cluster of inventions:** My favorite explanation of Hobart's Funnies is by the Imperial War Museum at https://www.iwm.org.uk/history/the-funny-tanks-of-d -day (accessed 2023).

279 **Many of the vessels in:** For example, Don Amer along with D. G. W. Slater and F. Phillips, who were all of COPP 6, led Team A of the DD tanks onto the beach from LCP(L) 189 in Group I. P. G. Wild, J. J. Watson, J. H. Bowden, and L. D. D. Hunt of COPP 6 traveled in LCP(L) 197 and led in the "E" squadron. Amer, Appendix V, 1944, 240; Talbot, "Operation Neptune," 1944, phase II, point 106.

279 **tanks would form a:** See maps in Office of Naval Commander, "Force 'S' Orders," S.7A, 1944.

279 **If the tank-launching process:** Office of Naval Commander, "Force 'S' Orders," S.7B, 1944, 2.

279 **At seven minutes past:** Honour, "Force 'S' Report," 1944.

279 **Their gyro compass had:** Lyne, "Force 'S' Report," 1944.

279 **did not feel confident enough:** Talbot, "Operation Neptune," 1944, phase II, point 41.

279 **The official name for:** Office of Naval Commander, "Force 'S' Orders," S.7B, 1944, 3.

279 **One of the X-men clambered:** Honour, "Force 'S' Report," 1944; Lyne, "Force 'S' Report," 1944.

280 **each of them got washed off:** Talbot, "Operation Neptune," 1944, phase II, point 44.

280 **return fire from the Wehrmacht . . . "events started to move swiftly.":** Talbot, "Operation Neptune," 1944, phase II, point 94.

280 **"streaked low across . . .":** Talbot, "Operation Neptune," 1944, phase II, point 95.

281 **Three snuck in . . . crew jumped into the water:** Talbot, "Operation Neptune," 1944, phase II, points 95–96.

281 **The freshly mangled pieces:** The torpedo attack that sank the *Svenner* was at 0530 and full sunrise was at 0557.

281 **tanks rolled smoothly out:** The launch of the DDs was successful per Talbot, "Operation Neptune," 1944, point 110.

281 **DD tanks of Group I:** Office of Naval Commander, "Force 'S' Orders," S.7B, 1944, appendix II, 2.

282 **Return fire flashed . . . "set [the house] well alight":** This happened sometime in the time window from 0620 to 0725, so its placement here in the sequence is approximate. Talbot, "Operation Neptune," 1944, phase II, point 123.

282 **When day broke fully:** Lyne, "Force 'S' Report," 1944, point 10.

282 **Royal Marine Donald B. King:** Wheeler, *D-Day*, 1994.

282 **Fellow Marine Bill Powell:** Wheeler, *D-Day*, 1994, time point 31:20.

282 **At 0633:** The schedule was earlier, with an H-hour of 0630 on Omaha, because the differing tides meant the clearance teams would have less time to work on the obstacles when they were in the dry (Commander Assault Force "O," Omaha Report, June 1944, 4-3 through 4-4). The Americans wanted to land even earlier on Omaha and Utah for obstacle clearance, but the Allies decided as a group to limit the spread of the landing times to within one hour. (Commander Assault Force "O," C. C. Mann, "Operation Instructions," 1942, 4-3.) The first personnel landed at Omaha at 0633 (C in C Pacific, "Naval Combat Demolition Units," 1944), and all NCDU/Army gap-clearance teams landed between 0633 and 0635, making them some of the earliest, if not the first, troops on foot onshore. These teams were included in waves O-1 and O-2, the first planned to land (C in C Pacific, "Naval Combat Demolition Units," 1944). The DD tanks were supposed to land starting at H-10, but because of the state of the sea, the landing order got mixed and the first landings overall occurred simultaneously in the 0635 time frame (Commander Assault Force "O," C. C. Mann, "Operation Instructions" 1942, 4-4).

282 **many either foundered:** Hall, Action Report, July 27, 1944, 7.

282 **At 0655, Sword beach's first:** Lyne, "Force 'S' Report," 1944, point 10.

282 **ten flat-bottomed, troop-carrying:** Office of Naval Commander, "Force 'S' Orders," S.7B, 1944, appendix III.

282 **Troops clinging:** The density with which troops were packed into many vessels is evident from photos and films of the landings, for example, the photograph printed at the end of chapter 17.

282 **"Many of you will be . . .":** Oakley, Interview, 1997.

282 **Gueritz rode behind the early lines:** Office of Naval Commander, "Force 'S' Orders," S.1, 1944, appendix I, 2, for Gueritz on LCH 185. LCH 185 was part of Group I per Talbot, "Operation Neptune," 1944, phase II, point 106.

283 **Two navigational ships:** Office of Naval Commander, "Force 'S' Orders," S.7B, 1944, appendix I.

283 **Three of the gunboats would:** Office of Naval Commander, "Force 'S' Orders," S.7B, 1944, appendix III.

283　**first DD line was supposed:** Office of Naval Commander, "Force 'S' Orders," S.7B, 1944, appendix II, 3.

283　**The tanks yawed heavily:** Talbot, "Operation Neptune," 1944, phase II, point 117.

283　**canvas waterproof coatings:** Talbot, "Operation Neptune," 1944, phase II, point 111.

283　**Thirty-four of the DDs:** Talbot, "Operation Neptune," 1944, phase II, point 110.

283　**Many of the tank crews had:** Quick and Shelford, "History of CCUBAs," 1970, 12. Images of tank crews with their DSEAs are preserved by the Imperial War Museum, e.g., War Office, Amphibious Tanks, 1944, https://www.iwm.org.uk/collections /item/object/205507085.

283　**Thirty-one tanks made it:** Talbot, "Operation Neptune," 1944, phase II, point 116.

283　**By 0730 the beach was:** Talbot, "Operation Neptune," 1944, phase II, point 130.

283　**The vessels offshore shifted:** Talbot, "Operation Neptune," 1944, phase II, point 124.

283　**More German gun emplacements:** Talbot, "Operation Neptune," 1944, phase II, points 118 and 137.

283　**Many of those gun sites:** Talbot, "Operation Neptune," 1944, phase II, points 118 and 144.

283　**To Gueritz, riding on the final:** Gueritz, Interview, 1997.

284　**As of Operation Postage Able:** Clogstoun-Willmott, COPP Postage Able Report, 1944, Part II, Section V, 2.

284　**shallows held ten-foot:** Talbot, "Operation Neptune," 1944, phase II, point 149.

284　**Element C was scarce on Sword:** Talbot, "Operation Neptune," 1944, phase II, point 151.

284　**"nasty waves":** Gueritz, Interview, 1997.

284　**bodyguard, Kenneth George Oakley:** Oakley told his own story in an oral history preserved at the Imperial War Museum. Oakley, Interview, 1997. His details and quotes are from that recording unless otherwise cited.

284　**took place amid what would:** Talbot, "Operation Neptune," 1944, phase II, point 135; Bush, Enclosure 1 to Report 6644, 1944, 170.

284　**Temporary Lieutenant J. O. Thomas had:** Talbot, "Operation Neptune," 1944, phase II, point 101.

284　**complete with the lyrics:** The song is available via YouTube.

284　**A bugler on another craft:** The song was "Cookhouse," not to be confused with the modern band. It contains the lines "Old soldiers never die, / They just fade away." Talbot, "Operation Neptune," 1944, phase II, point 101.

284　**personal bagpiper:** Oakley, Interview, 1997.

284　**Group II was hot on their tails:** Talbot, "Operation Neptune," 1944, phase II, points 128 and 135.

285　**Group II included landing craft:** Talbot, "Operation Neptune," 1944, phase II, point 128; Bush, Enclosure 1 to Report 6644, 1944, 191.

285　**number 947 was packed:** Bush, Enclosure 1 to Report 6644, 1944, 192.

285　**Crews sat inside their tanks:** Bellows, Interview, 1992.

285　**Before the departure from England:** Fielder, Interview, 1993.

285　**Only one tank crew:** Talbot, "Operation Neptune," 1944, phase II, point 132.

285　**It was hit in the turret:** The story of the explosion of LCT 947 and the death of Colonel Cocks is compiled from Talbot, "Operation Neptune," 1944, phase II, point 132; Bush, Enclosure 1 to Report 6644, 1944, 192; R. S. James, "Operation 'Neptune,'" 1944, 321; and Fielder, Interview, 1993.

285　**Edward Gueritz splashed off:** Unless otherwise cited, the details of Gueritz and Oakley's mad dash up the shore are compiled from their oral history recordings at the Imperial War Museum. Their narratives agreed and the details used were supported by fact-checking against contemporary records, but each oral history provided different, complementary information. Gueritz, Interview, 1997, and Oakley, Interview, 1997.

285 **tide was beginning to come:** Talbot, "Operation Neptune," 1944, phase II, point 154.

286 **"The horror of it was . . .":** Wheeler, *D-Day*, 1994, quote spoken by Siggy Ehinger, quartermaster, Royal Winnipeg Rifles, time stamp 1:15:00.

286 **"get off the bloody beach.":** Bellows, Interview, 1992.

286 **Frederic Ashcroft was moving:** Ashcroft, Interview, 1994.

286 **Private Bill Gray saw a man:** Wheeler, *D-Day*, 1994, unidentified veteran speaker, time stamp 36:00. The voice speaking this story matches the voice of Bill Gray, named elsewhere in the film.

286 **"the important thing [was] . . .":** Wheeler, *D-Day*, 1994, Charles Wheeler, time stamp 33:20.

286 **The rich smell of earth:** Wheeler, *D-Day*, 1994, Sergeant Leonard Daniels, the Parachute Regiment, time stamp 25:30.

287 **As landing craft continued:** Talbot, "Operation Neptune," 1944, phase II, points 170 and 173.

287 **scheduled touchdown time:** Office of Naval Commander, "Force 'S' Orders," S.7B, 1944, appendix I.

287 **had been distributed:** Office of Naval Commander, "Force 'S' Orders," S.7B, 1944, multiple locations.

287 **Some of the vessels carrying:** Talbot, "Operation Neptune," 1944, phase II, point 100.

287 **LCOCU bolted ashore:** Talbot, "Operation Neptune," 1944, phase II, points 161, 162, 166a, and 173.

287 **Briggs, the seasick British:** Unless otherwise cited, quotes and story of Briggs are taken from his oral history at the Imperial War Museum. Briggs, Interview, 1996.

287 **Once the density:** C in C Pacific, Lecture Notes, September 17, 1944; Hall, Action Report, July 27, 1944.

288 **"mechanical means":** Briggs, Interview, 1996.

288 **Lieutenant Harold Hargreaves:** Waldron and Gleeson, *The Frogmen*, 1955, 97–99.

288 **With him were bank clerks:** Waldron and Gleeson, *The Frogmen*, 1955, 95.

288 **"the obstacles were being . . .":** Waldron and Gleeson, *The Frogmen*, 1955, 95.

289 **kapok blast jackets developed:** H. C. Wright, "Review of Underwater Blast Research," 1951; H. C. Wright, "Investigations by RNPL," 1947.

289 **One petty officer was working:** Waldron and Gleeson, *The Frogmen*, 1955, 98–99.

289 **it is likely that the kapok:** While this incident has become something of a legend, and I typically prefer to trace legends back to multiple reliable sources before repeating them, experiments with these vests and suits showed a marked reduction in the risk of underwater blast trauma and fatality to animals. F. P. Ellis, "Present Position," 1944, 3. Another, similar incident also occurred and is better documented. A sailor wearing a diving kapok vest was talking to three other sailors wearing only smaller life belts when a blast went off. The vest-wearing sailor survived while the other three died instantaneously. F. P. Ellis, "Present Position," 1944, 6–7.

289 **Hugh Michael Irwin:** Irwin, Interview, 1987.

289 **Infantryman George Nicholson:** Nicholson, Interview, 1999.

289 **The LCOCU divers were not:** Talbot, "Operation Neptune," 1944, phase II, point 161; Bush, Enclosure 1 to Report 6644, 1944, 209.

289 **the X-craft crew was finished:** Honour, "Force 'S' Report," 1944, and Interview, 1987.

290 **By 0900, the rate:** Talbot, "Operation Neptune," 1944, phase II, points 161 and 170; Bush, Enclosure 1 to Report 6644, 1944, 170.

290 **British Number 4 Commando, many:** Number 4 Commando was part of the 4 SS Brigade, which was scheduled to go ashore with the British 8th Infantry Brigade. Office of Naval Commander, "Force 'S' Orders," S.7B, 1944, 3. Gordon Geoffrey Henry Webb was at both Dieppe and Normandy on D-Day (Webb, Interview, 1989), as were Peter Lincoln Fussell (Fussell, Interview, 1991), Bernard Kenneth Davies (Davies, Interview, 1989), and others.

290 **By 0906, the congested beach:** Talbot, "Operation Neptune," 1944, phase II, point 173; Gueritz, Interview, 1997.

290 **"runnels exist; near low . . .":** Clogstoun-Willmott, COPP Postage Able Report, 1944, Part I, Condensed Report.

291 **They had described at length:** Clogstoun-Willmott, COPP Postage Able Report, 1944, Part II, Section IV, 4–5.

291 **"featureless to a stranger . . .":** Clogstoun-Willmott, COPP Postage Able Report, 1944, Part II, Section VI, 1.

291 **troops at Omaha had deviated:** The navigational errors and causes are discussed in detail in Kilvert-Jones, *Omaha Beach*, 1999.

291 **scheduled bombing onshore:** Hall, World Today, July 27, 1944, 5.

291 **"the order of landing . . .":** Hall, World Today, July 27, 1944, 8.

291 **The order was supposed:** Hall, World Today, July 27, 1944, 7.

291 **The NCDUs landed:** Gibbons, World Today, June 18, 1944, 2–4.

292 **Gueritz and the other beachmasters:** Talbot, "Operation Neptune," 1944, phase II, point 167; Leggatt, "Report of NOIC," 1944, 25–53.

292 **sometimes through hand-to-hand:** *Wheeler*, D-Day, 1994, Raimund Steiner, time stamp 24:30.

292 **By 1300 that day:** Featherstone, *Wargaming*, 1979, 88.

292 **By 1555, Sword beach was:** Talbot, "Operation Neptune," 1944, phase II, point 173.

292 **Flail tanks with rollers . . . more Allies flood in:** Nicholson, Interview, 1999.

293 **had planned from the start:** Commander Assault Force "O," Omaha Report, June 1944, 4-2.

293 **paratrooper boots because:** Gawne, Naval Beach Battalions, 1991, 2.

293 **They brought their diving equipment:** C in C Pacific, Lecture Notes, September 17, 1944, 8–9.

293 **amphibious masks nestled:** Gawne, Naval Beach Battalions, 1991, 2.

293 **"far lighter than in the . . .":** C in C Pacific, "Naval Combat Demolition Units," 1944, 12.

293 **Around 2250, the Allies:** Talbot, "Operation Neptune," 1944, phase II, point 181.

293 **At 0730 the next morning:** Honour, "Force 'S' Report," 1944.

293 **the people of Portsmouth:** Gueritz, Interview, 1997.

294 **John Bridge had cleared:** Bridge's story, except where otherwise cited, is from his oral history records at the Imperial War Museum. Bridge, Interview, 1995.

294 **carried two gas mixtures:** F. P. Ellis, "Personnel Research," 1950, 15.

294 **75 percent oxygen blend:** Donald, "Report on the Use of Mixtures," 1943, 1.

294 **in a secret rush:** The proposed scheme for the P-parties was outlined November 6, 1943, leaving little time for training. Chief of Staff to Naval Commander Expeditionary Force, Proposed Scheme, 1943. All of the related documents were classified, indicating the secrecy of the program.

295 **Young divers had been picked:** Grosvenor and Bates, *Open the Ports*, 1956, 39.

295 **divers laid out:** Grosvenor and Bates, *Open the Ports*, 1956, 72.

295 **gave himself credit:** Donald's credit grabbing is grotesque and obvious once the documents related to diving are laid out together. Whenever a document was inter-

nal to the government, Donald would take full credit for all the innovation. For example, Committee, Admiralty Diving, Admiralty Fleet Order, 1944, uses a UCL lab report with the implication that it was generated by Admiralty Diving personnel, meaning Donald. In Donald, "Comparison of Wet," 1943, Donald claims that his investigations, rather than those of the UCL lab and Elizabeth Jermyn, discovered that there was a difference in oxygen toxicity when diving between the wet and dry. Meanwhile, only five months earlier, but in a document that Haldane would have been more likely to see, Donald gave the UCL lab and Jermyn credit for the same discovery (Donald, "Continued Report," 1942). Donald is on record as protesting when Haldane gave credit for diving discoveries to the UCL lab, to the point that one intermediary instead suggested they remove all attributions for any discoveries to avoid the conflict (Brown, Letter to Haldane, 1944). When documents were written by other people who were present for the work, they included the UCL lab, e.g., F. P. Ellis, "Personnel Research," 1950, 12–15. *Oxygen and the Diver* is the most egregious example of Donald's attempt to take credit for everything during the time period, and he could do so without pushback since he published the book after the deaths of most of the UCL lab members but while most of the reports were still classified.

295   **Helen Spurway made their risk curves . . . he took the credit:** Donald's original report, with the disordered heap of tables and numbers, is Donald, "Comparison of Wet," 1943. The compilation of his raw data into meaningful, mathematically based risk curves can be traced back to Helen Spurway in Donald, "Variation in the Oxygen Tolerance," 1944, where they are clearly labeled as the creations of Spurway and Haldane in the appendix. While both Spurway and Haldane worked on the statistics, the majority of the calculations are in Spurway's handwriting (J. B. S. Haldane and Spurway, "Physiology of Diving," 1944). Donald republished her work as his own in his 1992 book, *Oxygen and the Diver,* with no citations or indications that he was plagiarizing her descriptions of the analysis verbatim, in addition to her results in the form of the curves. It is important to emphasize that the conversion of raw data to risk curves is neither a small nor an insignificant step, as it requires the use of advanced statistical methods. In the time before computers, that conversion required a massive amount of work, and without it, the raw data provided at best a loose ballpark of the levels of danger. Today, the creation of risk curves from preexisting data sets is considered to be a contribution worthy of separate academic publication, so what Donald did is inarguably plagiarism of Spurway's results, and his actions should not be minimized because she began with his reported data.

295   **John Bridge of Messina's harbor:** Bridge, Interview, 1995.

CHAPTER 18: The Unremembered

299   **Acting Temporary Lieutenant Ronald Henry Saull:** Except where otherwise cited, Saull's story is compiled from the documents in his military records (Royal Navy, Saull Military Records, 1940–45) and the paperwork for his commendation (Hayes, "Recommendation for Decoration," 1944).

299   **Saull was a gangly, homely:** Photograph clipping provided by Hampstead, England, historical research group.

299   **His partner-in-munitions:** Hayes, "Recommendation for Decoration," 1944; (signature illegible), Work of the "P" Parties, 1945.

299   **Saull was an expert in:** This method of dismantling the obstacles, as used by the Royal Navy, is described in Admiralty, "Part III—Clearance of Obstacles," 1943.

300 **one of so many who:** In the Force "S" orders, the officers are listed, and a large number have elevated Acting positions. Office of Naval Commander, "Force 'S' Orders," S.1, 1944.

300 **recycled oxygen-rich gas:** F. P. Ellis, "Personnel Research," 1950.

300 **"the saving of the lock . . .":** Rear Admiral James, "Remarks," 1944.

300 **"As each port was occupied . . .":** (signature illegible), Work of the "P" Parties, 1945.

300 **awards for Kirkland and Saull's dangerous work:** Authority, "Central Chancery," 1945.

300 **three days before the announcement:** Admiralty, "Admiralty and Predecessors," 1945.

301 **buried together in the Rotterdam:** Commonwealth War Graves, "Sub-Lieutenant Donald Kirkland" and "Lieutenant Ronald Henry Saull."

301 **The troops storming the beaches:** This story is in Grünerberg's oral history (H. Grüneberg, Interview, 1979), and it is supported by his archived Admiralty reports regarding shrapnel distribution; they confirm that he was their expert in that area, e.g., H. Grüneberg et al., "Relations Between Antipersonnel," 1943–44.

301 **weaponry into London telephone:** Lewis and Hunt, "Hans Grüneberg," 1984, 240.

301 **used British Army case reports . . . ". . . body armor was stopped":** H. Grüneberg, Interview, 1979.

302 **nickname "Super Leech":** Lewis and Hunt, "Hans Grüneberg," 1984.

302 **more than fifteen thousand pints:** H. Grüneberg, Interview, 1979.

302 **been denied refugee status:** Zamet, "Refugee Dentists," 2007, 277.

302 **exempted from internment:** Home Office, "Exemption (Elsbeth Grüneberg)," 1939, and "Exemption (Hans Grüneberg)," 1939.

302 **suicide in April 1944:** High Court of Justice in England, Elsbeth Grüneberg Probate, 1944, 661; R. Grüneberg, Interview, 2020. These details have been included only with the express permission of the surviving family members.

302 **On June 26, 1944, only:** H. Grüneberg, Letter to Haldane, 1944.

302 **wrote several books about:** e.g., H. Grüneberg, *Genetics of the Mouse*, 1943; H. Grüneberg and Werner, *Moderne Biologie*, 1950.

302 **establish the humble mouse:** Lewis and Hunt, "Hans Grüneberg," 1984.

302 **after his seventy-fifth birthday:** Levine, *Strategic Bombing*, 1992.

303 **Reuben remembers:** R. Grüneberg, Interview, 2020.

303 **brothers and her parents:** C. Wright, Interview, 2020.

303 **younger brother Manfred was:** Stoller, Email, 2020.

303 **turned eighteen in late:** Selective Service System, Registration Card, 1943.

303 **Ursula Philip—the warm:** This story and the quotes are from Ursula Philip's daughter, Cathie Wright. C. Wright, Interview, 2020.

303 **womanizing, pro-nudist, free-love-advocating Minchin:** Minchin, *My Footsteps*, 1997.

304 **Ursula Philip became:** Her naturalization was granted on November 4, 1946. Philip, Letter to Ms. Ursell, 1946.

304 **pregnancy-related medical issues:** C. Wright, Interview, 2020.

304 **calling her lazy:** C. Wright, Interview, 2020; his general abuse of Philip is also documented in Keyfitz, *Notes of a Wayfarer*, 2004, 96.

304 **downgrading her position:** J. B. S. Haldane, Letter to the Provost, 1946, and Staff Form 3, 1946; Pye, Letter to Kalmus, 1946.

304 **nasty letters:** J. B. S. Haldane, Letter to the Provost, 1946.

304 **Grüneberg a raise in 1934:** H. Grüneberg, Interview, 1979.

304 **he fought for Grüneberg:** J. B. S. Haldane, Letter to Grüneberg, 1945.

304   **Philip and her young daughter:** C. Wright, Interview, 2020.

304   **scientific lecturer at Newcastle:** Philip, Letter to Ms. Ursell, 1946.

304   **words like "giggilation":** C. Wright, Interview, 2020.

304   **came up with a statistical test:** Philip and Haldane, "Relative Sexuality," 1939.

304   **her methods are still:** Loadsman and McCulloch, "Wielding the Search," 2017.

305   **discovery of concentration camps forced:** Ursula Philip, who certainly would have been imprisoned in a concentration camp had she not been granted academic immigration status in England, wrote a discussion of the German attitude to eugenics and its evolution as of twenty years after the war. Philip, "Genetics, and Eugenics," 1964.

305   **refugees of the UCL lab:** H. Grüneberg, Interview, 1979.

306   **Nussy Kalmus's brother:** G. Kalmus, Interview, 2020a.

306   **Hans and Nussy Kalmus did not know:** H. Kalmus, *Odyssey of a Scientist*, 1991, 55.

306   **Eventually, he did the same:** M. Kalmus, Interview, 2021.

306   **thanks in part to encouragement:** Roe, Interview, 2020.

307   **adult Peter reported that:** P. Kalmus, Interview, 2020b.

307   **immigrant families, Hans:** Home Office, Naturalisation Certificate, 1946. The naturalization certificates are filed in sequential order by date, so the number of families on that day was counted from the records in that file.

307   **Haldane had written in:** J. B. S. Haldane, Letter to the Secretery, SPSL, 1946.

307   **"I remember Prof. Haldane . . .":** (signature illegible), Note on inside cover, 1946.

307   **textbook about genetics:** H. Kalmus, *Genetics*, 1952.

307   **atypically lost:** M. Kalmus, Interview, 2021.

307   **His granddaughter fondly remembered:** M. Kalmus, Interview, 2021.

307   **Jim and Joan Rendel also lived:** Franklin et al., "James Meadows Rendel," 2004.

307   **would respond with shame:** J. Robertson, Interview, 2020.

308   **he ran the genetics unit:** Franklin et al., "James Meadows Rendel," 2004.

308   **"I still keep thinking . . .":** Rendel, Letter to Mitchison, April 19, 1941.

308   **glimmers of schizophrenia:** Joan Rendel's granddaughter, Celia Robertson, chronicled Joan's struggle with schizophrenia in a more complete, sensitive, and loving fashion than I could ever replicate in summary in this book. C. Robertson, *Who Was Sophie?*, 2008.

308   **put together roughly a half dozen:** To provide an incomplete list: H. C. Wright, "Subjective Effects," 1947; "Investigations by RNPL," 1947; "Review of Underwater Blast Research," 1951; "Review of Underwater Blast Research," 1957; "Injurious Effects," ca. 1950; "Underwater Blast Trials," 1949.

308   **most dramatic experimental series:** H. C. Wright, "Subjective Effects," 1947.

309   **lack of a medical degree:** Amoroso, "Memorial," 1979.

309   **His former colleagues at:** Bennett, Interview, 2021.

309   **At one point, the brusque:** Amoroso, "Memorial," 1979.

309   **his estate was distributed:** Nowlan, Interview, 2020.

309   **died on April Fools' Day:** Amoroso, "Memorial," 1979.

309   **Robert Alfred Martin Case ended up:** J. F. M. L. Anderson, Letter to Osborn, 1944; Director of Scientific Research, Letter to DNI, 1944.

309   **where, while using an oxygen rebreather:** His adventure was chronicled in multiple newspapers, e.g., Pincher, "Man to Swim Against Whale," 1947.

310   **Brother Ralph Case became a physician:** *British Medical Journal*, "Cases Previously Postponed," 1956, 356.

310   **one particularly notable mission:** Warren and Benson, *Above Us the Waves*, 1953, 228–33.

310   **"dropped Communism like 'a hot cake'":** J. E. Bamji, Letter #2 to Cramer, ca. 2007.

310   **she wanted to be a physician:** A. Bamji, Interview, 2020; J. E. Bamji, Letter #1 to
         Cramer, 2007.

310   **She had to get special permission:** J. E. Bamji, Letter #1 to Cramer, 2007; J. E. Jermyn,
         Letter to Manager, 1944; J. Morris, Letter to Jermyn, 1944.

311   **She planned a calendar:** J. E. Jermyn, Letter to the Manager, 1944.

311   **Ministry of Labour relented:** (signature illegible), Letter to Jermyn, 1944.

311   **"to be poisoned . . .":** J. B. S. Haldane, Letter to Killick, 1944.

311   **who had refused to write:** J. B. S. Haldane, Letter to Richardson, 1943.

311   **Dr. Joan Elizabeth Jermyn graduated:** London School of Medicine for Women, An-
         nual Reports, 1948, 79.

311   **got married to Bamji:** General Register Office, "England & Wales, Civil Registration
         Marriage," 1949.

311   **The job post that:** A. Bamji, Interview, 2020.

311   **most preestablished medical practices:** J. E. Bamji, Letter #1 to Cramer, 2007.

312   **Her spinal fracture caused:** A. Bamji, Interview, 2020.

312   **Joan Elizabeth Jermyn died:** A. Bamji, "Joan Elizabeth Bamji," 2011.

312   **textbook called *Oxygen and the Diver*:** Donald, *Oxygen and the Diver*, 1992. See cita-
         tions and explanation above regarding plagiarism.

313   **She and J. B. S. Haldane married:** J. B. S. Haldane, Letter to the Provost, 1945.

313   **then took a trip to Paris:** French, "Report from Port of Newhaven," 1945.

313   **"She is an extraordinary creature . . .":** Rendel, Letter to Mitchison, December 5,
         1940.

313   **the first adjective nearly every:** e.g., Marler, "Hark Ye to the Birds," 1985, 322, is the
         nicest, in calling her "well equipped vocally." Dronamraju, *Popularizing Science*, 2017,
         207, calls her "strident," and on page 212 says her voice "defied description—high
         pitched, but neither masculine nor feminine." Dronamraju's biography of Haldane
         is also problematic and reveals bias against Spurway through the gendered lan-
         guage it uses. For example, a chapter section titled "Spurway's Problems" dismisses
         her work and her as a person, and another called "Dress and Looks" mocks her for
         her masculine fashion style. D. Morris, *Watching*, 2006, 112, calls her "shrill" and says
         she "screeched" rather than talked. She was also described as "shrill" by two inter-
         viewees for this book, but I would prefer not to embarrass them by citing them by
         name because overall those two did not use sexist or gendered language.

313   **"like a circular saw":** Rendel, Letter to Mitchison, December 5, 1940.

313   **she privately confessed that:** Subramanian, *A Dominant Character*, 2020, 227.

313   **She was mortified about:** Rendel, Letter to Mitchison, December 5, 1940.

313   **pub called the Marlborough Arms:** G. Kalmus, Interview, 2020a, supported by N.
         A. Mitchison, "J. B. S. Haldane," 2017, which describes a favored pub "just outside
         UCL's backgate [*sic*]," consistent with the location of the Marlborough Arms.

313   **Haldane expressed in letters:** J. B. S. Haldane, Letter to *Fairplay*, 1945.

313   **those results remained cloaked:** Classification markings with years ranging from
         1972 to 2001 are evident on nearly all the records used from the United Kingdom
         National Archives. Per its markings, the documentation of the discussion to termi-
         nate J. B. S. Haldane from the underwater military research was originally supposed
         to stay classified until 2040. Director of Scientific Research, "Proposal to Dispense,"
         1943.

313   **Haldane and Spurway wrote a draft:** J. B. S. Haldane and Spurway, "Physiology of
         Diving," 1944.

313   **A year after the end:** (signature illegible) Manager, Letter to Haldane, 1946.

313   **He was granted:** Ministry of Labour, "Disabled Persons (Employment) Act," 1946.

313 **In 1956, while leaving ... jail cell:** *Times* (London), "Lecturer Stamped," 1956.

313 **described by some who:** N. A. Mitchison, "J. B. S. Haldane," 2017.

314 **they carried with them:** Subramanian, *A Dominant Character*, 2020, 273.

314 **By the time of J. B. S. Haldane's death:** *Times* (London), "Haldane," 1964.

314 **replaced Haldane as editor:** Lightman et al., *Circulation of Knowledge*, 2013, 76.

314 **she died of tetanus, which:** Emery, "Reader's Comment," 2004; *Guardian* (UK), "New Delhi," 1978; Wallace, *The Environment*, 1998, 206.

315 **After seeing how they enabled:** Quick and Shelford, "History of CCUBAs," 1970.

315 **dive teams still use:** Naval Sea Systems Command, *US Navy Diving Manual*, 2016.

315 **Cousteau asked for an introduction:** Shelford, Letter to Haldane, 1945.

315 **After the meeting, Jacques Cousteau:** Cousteau, Letter to Haldane, 1945.

316 **At worst, he was accused:** This is the premise of Tredoux, *Comrade Haldane*, 2018.

316 **he had express written permission:** J. B. S. Haldane, Letter to Pratt, 1943; Pratt, Letter to Haldane, 1943.

316 **In one of the most fascinating:** Operation Ivy Bells is documented fully in Sontag et al., *Blind Man's Bluff*, 2008.

317 **oxygen-filled Draeger Lar V:** Naval Sea Systems Command, *US Navy Diving Manual*, 2016.

Bibliography

Acott, Chris. 1999. "A Brief History of Diving and Decompression Illness." *Journal of South Pacific Underwater Medicine Society* 29 (2): 98–109.

Adjutant General Office. 1944. "'Jubilee': The Dieppe Raid Highlights," March. RG 407, Entry 427D, Box 19260. US National Archives and Records Administration, Washington, DC.

Admiral (Submarines). 1943. "Employment of X-craft X.20–X.23," November 3. No. 04351/846 from Admiral (Submarines). DEFE 2/1059 "X" Craft in Combined Operations: Type, Capabilities, etc. The National Archives, Kew, Richmond, United Kingdom.

Admiralty. (undated). "Comparison Between COPP and Chariot Methods of Beach Reconnaisance." DEFE 2/1116 Combined Operations Pilotage Parties: History and Reports. The National Archives, Kew, Richmond, United Kingdom.

———. 1939–41. "Methods of Saving Life in Sunken Submarines." ADM 116/4460. The National Archives, Kew, Richmond, United Kingdom.

———. 1942. Lieutenant P. C. A. Brownrigg, RNVR: Report of Death Whilst Diving, June 8. Reference Sheet 53/283. ADM 358/691. The National Archives, Kew, Richmond, United Kingdom.

———. 1942. Service Records. ADM 363/414/52. "Name: Redmond Vaughan Rowkins. Official Number: J. 70288. Date of Birth: 7 April . . ." The National Archives, Kew, Richmond, United Kingdom.

———. 1942. Translation. Instructions, October 8. HS 2/202 Operation Title: Abortive Chariot Attack on *Tirpitz*. The National Archives, Kew, Richmond, United Kingdom.

———. 1943. Enclosure to Portsmouth No. 6745/0/0302/45/16 of 31 December 1943: "Minutes of a Meeting Held at Fort Blockhouse, Gosport, on 30 December 1943," December 31. ADM 179/354 Operation "K.J.H.": Reference to Reconnaissance: Minutes of Meeting. The National Archives, Kew, Richmond, United Kingdom.

———. 1943. "Miniature Torpedoes and Two-Man Motor Boats," November 27. ADM 1/14789 Reports on Italian Miniature Torpedoes and Two-Man Motor Boats. The National Archives, Kew, Richmond, United Kingdom.

———. 1943. "Minutes of the Meeting Held at C.O.H.Q. on 5.5.43 to Discuss the Destruction of Underwater Obstacles," May 5. Ref. C.R.4180/43. DEFE 2/963 Clearance of

Underwater Obstacles: Organisation of Landing Craft Obstruction Clearance Units. The National Archives, Kew, Richmond, United Kingdom.

———. 1943. "Naval Cypher O.T.P. to Australian Commonwealth Naval Board 329," March 24. Message 0055A/24 March. ADM 1/12614 Definitions of Terms and Expressions (35): "X" Craft (Midget Submarines): Definitions of Terms. The National Archives, Kew, Richmond, United Kingdom.

———. 1943. "Part III—Clearance of Obstacles (Above and Below Water)." ADM 1/15494 Foreign Countries (52): Port Mine Clearance Parties for Clearance of Underwater Obstructions in Captured Ports. The National Archives, Kew, Richmond, United Kingdom.

———. 1944. Letter No. 2955 to Mrs. M. Sargent, May 20. ADM 358/1373 Temporary Lieutenant H. F. Cook, RNVR; Temporary Lieutenant C. E. Bonnell, RCNVR, etc. The National Archives, Kew, Richmond, United Kingdom.

———. 1945. "Admiralty and Predecessors: Office of the Director General of the Medical Department of the Navy and Predecessors: Service Registers and Registers of Deaths and Injuries." Registers of Reports of Deaths: Ship; Class: ADM 104; Piece: 117, entry Osprey, digital image s.v. "R. H. Saull," Ancestry.com.

Air Ministry (unsigned). 1941. Letter to Cameron Wright, April 2. Box 19, Folder 38, Institute of Naval Medicine, Gosport, United Kingdom.

Alexander, A. V. 1942. "Chariots, X-Craft, Etc.," September 27. ADM 1/25845 Training of Crews to Man X-Craft and Chariots (2-Man Human Torpedo). The National Archives, Kew, Richmond, United Kingdom.

———. 1942. Letter to Prime Minister Churchill, September 22. ADM 1/25845 Training of Crews to Man X-Craft and Chariots (2-Man Human Torpedo). The National Archives, Kew, Richmond, United Kingdom.

Alexander, William, P. Duff, John B. S. Haldane, G. Ives, and D. Renton. 1939. "After-Effects of Exposure of Men to Carbon Dioxide." The Lancet, August 19, 419–20.

Aliens War Service Department. 1943a. Letter to Professor J. B. S. Haldane re Miss Anna Ursula Philip, March 28. MS.20534 Correspondence of J. B. S. Haldane, Part 1. National Library of Scotland, Edinburgh.

———. 1943b. Letter to Professor J. B. S. Haldane re Mr. Hans Kalmus, March 28. MS.20534 Correspondence of J. B. S. Haldane, Part 1. National Library of Scotland, Edinburgh.

Amer, Don W. 1944. Appendix V to Naval Assault Group Commander's No. 6644 dated 30th June 1944, Enclosure No. 1. DEFE 2/420 Report by Naval Commander, Force "S," June 13. The National Archives, Kew, Richmond, United Kingdom.

Amoroso, Emmanuel C. 1979. "Memorial Address Given by Professor E. C. Amoroso at St. Luke's Church, RNH Haslar, on April 24, 1979."

Anderson, A. G. 1940. Letter from Mrs. A. G. Anderson to Sir John Anderson, Home Secretary, July 5. KV 2/1832 John Burdon Sanderson Haldane. The National Archives, Kew, Richmond, United Kingdom.

Anderson, J. F. Mac L. 1944. Letter to Osborn, October 7, Ref. No. SLO.14410/N. KV 2/1832 John Burdon Sanderson Haldane: "British. A well-known scientist, he announced his membership of the CPGB in 1942, having been an undisclosed member for some years previously." The National Archives, Kew, Richmond, United Kingdom.

Arts, Ilja C. W., Betty van de Putte, and Peter C. H. Hollman. 2000. "Catechin Contents of Foods Commonly Consumed in the Netherlands. 1. Fruits, Vegetables, Staple Foods, and Processed Foods." Journal of Agricultural and Food Chemicals 48 (5): 1746–51.

———. 2000. "Catechin Contents of Foods Commonly Consumed in the Netherlands. 2. Tea, Wine, Fruit Juices, and Chocolate Milk." Journal of Agricultural and Food Chem. 48 (5): 1752–57.

Ashcroft, Frederic Parker. 1994. Interview by Conrad Wood. Imperial War Museum, London. Catalogue number 12932. https://www.iwm.org.uk/collections/item/object /80013637. Accessed 2020.

Aspinall-Oglander, Brig. Gen. C. F. 2013. *Gallipoli,* Vol. 1: *Inception of the Campaign to May 1915.* Uckfield, UK: Naval & Military Press.

Assistant Controller (R&D). 1943. Letter to Rear Admiral R. B. Darke DSO, November 9. ADM 178/313 Royal Naval Personnel Research Committee: "proposed to terminate the appointment of Professor J. B. S. Haldane." The National Archives, Kew, Richmond, United Kingdom.

Authority. 1945. "Central Chancery of the Orders of Knighthood." *Third Supplement to the London Gazette of Friday, 11th of May 1945,* November 28, 2523.

———. 1950. "Central Chancery of the Orders of Knighthood." *Supplement to the London Gazette of Friday, 24th November 1950,* November 28, 5937 (preserved in Box 19, Folder 26, Institute of Naval Medicine).

Bagot, Milicent J. E. 1941. Letter, April 29. KV 2/1562 Hans Kahle: "German. An officer in the German army in the First World War, Kahle went to . . ." The National Archives, Kew, Richmond, United Kingdom.

———. 1941. Letter, May 21. KV 2/1562 Hans Kahle: "German. An officer in the German army in the First World War, Kahle went to . . ." The National Archives, Kew, Richmond, United Kingdom.

Bain, H. B. 1943. "Description of the 'Sladen' Dress—To Be Introduced into the Service as the 'Admiralty Shallow-Water Diving Dress,'" July 9. ADM 1/16669 Admiralty (5) and Inventions and Suggestions (59): "Admiralty diving equipment: research, development and supplies, etc." The National Archives, Kew, Richmond, United Kingdom.

Bamji, Andrew. 2011. "Joan Elizabeth Bamji (née Jermyn)." *British Medical Journal* 342: d659.

———. 2020. Interview by Rachel Lance.

Bamji, Joan Elizabeth. 2007. Letter #1 to Philip Cramer, June 26. Private collection of Andrew Bamji.

———. ca. 2007. Letter #2 to Philip Cramer, undated. Private collection of Andrew Bamji.

Bando, Mark. 2007. *101st Airborne: The Screaming Eagles in World War II.* McGregor, MN: Voyageur Press.

Banks, W. E. 1943. "Narrative of Exercise PBX4—Passage of X Craft from Port HHZ to Loch Kishorn and Return to Attack H.M.S. 'Bonaventure' at Port HHZ," April 8. ADM 1/12929 "Simulated Attack by 'X' Craft on HMS BONAVENTURE (Exercise 'PBX4') 2 April." The National Archives, Kew, Richmond, United Kingdom.

Barger, G., and H. H. Dale. 1910. "Chemical Structure and Sympathomimetic Action of Amines." *The Journal of Physiology* 41 (1–2): 19–59.

Barnes, J. S. 1943a. Letter to J. B. S. Haldane, December 21, Reference No. C.P.34570/43. MS.20534 Correspondence of J. B. S. Haldane, Part 1. National Library of Scotland, Edinburgh.

———. 1943b. Letter to J. B. S. Haldane, December 21, Reference No. C.P.34570/43. ADM 178/ 313 Royal Naval Personnel Research Committee: "proposed to terminate the appointment of Professor J. B. S. Haldane." The National Archives, Kew, Richmond, United Kingdom.

Barry, C. B. 1943. Letter No. 2245/SM 04351 to the Secretary of the Admiralty, October 23. DEFE 2/1059 "X" Craft in Combined Operations: Type, Capabilities, etc. The National Archives, Kew, Richmond, United Kingdom.

———. 1943. SM 04349/990 in Register No. P.D. 052/1943. ADM 1/18651 "Special Service Craft," March 17. The National Archives, Kew, Richmond, United Kingdom.

Bayne, Ronald Christopher. 1939. "HMS/M 'Thetis'—Detailed Statement of Staff Officer (Operations) to Commander-in-Chief, Plymouth," June 7. Private Papers of Captain R. C. Bayne RN, Documents.1213. Imperial War Museum, London.

Bean, J. W., and G. Rottschafer. 1939. "Reflexogenic and Central Structures in Oxygen Poisoning." *The Journal of Physiology* 94(3): 294–306.

Bedford College. 1937. "Bedford College for Women Report Card" (#6111). BC AR.200.80 Joan E. Jermyn student record. Royal Holloway, University of London.

Behnke, Lieutenant Commander A. R., and Lieutenant T. L. Wilmon. 1939. "USS *Squalus*, Medical Aspects of the Rescue and Salvage Operations and the Use of Oxygen in Deep-Sea Diving," November 25. Printed by Army and Navy Register. Preserved in Haldane's records and found in HALDANE/1/5/3/9 "Offprint of an Article on the USS *Squalus*." University College London Special Collection, College Archives, London.

Bellows, James Henry. 1992. Interview by Conrad Wood. Imperial War Museum. Catalogue number 12913. https://www.iwm.org.uk/collections/item/object/80012643.

Bennett, Peter. 2021. Interview by Rachel Lance.

Benzinger, Theodor. 1950. "Chapter IV-M Explosive Decompression." In *German Aviation Medicine World War II*, Vol. 1. Washington, DC: The Surgeon General, US Air Force.

Bett, W. R. 1946. "Benzedrine Sulphate in Clinical Medicine: A Survey of the Literature." *Post-Graduate Medical Journal* 10 (1136): 205–18.

Bigg-Wither, Richard Lovelace. 1991. Interview by Conrad Wood. Imperial War Museum. Catalogue number 12175. https://www.iwm.org.uk/collections/item/object /80011910.

Birindelli, Gino. 1994. Interview by Conrad Wood. Imperial War Museum, London. Catalogue number 14236. https://www.iwm.org.uk/collections/item/object/80013932.

Blackett, P. M. S. 1942. Letter to J. B. S. Haldane Labeled "Secret," July 4. MS.20534 Correspondence of J. B. S. Haldane, Part 1. National Library of Scotland, Edinburgh.

Booth, Tony. 2009. *Thetis Down: The Slow Death of a Submarine*. Barnsley, UK: Pen & Sword Books.

Boothby, W. M., and William Randolph Lovelace. 1938. "The Necessity for the Use of Oxygen and a Practical Apparatus for Its Administration to Both Pilots and Passengers." *The Journal of Aviation Medicine* 9: 172–98.

——, and O. O. Benson. 1940. "High Altitude and Its Effect on the Human Body." *Journal of the Aero Society of America* 7 (11): 461–68.

Bovill, Albert. 2002. Interview by Peter M. Hart. Imperial War Museum. Catalogue number 23786. https://www.iwm.org.uk/collections/item/object/80022014.

Box, K. (Diarist 5257). 1940. Diary. "Typist and Full-time Mass Observer, Female, DOB 1911, Living in London, SW6." Mass Observation Online.

Boycott, A. E., G. C. C. Damant, and John Scott Haldane. 1908. "Prevention of Compressed-Air Illness." *Epidemiology & Infection* 8 (3): 342–443.

Bozic, Kevin J., Joyce H. Keyak, Harry B. Skinner, H. Ulrich Bueff, and David S. Bradford. 1994. "Three-Dimensional Finite Element Modeling of a Cervical Vertebra: An Investigation of Burst Fracture Mechanism." *Journal of Spinal Disorders & Techniques* 7 (2): 102–10.

Bradshaw, William Harold. 1989. Interview by Conrad Wood. Imperial War Museum, London. Catalogue number 10798. https://www.iwm.org.uk/collections/item/object /80010575.

Brangwyn, Frank. 1954. Christmas Card to Cameron Wright, Addressed "My Dear Friend." Box 19, Folder 37, Institute of Naval Medicine, Gosport, United Kingdom.

Brickhill, Paul. 1951. *The Dam Busters: The Thrilling Exploits of the R.A.F.'s Most Famous Bomber Squadron*. New York: Bantam.

Bridge, John. 1995. Interview by Conrad Wood. Imperial War Museum, London. Catalogue number 15560. https://www.iwm.org.uk/collections/item/object/80015088.

Briggs, Keith May. 1996. Interview by Conrad Wood. Imperial War Museum, London. Catalogue number 16696. https://www.iwm.org.uk/collections/item/object/80016162.

British Medical Journal. 1956. "Cases Previously Postponed." *Supplement to the British Medical Journal* 2682 (Jun 16): 351–66.

Brooklyn Daily Eagle. 1872. "The Bridge: How the Great Work Goes On." June 22, 2.

———. 1872. "Frightful Accident: The Bridge of Death Claims Another Victim." January 27, 3.

———. 1901. "Rapid Transit Tunnel Builders at Last Find Means to Thwart Dangerous 'Caisson Disease.'" September 15, 24.

Brooklyn Times-Union. 1872. "Death in the Caisson." April 25, 4.

Brooklyn Union. 1867. "The River: An Ice Bridge from Brooklyn to New York." January 23, 4.

———. 1872. "Another Victim to the Caisson." May 16, 2.

———. 1872. "The Caisson Again: Another Victim to Compressed Air." April 7, 4.

———. 1872. "Life in the Caisson." May 3, 1.

Broome, Roy. 1993. Interview by Conrad Wood. Imperial War Museum. Catalogue number 13367. https://www.iwm.org.uk/collections/item/object/80013083.

Brown, G. L. 1943. Letter Addressed "Dear Haldane," December 10. MS.20569 Miscellany of Papers, Chiefly Undated, Chiefly of J. B. S. Haldane, Including Notes, Memoranda, Reports and Mathematical Calculations Concerning Diving. National Library of Scotland, Edinburgh.

———. 1943. "Minutes of a Special Meeting of the Sub-Committee on Underwater Physiology," November 18; UPS Reports; Sub-Committee on Underwater Physiology; Royal Naval Personnel Research Committee of the Medical Research Council. ADM 178/313 Royal Naval Personnel Research Committee: "proposed to terminate the appointment of Professor J. B. S. Haldane." The National Archives, Kew, Richmond, United Kingdom.

———. 1944. Letter to J. B. S. Haldane, November 24, RNP/U.79/44. MS.20569 Miscellany of Papers, Chiefly Undated, Chiefly of J. B. S. Haldane, Including Notes, Memoranda, Reports and Mathematical Calculations Concerning Diving. National Library of Scotland, Edinburgh.

Brubakk, Alf O., David Hallen Elliott, Peter B. Bennett, and Tom S. Neuman, eds. 2003. *Bennett & Elliott's Physiology and Medicine of Diving*. Philadelphia: Saunders.

Buckingham, J. 1939. Letter to J. B. S. Haldane, August 5. HALDANE/5/6/3/1/6 1 August 1939. University College London Special Collection, College Archives, London.

Buffalo Commercial. 1872. "The News." April 7, 2.

Buffalo Morning Express. 1872. "Strikes." May 7, 4.

Burt, Stephen. 1992. "The Exceptional Hot Spell of Early August 1990 in the United Kingdom." *International Journal of Climatology* 12(6): 547–67.

Bush, Eric. 1944. Enclosure (No. 1) to Naval Assault Group Commander's No. 6644 Dated 30th June 1944. "The Narrative: Operation 'Neptune'—5th and 6th June." DEFE 2/420 Report by Naval Commander, Force "S." The National Archives, Kew, Richmond, United Kingdom.

Butler, W. P. 2004. "Caisson Disease During the Construction of the Eads and Brooklyn Bridges: A Review." *Undersea & Hyperbaric Medicine* 31 (4): 445–59.

C in C Levant. 1943. Message 101031C/April Regarding Naval Code X by W/T. ADM 1/12614 Definitions of Terms and Expressions (35), April 10: "'X' Craft (Midget Submarines): Definitions of Terms." The National Archives, Kew, Richmond, United Kingdom.

C in C Pacific. 1944. "Naval Combat Demolition Units—Report on Operation Neptune June 1944," September 17, COMFIFTHPHIB File No A 16-3(3). RG 331, Entry 23, Box 16. US National Archives and Records Administration, Washington, DC.

C in C Portsmouth. 1943. Operation "K.J.H.," December 30. Ref. No. 0850/19. ADM 179/354 Operation "K.J.H.": Reference to Reconnaissance: Minutes of Meeting. The National Archives, Kew, Richmond, United Kingdom.

Cabot, Hugh. 1918. *General Considerations: Diseases of Penis and Urethra, Diseases of Scrotum and Testicle, Diseases of Prostate and Seminal Vesicles.* Philadelphia: Lea & Febiger.

Calder, I. M., A. C. Palmer, J. T. Hughes, J. F. Bolt, and J. D. Buchanan. 1989. "Spinal Cord Degeneration Associated with Type II Decompression Sickness: Case Report." *Paraplegia* 27(1): 51–57.

Calder, Jenni. 2019. *The Burning Glass: The Life of Naomi Mitchison.* Dingwall, Scotland: Sandstone Press Ltd.

Cameron, A. J. D., and Horace Cameron Wright. 1935. "Correspondence: Short Wave Therapy." *The British Journal of Physical Medicine*, April: 239–40.

Cameron, Roy. 1964. Testimonial, December 3. Box 19, Folder 14, Institute of Naval Medicine, Gosport, United Kingdom.

Car Records. 1940–52. Miscellaneous documents. HALDANE/6/2/4/2 "Financial Records: Records Relating to His Cars." University College London Special Collection, College Archives, London.

Case, Edwin Martin. 1940. Lab Notebook. MS.20570 Notebook Containing Records of Experiments on Respiratory Problems in Diving Carried Out by J. B. S. Haldane and E. M. Case. National Library of Scotland, Edinburgh.

———. 1940. Letter to J. B. S. Haldane, January 19. HALDANE/1/5/3/45 Letter from E. M. Case to J. B. S. Haldane. University College London Special Collection, College Archives, London.

———.1941. "Human Physiology Under High Pressure. Effects of Nitrogen, Carbon Dioxide, and Cold." *Journal of Hygiene* 41 (3): 225–49.

Case, Edwin Martin, and John B. S. Haldane. 1940. Report, July 22. HALDANE/1/5/3/25 "Report on Effects of High Pressure, Carbon Dioxide and Cold." University College London Special Collection, College Archives, London.

Chief Inspector. 1944. Letter to Metropolitan Police Aliens Registration Office, Reference 79/K/2760, July 31. KV 2/1565 Hans Kahle: "German. An officer in the German army in the First World War, Kahle went to . . ." The National Archives, Kew, Richmond, United Kingdom.

Chief of Combined Operations. 1943. Enclosure to CCO's SR 380/43, Dated 11th February 1943. ADM 1/15583 Combined Operations (47): Supply of Stores to Combined Operations Pilotage Parties. The National Archives, Kew, Richmond, United Kingdom.

Chief of Staff (X). 1943. "Note of a Meeting Held in COS(X)'s Room, Norfolk House on Friday, 17th September 1943," September 18. No. X/0560/6/15. DEFE 2/1101 Combined Operations Pilotage Parties: Composition and Movement of Teams, Complement, Requirements for Beach Reconnaissance, etc. The National Archives, Kew, Richmond, United Kingdom.

Chief of Staff to Naval Commander Expeditionary Force. 1943. Proposed Scheme for "P" Party, Enclosure No. 4 to Letter No. X/0940/2 of 6/11/43, November 6. ADM 1/15494 Foreign Countries (52): Port Mine Clearance Parties for Clearance of Underwater Obstructions in Captured Ports. The National Archives, Kew, Richmond, United Kingdom.

Churchill, Winston. 2014. *The Hinge of Fate.* New York: Rosetta Books.

Clark, P. W. 1944. C.O.P.P. 1 Progress Report No. 1, March 13. DEFE 2/1152 Combined Operations Pilotage Parties: Progress Reports. The National Archives, Kew, Richmond, United Kingdom.

Clark, Ronald W. 1969. *JBS: The Life and Work of JBS Haldane*. New York: Coward-McCann, Inc.

Clogstoun-Willmott, Nigel H. 1943. C.O.P.P.—Command Liaison, Letter Ref. No. 26/1 from C.O.P.P. Depot c/o G.P.O., July 30. DEFE 2/971 Combined Operations Pilotage Parties: Brief on Information Required, Establishment, Function and Use. The National Archives, Kew, Richmond, United Kingdom.

Clyde Cruising Club. 2020. *Firth of Clyde, Including Solway Firth and North Channel*. Cambridgeshire, UK: Imray, Laurie, Norie and Wilson Limited.

Cole, Robert Hugh. 1948. *Underwater Explosions*. Princeton, NJ: Princeton University Press.

Combined Operations Headquarters. 1944. Chapter IV Ship to Shore Movement—Omaha Beach, June. Amphibious Operations Invasion of Northern France Western Task Force June 1944 COMINCH P-006 (ARC ID 12005066). US National Archives and Records Administration, Washington, DC.

———. 1944. "Minutes of Meeting Held on 18th January 1944 to Discuss Points Arising from Admiralty Letter M.053669/43," January 18. DEFE 2/963 Clearance of Underwater Obstacles: Organisation of Landing Craft Obstruction Clearance Units. The National Archives, Kew, Richmond, United Kingdom.

———. 1945. Bulletin T/18 COPP Combined Operations Pilotage Parties, June. DEFE 2/2084 Combined Operations Pilotage Parties. The National Archives, Kew, Richmond, United Kingdom.

Commander Assault Force "O" (Commander Eleventh Amphibious Force). 1944 Chapter IV Ship to Shore Movement—Omaha Beach, August 5. Amphibious Operations Invasion of Northern France Western Task Force June 1944 COMINCH P-006 (ARC ID 12005066). US National Archives and Records Administration, Washington, DC.

Commander R.N. for Chief of Combined Operations. 1943. Letter to D. of P. (Q) with Copy to Messrs. Siebe Gorman & Co., with Attached Spreadsheet of Measurements, April 15. ADM 1/15583 Combined Operations (47): Supply of Stores to Combined Operations Pilotage Parties. The National Archives, Kew, Richmond, United Kingdom.

Committee, Admiralty Diving. 1944. Admiralty Fleet Order: Diving- Effects of Breathing Oxygen Under Pressure AFO 4565/44, August 24. ADM 1/16669 Admiralty (5) and Inventions and Suggestions (59): Admiralty Diving Equipment: Research, Development, and Supplies, etc. The National Archives, Kew, Richmond, United Kingdom.

Commonwealth War Graves. "Lieutenant Ronald Henry Saull, entry 2635018, Rotterdam (Crooswijk) General Cemetery." https://www.cwgc.org/find-records/find-war-dead/casualty-details/2635018/ronald-henry-saull/.

———. "Sub-Lieutenant Donald Kirkland, entry 2634981, Rotterdam (Crooswijk) General Cemetery." https://www.cwgc.org/find-records/find-war-dead/casualty-details/2634981/donald-kirkland/.

COMSUBRON 4—Japanese Midget Submarine, No. 19—Navigational Charts. 1941. RG 38, Series P 17, UD-WW 46, NAID: 133891335, December 30. US National Archives and Records Administration, Washington, DC.

Connolly, Doris Eileen. 1999. Interview by Conrad Wood. Imperial War Museum. Catalogue number 19050. https://www.iwm.org.uk/collections/item/object/80018059.

Cook, George. 1987. Interview by Conrad Wood. Imperial War Museum, London. Catalogue number 9977. https://www.iwm.org.uk/collections/item/object/80009760.

Copeman, Frederick Bayes. 1980. Interview by Alan Ereira. Imperial War Museum, London. Catalogue number 5829. https://www.iwm.org.uk/collections/item/object/80005785.

Cousteau, Jacques-Yves. 1945. Letter to J. B. S. Haldane, March 16. HALDANE/5/6/3/1/65 4 March 1945. University College London Special Collection, College Archives, London.

Cowan, S. L., W. Wilson-Dickson, and Horace Cameron Wright. 1947. "Respiratory and Other Responses of Human Subjects During Rapid Decompression—Preliminary Experiments Using a Hot-Wire Recoder" (R.N.P.L 4/47), July 1. ADM 218/172 Royal Naval Personnel Research Committee: Minutes of Meetings and Results of Experiments, etc. The National Archives, Kew, Richmond, United Kingdom.

Cowell, Frederick, Xuan Goh, James Cambrook, and David Bulley. 2007. "Chlorine as the First Major Chemical Weapon." In *An Element of Controversy: The Life of Chlorine in Science, Medicine, Technology and War*, edited by Hasok Chang and Catherine Jackson, 220–264. London: British Society for the History of Science.

Crow, Alwyn D. 1948. "The Rocket as a Weapon of War in the British Forces." *Proceedings of the Institution of Mechanical Engineers* 158 (1): 15–21.

Daily Worker. 1939. "J. B. S. Haldane and Four Spain Brigadiers in Thetis Test Ordeal" (July 21 newspaper clipping inserted into MI5 file). KV 2/1832 John Burdon Sanderson Haldane: "British. A well-known scientist, he announced his membership of the CPGB in 1942, having been an undisclosed member for some years previously. The National Archives, Kew, Richmond, United Kingdom.

———. 1939. "100 Scientists at Historic Meeting" (August 15 newspaper clipping inserted into MI5 file). KV 2/1832 John Burdon Sanderson Haldane: "British. A well-known scientist, he announced his membership of the CPGB in 1942, having been an undisclosed member for some years previously." The National Archives, Kew, Richmond, United Kingdom.

Damant, G. C. C. 1937. "Deep Sea Diving." *Royal United Services Institution Journal* 82 (526): 315–23.

Dansey, C. 1941. Letter to Cameron Wright, March 6. Box 19, Folder 27, Institute of Naval Medicine, Gosport, United Kingdom.

Dansey, C. P. 1941. Letter to Cameron Wright, January 19. Box 19, Folder 27, Institute of Naval Medicine, Gosport, United Kingdom.

Danto, Arthur C. 1987. "Gettysburg." *Grand Street* 6 (3): 98–116.

Daubin, F. A. December 30, 1941. COMSUBRON 4—Japanese Midget Submarine, No 19—Navigational Charts. RG 38, Series P 17, UD-WW 46, NAID: 133891335. US National Archives and Records Administration, Washington, DC, USA.

Davies, Bernard Kenneth. 1989. Interview by Conrad Wood. Imperial War Museum, London. Catalogue number 10783. https://www.iwm.org.uk/collections/item/object/80010560.

Davis, F. Michael. 2003. "An Element of Doubt: Four Divers with Acute Neurological Problems." *Journal of South Pacific Underwater Medicine Society* 33 (4): 187–91.

Davis, Robert H. 1939. Letter to Commander Cumming, March 1. HALDANE/1/5/3/10 Letter from Robert Henry Davis to A. S. Cumming. University College London Special Collection, College Archives, London.

———. 1939. Letter to J. B. S. Haldane, June 6. HALDANE/5/6/3/1/3 6 June 1939. University College London Special Collection, College Archives, London.

———. 1939. Letter to J. B. S. Haldane, July 7. HALDANE/5/6/3/1/3 6 June 1939. University College London Special Collection, College Archives, London.

———. 1939. Letter to Miss Jermyn, August 15. HALDANE/5/6/3/1/3 6 June 1939. University College London Special Collection, College Archives.

———. 1940. Letter to J. B. S. Haldane, January 11. HALDANE/5/6/3/1/8 9 January 1940. University College London Special Collection, College Archives, London.

———. 1940. Letter to J. B. S. Haldane, January 23. HALDANE/5/6/3/1/10 23 January 1940. University College London Special Collection, College Archives, London.

———. 1942. Letter to the Admiralty Regarding Employment of Helen Spurway, March 31. HALDANE/5/6/3/1/49 4 February 1942. University College London Special Collection, College Archives, London.

———. 1942. Letter to the Admiralty Regarding Employment of Joan Elizabeth Jermyn, March 31. Private collection of Andrew Bamji.

———. 1943. Letter to J. B. S. Haldane, March 31. MS.20534 Correspondence of J. B. S. Haldane, Part 1. National Library of Scotland, Edinburgh.

———. 1943. Letter to Mr. Dale Bussell, December 14, second with reference RHD/FW. ADM 178/313 Royal Naval Personnel Research Committee: "proposed to terminate the appointment of Professor J. B. S. Haldane." The National Archives, Kew, Richmond, United Kingdom.

———. 1944. Letter to J. B. S. Haldane, October 26. HALDANE/5/6/3/1/61 13 May 1944. University College London Special Collection, College Archives.

———, R. W. Gorman Davis, and W. Eric Davis. 1946. *A Record of War and Peace.* "Neptune" Works, Tolworth, Surrey, UK: Siebe, Gorman & Company Limited.

Davy, Sir Humphry. 1839. *The Collected Works of Sir Humphry Davy.* Sacramento, CA: Creative Media Partners, LLC.

Demchenko, Ivan T., Karen E. Welty-Wolf, Barry E. Allen, and Claude A. Piantadosi. 2007. "Similar but Not the Same: Normobaric and Hyperbaric Pulmonary Oxygen Toxicity, the Role of Nitric Oxide." *American Journal of Physiology—Lung Cellular and Molecular Physiology* 293 (1): L229–38.

Department of Defense Office of Public Information. 1955. "Press Release: Air Force Presents Exceptional Service Awards to Former Scientific Advisory Board Members." No. 1010-55 LI5-6700 Ext 75131, October 19.

Diarist 5094. 1940. Diary. "Policeman, Male, Living in London, W11." Mass Observation Online.

Diarist 5142. 1940. Diary. "Student (Medical), Male, DOB 1917, Living in Reigate, Surrey." Mass Observation Online.

Diarist 5150. 1940. Diary. "Commercial Traveller, Male, DOB 1902, Living in Wembley, Middlesex." Mass Observation Online.

Diarist 5166. 1940. Diary. "Student, Male, DOB 1920, Living in Cambridge, Cambridgeshire." Mass Observation Online.

Diarist 5245. 1940. Diary. "Tax Inspector and Mother, Female, DOB 1908, Living in Purley, Surrey, and Belfast, Northern Ireland." Mass Observation Online.

Diarist 5250. 1940. Diary. "Actress, Female, DOB 1909, Living in London, NW3." Mass Observation Online.

Diarist 5280. 1940. Diary. "Psychiatrist, Female, DOB 1902, Living in London, NW8, and London, NW1." Mass Observation Online.

Diarist 5368. 1940. Diary. "Housewife, Female, DOB 1920, Living in Manchester, Lancashire." Mass Observation Online.

Diarist 5438. 1940. Diary. "Teacher, Female, DOB 1887, Living in Crouch End, London, N8." Mass Observation Online.

Director of Plans. 1943. Letter PLS/PBA, March 3. ADM 1/18651 Special Service Craft. The National Archives, Kew, Richmond, United Kingdom.

Director of Scientific Research. 1943. "Proposal to Dispense with the Services of Professor Haldane and His Assistants on Work for the Admiralty Diving Committee at the Works of Messrs. Siebe, Gorman & Co.," November 14. ADM 178/313 Royal Naval Personnel Research Committee: "proposed to terminate the appointment of Professor J. B. S. Haldane." The National Archives, Kew, Richmond, United Kingdom.

———. 1944. Letter to Director of Naval Intelligence, September 26. KV 2/1832 John Burdon Sanderson Haldane: "British. A well-known scientist, he announced his membership of the CPGB in 1942, having been an undisclosed member for some years previously." The National Archives, Kew, Richmond, United Kingdom.

Director of Victualling. 1943. Enclosure to D. of V.'s memo, June 11. V.08/43 dated 11th June 1943. ADM 1/15583 Combined Operations (47): Supply of Stores to Combined Operations Pilotage Parties. The National Archives, Kew, Richmond, United Kingdom.

DM. 1943. Notes on M.E. C.O.P.P.'s. DEFE 2/1116 Combined Operations Pilotage Parties: History and Reports. The National Archives, Kew, Richmond, United Kingdom.

Donald, Kenneth. 1942. "Continued Report on Experimental Work at Siebe Gorman (up till 12/12/1942)," December 12. ADM 315/9 Diving Suits for Chariots. The National Archives, Kew, Richmond, United Kingdom.

———. 1942. "Time-Pressure Curves," December 12. ADM 315/15. The National Archives, Kew, Richmond, United Kingdom.

———. 1943. "Comparison of Wet and Dry Performances," April 6. ADM 315/12 Comparison of Wet and Dry Performance of Diving Suits. The National Archives, Kew, Richmond, United Kingdom.

———. 1943. "Report Concerning the Use of the M.R.S. Suit at Greater Depths," May 8. ADM 315/14 Use of MRS Suit at Greater Depths. The National Archives, Kew, Richmond, United Kingdom.

———. 1943. "Report on the Use of Mixtures in Counterlung Breathing Apparatus," September 24. ADM 315/18 Use of Mixtures in Counter Lung Breathing Apparatus. The National Archives, Kew, Richmond, United Kingdom.

———. 1944. Letter to the Secretary, Underwater Physiology Sub-Committee, Subject: The Future Programme of Research in Diving and Submarine Escape, January 24. MS.20569 Miscellany of Papers, Chiefly Undated, Chiefly of J. B. S. Haldane, Including Notes, Memoranda, Reports and Mathematical Calculations Concerning Diving. National Library of Scotland, Edinburgh.

———. 1944. "Variation in the Oxygen Tolerance of Human Subjects in the Wet and in the Dry with Statistical Appendix by J. B. S. Haldane, FRS, & Helen Spurway," March. ADM 315/19 Oxygen Tolerance of Human Subjects in Wet and Dry Conditions. The National Archives, Kew, Richmond, United Kingdom.

———. 1992. *Oxygen and the Diver.* Hanley Swan, Worcestershire, England: The SPA Ltd.

Draeger, R. Harold, Joseph S. Barr, and W. W. Sager. 1946. "Blast Injury." *The Journal of the American Medical Association* 132 (13): 762–67.

Dronamraju, Krishna. 2017. *Popularizing Science: The Life and Work of JBS Haldane.* New York: Oxford University Press.

Dunbar-Nasmith, Admiral Martin Eric. 1939. Letter to Professor Haldane, July 24. HALDANE/5/6/3/1/4 24 July 1939. University College London Special Collection, College Archives, London.

Duncan, D. 1940. Letter from Duncan on Behalf of Medical Director General to the Secretary, Dunbar-Nasmith, Physiological Subcommittee, Royal Naval Medical School, August 22. HALDANE/5/6/3/1/48 26 January 1942. University College London Special Collection, College Archives, London.

Dunn, Leonard Alfred. 2000. Interview by Conrad Wood. Imperial War Museum. Catalogue number 20473. https://www.iwm.org.uk/collections/item/object/80019930.

Easthope, William G. 2010. *Smitten City: The Story of Portsmouth in the Air Raids 1940–1944.* Portsmouth, UK: Portsmouth Publishing and Printing Limited.

Edgar, Mia, Michael A. Franco, and Hugh M. Dainer. 2021. "Case Series of Arterial Gas Embolism Incidents in U.S. Navy Pressurized Submarine Escape Training from 2018 to 2019." *Military Medicine* 186 (5–6): e613–18.

Edwards, A. W. F. 2017. "Haldane and Fisher—Scientific Interactions." *Journal of Genetics* 96 (5): 747–52.

Ellis, F. P. 1944. "The Present Position of Our Knowledge of the Injurious Effect of the 'Blast' of Underwater Explosions." ADM 298/464 The Injurious Effect of Blast of Underwater Explosions. The National Archives, Kew, Richmond, United Kingdom.

———. 1950. "Personnel Research in the Royal Navy 1939–1945 with Special Reference to the Activities of the Royal Naval Personnel Research Committee." ADM 298/78 Personnel Research in the RN (1939–1945). The National Archives, Kew, Richmond, United Kingdom.

Ellis, R. M. 1942. "Boom Commando Party Proposed Issue of Battle Dress to Personnel," November 25 B.D.01319/42. DEFE 2/963 Clearance of Underwater Obstacles: Organisation of Landing Craft Obstruction Clearance Units. The National Archives, Kew, Richmond, United Kingdom.

Emery, Alan E. 2004. "Reader's Comment." *Wellcome History*, summer (26): 20.

Erskine, Ralph, and Frode Weierud. 2010. "Naval Enigma: M4 and Its Rotors." *Cryptologia* 11 (4): 235–44.

Eugene Guard (OR). 1949. "Feeling Old? You Are Younger Than You Think!" December 3, 36.

Extended Defence Officer Gibraltar. 1944. "Report on A/S Trials with Human Torpedo at Gibraltar December 1943," January 26. ADM 1/16490 Armaments (11): Report on Anti-submarine Trials with Human Torpedo at Gibraltar and Comparison with British Chariot. The National Archives, Kew, Richmond, United Kingdom.

FDA. 1974. "Notices of Judgment." *FDA Consumer*, May 8 (4): 42.

Featherstone, Donald F. 1979. *Wargaming Airborne Operations*. New York: A. S. Barnes.

Ferris, Eugene B., and George L. Engel. 1951. "The Clinical Nature of High Altitude Decompression Sickness." In *Decompression Sickness: Caisson Sickness, Diver's and Flier's Bends and Related Syndromes*. Philadelphia: W. B. Saunders Company.

Fielder, Harold Sidney. 1993. Interview by Conrad Wood. Imperial War Museum. Catalogue number 12986. https://www.iwm.org.uk/collections/item/object/80012714.

Flag Officer Gibraltar. ca. 1945. Pack No. 73/4 Midget Submarines—General. ADM 199/1812 Flag Officer, Gibraltar: Midget Submarine and Human Torpedo Attacks. Includes 37, etc. The National Archives, Kew, Richmond, United Kingdom.

Fosdick, Raymond B. 1989. *The Story of the Rockefeller Foundation*. New York: Routledge.

Fotali. 1943. Naval Cypher XD by W/T 2419431/November to F.O.C. Gibraltar, November 25. ADM 1/14789 Reports on Italian Miniature Torpedoes and Two-Man Motorboats. The National Archives, Kew, Richmond, United Kingdom.

Frankland, George James. 1999. Interview by Conrad Wood. Imperial War Museum. Catalogue number 19041. https://www.iwm.org.uk/collections/item/object/80018060.

Franklin, Ian, Geoff Grigg, and Oliver Mayo. 2004. "James Meadows Rendel 1915–2001." *Historical Records of Australian Science* 15(2): 269–84.

Fraser, Ian Edward. 1987. Interview by Conrad Wood. Imperial War Museum. Catalogue number 9822. https://www.iwm.org.uk/collections/item/objec/80009605.

Frederick, Reeks, and Goode. 1940. "Inventory of Household Furniture and Effects at the Crossways, Salisbury Avenue, Harpenden, 15 Nov 1940." HALDANE/6/2/4/1 and 2 Financial Records. University College London Special Collection, College Archives, London.

Frederick, Robert C. 1940. Letter to J. B. S. Haldane, January 9. HALDANE/5/6/3/1/8 9 January 1940. University College London Special Collection, College Archives, London.

——. 1940. Letter to J. B. S. Haldane, June 17. HALDANE/5/6/3/1/15 17 June 1940. University College London Special Collection, College Archives, London.

——. 1940. Researches by Professor J. B. S. Haldane, FRS, July 30. Reference 6/1B/DNA9703/40. ADM 116/4460 Methods of Lifesaving in Sunken Submarines: Investigations by Professor R. R. MacIntosh and Sir Robert Davies, Reports of Admiral Dunbar-Nasmith Committee, etc. The National Archives, Kew, Richmond, United Kingdom.

——. 1941. Letter to Divisional Petroleum Officer, Mines Department, February 25, as Copied to J. B. S. Haldane. HALDANE/5/6/3/1/30 13 February 1941. University College London Special Collection, College Archives, London.

Fremlin, Celia. 1939. "Air Raid Shelters." File Report A14. (Note: the date for this report is listed in the database as March 1939, but based on the survey information it contains of people sleeping in the Tube for shelter from bombs, that seems incorrect.) Mass Observation Online.

French, G. 1945. "Report from Port of Newhaven to Metropolitan Police (Special Branch)," December 18. KV 2/1832 John Burdon Sanderson Haldane: "British. A well-known scientist, he announced his membership of the CPGB in 1942, having been an undisclosed member for some years previously." The National Archives, Kew, Richmond, United Kingdom.

Fussell, Peter Lincoln. 1991. Interview by Conrad Wood. Imperial War Museum, London. Catalogue number 10242. https://www.iwm.org.uk/collections/item/object/80010022.

Fye, W. Bruce. 1986. "T. Lauder Brunton and Amyl Nitrite: A Victorian Vasodilator." *Circulation* 74 (2): 222–29.

Gallagher, Thomas. 1971. *The X-Craft Raid*. San Diego: Harcourt Brace Jovanovich.

Galwey, Geoff V. 1992. "Geoff's Opus: A Record of Survival, One Way and Another. Part III: Combined Operations Pilotage Parties." London Metropolitan Archives, London, LMA/4462/P/04/001 G V Galway: "Geoff's Opus."

General Register Office. 1892. "England & Wales, Civil Registration Birth Index, 1837–1915, London, England," digital image s.v. "John Burdon Sanderson Haldane," Ancestry.com.

——. 1900. "Marriage Registration of Edward Cameron James Wright and Laura Elizabeth White, September 30, 1900. England, Select Marriages, 1538–1973," digital image s.v. "Laura Elizabeth White," Ancestry.com.

——. 1901. "Class: RG13; Piece: 159; Folio: 79; Page: 13; Census Returns of England and Wales, 1901, London, England," digital image s.v. "Edward Wright," Ancestry.com.

——. 1902. "Horace Cameron Wright birth certificate."

——. 1915. "England & Wales, Civil Registration Birth Index, 1837–1915, London, England," digital image s.v. "Helen Spurway," Ancestry.com.

——. 1916. "England and Wales Marriage Registration Index, 1837–2005," database Family Search (https://familysearch.org/ark:/61903/1:1:26XG-GF7: 13 December 2014), "Naomi M. M. Haldane and Null," 1916; from "England & Wales Marriages, 1837–2005," database Find My Past (http://www.findmypast.com: 2012); citing 1916, quarter 1, vol. 3A, p. 1955, Headington, Oxfordshire, England, General Register Office, Southport, England.

——. 1918. "London, England, Electoral Registers, 1832–1965, London Metropolitan Archives, London, England," digital image s.v. "Edward C. J. Wright," Ancestry.com.

——. 1921. "Census Returns of England and Wales, 1921, London, England," digital image s.v. "Horace Cameron Wright," findmypast.co.uk.

——. 1932. "England and Wales, National Index of Wills and Administrations, 1858–1957," database Family Search (https://familysearch.org/ark:/61903/1:1:7X6Z-9VPZ: 27 August 2019), "Laura Elizabeth Wright," 29 July 1932; citing Probate, City of London,

Middlesex, England, United Kingdom, Her Majesty's Stationery Office, Great Britain; FHL microfilm.

———. 1936. "England and Wales Birth Registration Index, 1837–2008," database Family Search (https://familysearch.org/ark:/61903/1:1:QVQ6-VH2Z: 1 October 2014), "Reuben N. Grüneberg," 1936; from "England & Wales Births, 1837–2006," database Find My Past (http://www.findmypast.com: 2012); citing Birth Registration, Hampstead, London, England, citing General Register Office, Southport, England.

———. 1938. "Marriage Registration of Dorris E. Butt and Roger J. A. Courtney. England & Wales, Civil Registration Marriage Index, 1916–2005," digital image s.v. "Roger Courtney and Dorris Butt," Ancestry.com.

———. 1940. "England and Wales Birth Registration Index, 1837–2008," database Family Search (https://familysearch.org/ark:/61903/1:1:QVQK-DBHT: 1 October 2014), "Daniel S. Grüneberg," 1940; from "England & Wales Births, 1837–2006," database Find My Past (http://www.findmypast.com: 2012); citing Birth Registration, Hendon, Middlesex, England, citing General Register Office, Southport, England.

———. 1948. "GRO Reference: 1948 M Quarter in BATTLE Volume 05H Page 25."

———. 1949. "England & Wales, Civil Registration Marriage Index, 1916–2005, London Metropolitan Archives, London, England," digital image s.v. "Joan Elizabeth Jermyn," spouse "Nariman Bamji," Ancestry.com.

———. 1978. "England and Wales Death Registration Index 1837–2007," database Family Search (https://www.familysearch.org/ark:/61903/1:1:QVZ3-DCJR: 22 October 2021), "Edwin Martin Case."

Goedert, James J., William C. Wallen, Dean L. Mann, Douglas M. Strong, Carolyn Y. Neuland, Mark H. Greene, Christine Murray, Joseph F. Fraumeni Jr., and William A. Blattner. 1982. "Amyl Nitrite May Alter T Lymphocytes in Homosexual Men." *The Lancet* 319 (8269): 412–15.

Goodman, Martin. 2008. *Suffer and Survive: Gas Attacks, Miners' Canaries, Spacesuits and the Bends: The Extreme Life of J. S. Haldane.* New York: Pocket Books.

Gordon, Cecil, Helen Spurway, and P. A. R. Street. 1939. "An Analysis of Three Wild Populations of *Drosophila Subobscura.*" *Journal of Genetics* 38 (1): 37–90.

Gordon-Taylor, G. 1943. "Injuries of the Large Intestine in War." *The Medical Press and Circular,* March 24, 180–83 (preserved in Box 19, Folder 31, Institute of Naval Medicine, Gosport, United Kingdom).

Government of Canada. "The Dieppe Raid, 19 August 1942." Veterans Affairs Canada. https://www.veterans.gc.ca/eng/remembrance/wars-and-conflicts/second-world -war/battle-of-dieppe. Accessed 2021.

Granatstein, J. L. 2016. *The Weight of Command: Voices of Canada's Second World War Generals and Those Who Knew Them.* Vancouver, BC: UBS Press.

Green, Samuel W. 1883. *A Complete History of the New York and Brooklyn Bridge, from Its Conception in 1866 to Its Completion in 1883.* New York: S. W. Green's Son.

Grehan, John, and Martin Mace. 2014. *Capital Ships at War, 1939–1945.* Barnsley, UK: Pen & Sword Books.

Gribble, M. de G. 1960. "A Comparison of the 'High-Altitude' and 'High-Pressure' Syndromes of Decompression Sickness." *British Journal of Industrial Medicine* 17: 181–86.

Griffiths, C. V. 1940. Letter from C. V. Griffiths for the Medical Director-General to the Secretary, Dunbar-Nasmith Sub-Committee, Royal Naval Medical School, July 23. HALDANE/5/6/3/1/16 19 June 1940. University College London Special Collection, College Archives, London.

Grosvenor, Joan, and L. M. Bates. 1956. *Open the Ports: The Story of Human Minesweepers.* London: William Kimber.

Grüneberg, Hans. 1943. *The Genetics of the Mouse*. London: Cambridge University Press.

———. 1944. Letter to J. B. S. Haldane, November 19. HALDANE/3/5/1/2/34 Letter from Hans Grüneberg to J. B. S. Haldane. University College London Special Collection, College Archives, London.

———. 1979. Interview by Margaret A. Brooks. Imperial War Museum, London. Catalogue number 4478. https://www.iwm.org.uk/collections/item/object/80004438.

———, R. Powell, and C. C. Spicer. 1943–44. "The Relations Between Antipersonnel Effect and Size of Projectile." HO 195/14/406 The Relations Between Antipersonnel Effect and Size of Projectile by H. Grüneberg, R. Powell and C. C. Spicer. Rotaprinted. The National Archives, Kew, Richmond, United Kingdom.

Grüneberg, Hans, and Werner Ulrich. 1950. *Moderne Biologie; Festschrift zum 60. Geburtstag von Hans Nachtsheim*. Berlin: F. W. Peters.

Grüneberg, Reuben. 2020. Interview by Rachel Lance.

Grüneberg-Capell, Elsbeth. 1934. *Systeme und Methoden der Orthodontie. Amerika, der Europäische Kontinent und England*. Inaugural-Dissertation zur Erlangung der Zahnarztlichen Doktowürde der Hohen Medizinischen Fakultät der Rheinischen Friedrich Wilhelms-Universität zu Bonn. Vorgelegt am 28 Februar 1934.

Guardian (UK). 1978. "New Delhi." February 18, 4.

Gueritz, Edward Findlay. 1997. Interview by Conrad Wood. Imperial War Museum, London. Catalogue number 17394. https://www.iwm.org.uk/collections/item/object/80016848

Haldane, Charlotte. 1949. *Truth Will Out*. London: G. Weidenfeld & Nicolson.

Haldane, John B. S. (undated). "Cats F. Domesticus Gatti Italiani." HALDANE/6/3/2 Personal, Family Correspondence. University College London Special Collection, College Archives, London.

———. (undated). MS.20548 Papers of J. B. S. Haldane, Part 1. National Library of Scotland, Edinburgh.

———. (undated). MS.20548 Papers of J. B. S. Haldane, Part 2. National Library of Scotland, Edinburgh.

———. 1914. Disciplinary Form Against Robert M. Murray, October 24. Private Papers of Naomi Mitchison, Documents.681. Imperial War Museum, London.

———. ca. 1914. Letter to Naomi Haldane (later Mitchison). Private Papers of Naomi Mitchison, Documents.681. Imperial War Museum, London.

———. 1914–18. MS.20582 WWI Field Service Notebook. National Library of Scotland, Edinburgh.

———. 1915. Letter to Naomi Haldane (later Mitchison), February 24. Private Papers of Naomi Mitchison, Documents.681. Imperial War Museum, London.

———. 1917. Letter to Nou (Naomi Mitchison, then Haldane), February 1. Private Papers of Naomi Mitchison, Documents.681. Imperial War Museum, London.

———. 1925. *Callinicus; a Defence of Chemical Warfare*. London: Kegan Paul, Trench, Trubner & Co. Ltd.

———. 1934. Application for Grant of Exit Permit, December 14. KV 2/1832 John Burdon Sanderson Haldane: "British. A well-known scientist, he announced his membership of the CPGB in 1942, having been an undisclosed member for some years previously." The National Archives, Kew, Richmond, United Kingdom.

———. 1938. *A.R.P. [Air Raid Precautions]*. London: Victor Gollancz Ltd.

———. 1938. *Heredity & Politics*. London: George Allen & Unwin Ltd.

———. 1938. Letter to Michael Pease, June 29. HALDANE/3/5/1/3/1 Letters between J. B. S. Haldane and Michael Pease. University College London Special Collection, College Archives, London.

——. 1938. Letter to Mr. Tanner, Secretary of University College, December 21. HAL-DANE/3/5/2/1/7 Letters between UCL and J. B. S. Haldane. University College London Special Collection, College Archives, London.

——. 1938. Multiple Letters and Articles. HALDANE/5/6/1/12 Haldane, John Burdon Sanderson. University College London Special Collection, College Archives, London.

——. 1939. Letter to J. Buckingham, Deputy Director, Department of Scientific Research & Experiment, Admiralty, August 12. HALDANE/5/6/3/1/3 6 June 1939. University College London Special Collection, College Archives, London.

——. 1939. Letter to Lieutenant Honorable Leslie Hore-Belisha, MP, the War Office, September 7. HALDANE/5/6/2/1 War Office. University College London Special Collection, College Archives, London.

——. 1939. Letter to "Sir," September 28. HALDANE/3/5/1/5/3 Letter from J. B. S. Haldane to an Unknown Addressee. University College London Special Collection, College Archives, London.

——. 1939. Letter to "To Whom It May Concern," June 15. HALDANE/3/5/1/5/2 Letters from J. B. S. Haldane to Ursula Philip and the UCL Provost. University College London Special Collection, College Archives, London.

——. 1939. Letter to the UCL Provost, March 25. HALDANE/3/5/1/5/2 Letters from J. B. S. Haldane to Ursula Philip and the UCL Provost. University College London Special Collection, College Archives, London.

——. 1940. Letter to Commander Fletcher, MP, House of Commons, July 25. HALDANE/1/5/3/44 Letter from J. B. S. Haldane to Commander Fletcher. University College London Special Collection, College Archives, London.

——. 1940. Letter to Edwin Martin Case, January 17. HALDANE/1/5/3/47 Letter from J. B. S. Haldane to the Undersecretary of State at the War Office. University College London Special Collection, College Archives, London.

——. 1940. Letter to Edwin Martin Case, February 7. HALDANE/1/5/3/54 Letter from J. B. S. Haldane to E. M. Case. University College London Special Collection, College Archives, London.

——. 1940. Letter to Hans Grüneberg, October 14. HALDANE/3/5/1/2/13 Letters between J. B. S. Haldane and Hans Grüneberg. University College London Special Collection, College Archives, London.

——. 1940. Letter to the Manager, Local Office of the Ministry of Labour and National Service, May 9. HALDANE/5/6/3/1/13 7 May 1940. University College London Special Collection, College Archives, London.

——. 1940. Letter to Messrs. N. A. C. Salvesen & Co., October 21. HALDANE/6/2/4/2 Papers re Rented Homes in Harpenden. University College London Special Collection, College Archives, London, United Kingdom.

——. 1940. Letter to Michael Fry Esq., February 19. KV 2/1832 John Burdon Sanderson Haldane: "British. A well-known scientist, he announced his membership of the CPGB in 1942, having been an undisclosed member for some years previously." The National Archives, Kew, Richmond, United Kingdom.

——. 1940. Letter to Professor E. D. Adrian, Trinity College, May 3. HALDANE/5/6/3/1/13 7 May 1940. University College London Special Collection, College Archives, London.

——. 1940. Letter to Robert C. Frederick, Royal Naval Medical School, January 10. HALDANE/5/6/3/1/8 9 January 1940. University College London Special Collection, College Archives, London.

——. 1940. Letter to Robert C. Frederick, Royal Naval Medical School, February 16. HALDANE/1/5/3/52 Letters from J. B. S. Haldane to Robert Frederick. University College London Special Collection, College Archives, London.

———. 1940. Letter to Robert C. Frederick, Royal Naval Medical School, April 13. HAL-DANE/1/5/3/66 Letters from J. B. S. Haldane to Robert Frederick. University College London Special Collection, College Archives, London.

———. 1940. Letter to Robert C. Frederick, Royal Naval Medical School, April 22. HAL-DANE/1/5/3/63 Letters from J. B. S. Haldane to Robert Frederick. University College London Special Collection, College Archives, London.

———. 1940. Letter to Robert C. Frederick, Royal Naval Medical School, May 7. HALDANE/1/5/3/64 Letters from J. B. S. Haldane to Robert Frederick. University College London Special Collection, College Archives, London.

———. 1940. Letter to Robert C. Frederick, Royal Naval Medical School, May 17. HALDANE/1/5/3/60 Letters from J. B. S. Haldane to Robert Frederick. University College London Special Collection, College Archives, London.

———. 1940. Letter to Robert C. Frederick, Royal Naval Medical School, June 18. HAL-DANE/5/6/3/1/15 Letters from J. B. S. Haldane to Robert Frederick. University College London Special Collection, College Archives, London.

———. 1940. Letter to Robert C. Frederick, Royal Naval Medical School, August 1. HAL-DANE/1/5/3/42 Letters from J. B. S. Haldane to Robert Frederick. University College London Special Collection, College Archives, London.

———. 1940. Letter to Robert C. Frederick, Royal Naval Medical School, October 8. HAL-DANE/5/6/3/1/17 Letters from J. B. S. Haldane to Robert Frederick. University College London Special Collection, College Archives, London.

———. 1940. Letter to Robert Davis, January 11. HALDANE/5/6/3/1/8 9 January 1940. University College London Special Collection, College Archives, London.

———. 1940. Letter to R. P. Winfrey, Esq., Secretary, May 15. HALDANE/5/6/3/1/13 7 May 1940. Letters between J. B. S. Haldane and the Sir Halley Stewart Trust. University College London Special Collection, College Archives, London.

———. 1940. Letter to R. P. Winfrey, Esq,, Secretary, June 20. HALDANE/3/5/1/3/5 Letters between J. B. S. Haldane and the Sir Halley Stewart Trust. University College London Special Collection, College Archives, London.

———. 1940. Letter to the Secretary of the Admiralty, June 29. HALDANE/3/5/1/3/6 Letters between J. B. S. Haldane and the Secretary of the Admiralty. University College London Special Collection, College Archives, London.

———. 1940. Letter to Surgeon Commander Seymour Rainsford, February 7. HALDANE/1/5/3/53 Letters from J. B. S. Haldane to Surgeon Commander Rainsford. University College London Special Collection, College Archives, London.

———. 1940. Letter to Surgeon Commander Seymour Rainsford, June 3. HALDANE/1/5/3/57 Correspondence between Seymour Rainsford and J. B. S. Haldane. University College London Special Collection, College Archives, London.

———. 1940. Letter to Surgeon Commander Seymour Rainsford, July 1. HALDANE/5/6/3/1/15 17 June 1940. University College London Special Collection, College Archives, London.

———. 1940. Letter to the Undersecretary of State, the War Office, January 17. HALDANE/1/5/3/47 Letter from J. B. S. Haldane to the Undersecretary of State at the War Office. University College London Special Collection, College Archives, London.

———. 1940. Letter to Walsham Brothers & Co. Ltd., June 11. HALDANE/5/6/3/1/14 11 June 1940. University College London Special Collection, College Archives, London.

———. 1940. Undated Notes and Bills. HALDANE/5/6/3/1/17 2 July 1940. University College London Special Collection, College Archives, London.

———. 1941. Letter to Dr. Andrew J. Morland, December 9. HALDANE/5/6/3/1/44 7 December 1941. University College London Special Collection, College Archives, London.

———. 1941. Letter to Messrs. N. A. C. Salvesen & Co., April 10. HALDANE/6/2/4/2 Papers re Rented Homes in Harpenden. University College London Special Collection, College Archives, London.

———. 1941. Letter to Robert C. Frederick, Royal Navy Medical School, March 14. HALDANE/5/6/3/1/32 24 February 1941. University College London Special Collection, College Archives, London.

———. 1941. Letter to Robert C. Frederick, Royal Naval Medical School, May 21. HALDANE/5/6/3/1/35 28 March 1941. University College London Special Collection, College Archives, London.

———. 1941. Letter to R. P. Winfrey, Esq., Secretary, January 29. HALDANE/3/5/1/3/5 Letters between J. B. S. Haldane and the Sir Halley Stewart Trust. University College London Special Collection, College Archives, London.

———. 1941. Compiled Report Beginning with Extracts from *Journal of Hygiene* XLI (3). FD 1/6068 Royal Naval Personnel Research Committee: Underwater Physiology. The National Archives, Kew, Richmond, United Kingdom.

———. ca. 1942. "Contributions to the Physiology of Diving." MS.20548 Papers of J. B. S. Haldane, Part 1. National Library of Scotland, Edinburgh.

———. ca. 1942. "Effects of High-Pressure Oxygen on Vision." MS.20548 Papers of J. B. S. Haldane, Part 1. National Library of Scotland, Edinburgh.

———. 1942. Letter to Commander Sladen at Fort Blockhouse, April 20. HALDANE/5/6/3/1/54 20 April 1942. University College London Special Collection, College Archives, London.

———. 1942. Letter to Edwin Martin Case, April 9. HALDANE/5/6/3/1/49 4 February 1942. University College London Special Collection, College Archives, London.

———. 1942. Letter to Joan Elizabeth Jermyn, August 18. Private collection of Andrew Bamji.

———. 1942. Letter to Mr. S. F. Bywaters Esq., April 8. HALDANE/6/2/4/1 Papers re Rented Homes in Harpenden. University College London Special Collection, College Archives, London.

———. 1943. Letter Addressed to "Dear Sir," April 29. Private collection of Andrew Bamji.

———. 1943. Letter to James Rendel, September 29. HALDANE/5/6/3/1/70 29 September 1943. University College London Special Collection, College Archives, London.

———. 1943. Letter to J. S. Barnes in Response to His Letter, December 23, C.P.34570/43. ADM 178/313 Royal Naval Personnel Research Committee: "proposed to terminate the appointment of Professor J. B. S. Haldane." The National Archives, Kew, Richmond, United Kingdom.

———. 1943. Letter to Lewis F. Richardson, November. HALDANE/3/5/2/2/67 Letters between L. Richardson and J. B. S. Haldane. University College London Special Collection, College Archives, London.

———. 1943. Letter to Pratt, February 17. HALDANE/1/5/3/31 Letter from J. B. S. Haldane to C. L. G. Pratt. University College London Special Collection, College Archives, London.

———. 1943. Letter to Professor Frederick, January 26. ADM 1/17701 Admiralty (5) and Compensation (9) and Civil Power and Legal Matters (19) and Estimates. The National Archives, Kew, Richmond, United Kingdom.

———. 1944. Letter to Professor Killick, April 17. HALDANE/5/2/2/182 Esther M. Killick. University College London Special Collection, College Archives, London.

———. 1944. Letter to "Sir," February 21. HALDANE/6/2/4/4 Financial Matters. University College London Special Collection, College Archives, London.

———. 1945. Letter to the Editor of *Fairplay*, January 9. HALDANE/5/6/3/1/62 1 November 1944. University College London Special Collection, College Archives, London.

———. 1945. Letter to Frank Jackson, April 25. HALDANE/3/5/2/1/33 Letters between J. B. S. Haldane and the Communist Party. University College London Special Collection, College Archives, London.

———. 1945. Letter to Hans Grüneberg, February 8. HALDANE/3/5/1/2/37 Letters between J. B. S. Haldane and Hans Grüneberg. University College London Special Collection, College Archives, London.

———. 1945. Letter to Harry Pollitt, August 8. KV 2/1832 John Burdon Sanderson Haldane: "British. A well-known scientist, he announced his membership of the CPGB in 1942, having been an undisclosed member for some years previously." The National Archives, Kew, Richmond, United Kingdom.

———. 1945. Letter to the Provost, University College London, December 7. HALDANE/3/5/1/4/16 Letters from J. B. S. Haldane to the UCL Provost. University College London Special Collection, College Archives, London.

———. 1946. Letter to the Provost, University College London, ca. April. HALDANE/3/5/1/1/6 Letters from J. B. S. Haldane to the UCL Provost. University College London Special Collection, College Archives, London.

———. 1946. Letter to the Secretary, Society for the Protection of Science and Learning, April 10. HO 405/28087 Kalmus, H. Date of birth: 11/01/1906. The National Archives, Kew, Richmond, United Kingdom.

———. 1946. University College London Non-Professorial Staff Form 3, ca. April. HALDANE/3/5/1/1/6 Letters between J. B. S. Haldane and UCL. University College London Special Collection, College Archives, London.

———. 1960. "The Scientific Work of J. S. Haldane." *Nature* 187 (Jul 9): 102–5.

———. 1968. "The Dark Religions." London: Pemberton Publishing Co. Ltd.

———. 1968. *Science and Life: Essays of a Rationalist*. London: Pemberton Publishing Co. Ltd.

———. 2009. *What I Require from Life*. New York: Oxford University Press.

———, and Edwin Martin Case. 1940. Untitled report labeled "SECRET." MS.20547 Papers of J. B. S. Haldane. National Library of Scotland, Edinburgh.

Haldane, John B. S., A. D. Sprunt, and Naomi Haldane. 1915. "Reduplication in Mice." *Journal of Genetics* 5: 133–35.

Haldane, John B. S., and Helen Spurway. ca. 1943. "Contributions to the Physiology of Diving." HALDANE/1/5/3/32 Summary of Physiological Results. University College London Special Collection, College Archives, London, United Kingdom.

———. 1944. "The Physiology of Diving with Oxygen-Air Mixtures," January. MS.20547 Papers of J. B. S. Haldane. National Library of Scotland, Edinburgh.

Haldane, John Scott. 1915. "War with Poisonous Gases: The Gap at Ypres Made by German Chlorine Vapor Bombs. Reports by the Official 'Eyewitness' and Dr. J. S. Haldane." *Current History* 2 (3): 458–63.

———. 1922. *Respiration*. New Haven: Yale University Press.

Haldane, Louisa Kathleen. 1961. *Friends and Kindred*. London: Faber and Faber Limited.

Hall, J. L. 1944. Action Report—Assault on Vierville-Colleville Sector, Coast of Normandy, Eleventh Amphibious Force, July 27. COM 11TH PHOBFOR—Report of Ops Period 6/4-29/44—Assault on Vierville-Colleville Sector, Coast of Normandy, France. US National Archives and Records Administration, Washington, DC.

Hallett, T. J. 1942. "Underwater Obstacles," June 17. Ref.2613/TOU/3 to Chief of Combined Operations. DEFE 2/963 Clearance of Underwater Obstacles: Organisation of Landing Craft Obstruction Clearance Units. The National Archives, Kew, Richmond, United Kingdom.

Hanley, M. B. 1949. B.2.b. Note on Hans Georg Kahle, September 16. KV 2/1566 Hans Kahle: "German. An officer in the German army in the First World War, Kahle went to …" The National Archives, Kew, Richmond, United Kingdom.

Harper's Weekly. 1870. Images of the construction of the Brooklyn Bridge (East River Bridge), New York City. Illustration in: *Harper's Weekly* 14 (Dec. 17, 1870), p. 812, Library of Congress Control Number 99471867. Library of Congress, Washington, DC, USA.

Harris, Peter. 1950. "He Risks Life Escaping from 'Submarines' in a Cottage." *The Daily Mirror*, November 29, 3 (preserved in Box 19, Folder 43, Institute of Naval Medicine, Gosport, United Kingdom).

Harvey, Mabel. 1947. Letter to J. B. S. Haldane, March 16. HALDANE/3/5/2/1/75 Letters from Mabel Harvey to J. B. S. Haldane. University College London Special Collection, College Archives, London.

Hastings, Nicholas. 1943. C.O.P.P. Depot Progress Report Number 33 (Week Ending 18th September 1943), September 22. DEFE 2/1111 Combined Operations Pilotage Party: Progress Reports and Honours and Awards. The National Archives, Kew, Richmond, United Kingdom.

———. 1944. C.O.P.P. Depot Progress Report Number 58 (Week Ending 12th March 1944), March 14. DEFE 2/1111 Combined Operations Pilotage Party: Progress Reports and Honours and Awards. The National Archives, Kew, Richmond, United Kingdom.

———. 1944. C.O.P.P. Progress Report Number 51 (Week Ending 23rd January 1944), January 25. DEFE 2/1111 Combined Operations Pilotage Party: Progress Reports and Honours and Awards. The National Archives, Kew, Richmond, United Kingdom.

Hayes, J. D. 1944. "Recommendation for Decoration, Ship Naval Party 1502C," August 11. ADM 1/15494 Foreign Countries (52): Port Mine Clearance Parties for Clearance of Underwater Obstructions in Captured Ports. The National Archives, Kew, Richmond, United Kingdom.

Henry, Hugh G. 1995. "The Calgary Tanks at Dieppe." *Canadian Military History* 4 (1): 61–74.

High Court of Justice in England. 1944. "England & Wales, National Probate Calendar (Index of Wills and Administration), 1958–95, London, England," digital image s.v. "Elsbeth Grüneberg," Ancestry.com.

Hilgard, J. E. 1875. "On Tidal Waves and Currents Along Portions of the Atlantic Coast of the United States." *American Journal of Science and Arts* 10 (56): 117–27.

Hills, Eric Arthur. 1998. Interview by Conrad Wood. Imperial War Museum. Catalogue number 18504. https://www.iwm.org.uk/collections/item/object/80017719.

Hobbs, D. H. 1942. Government Telegram. ADM 358/691 Lieutenant P. C. A. Brownrigg, RNVR: Report of Death Whilst Diving. The National Archives, Kew, Richmond, United Kingdom.

Holloway, Adrian George. 2005. Interview by Lindsay Baker. Imperial War Museum. Catalogue number 27767. https://www.iwm.org.uk/collections/item/object/80024996.

Home Office. 1939. "Female Enemy Alien—Exemption from Internment—Refugee Card (Elsbeth Grüneberg)," October 20. The National Archives, Kew, Richmond, United Kingdom; HO 396 WW2 Internees (Aliens) Index Cards 1939–47; Reference Number HO 396/227. Ancestry.com. UK, World War II Alien Internees, 1939–45 [database online]. Lehi, UT: Ancestry.com Operations, Inc., 2019.

———. 1939. "Female Enemy Alien—Exemption from Internment—Refugee Card (Ursula Philip)," November 24. The National Archives, Kew, Richmond, United Kingdom; HO 396 WW2 Internees (Aliens) Index Cards 1939–47; Reference Number HO 396/

68. Ancestry.com. UK, World War II Alien Internees, 1939–45 [database online]. Lehi, UT: Ancestry.com Operations, Inc., 2019.

———. 1939. "Male Enemy Alien—Exemption from Internment—Refugee Card (Hans Grüneberg)," October 6. The National Archives, Kew, Richmond, United Kingdom; HO 396 WW2 Internees (Aliens) Index Cards 1939–47; Reference Number HO 396/ 227. Ancestry.com. UK, World War II Alien Internees, 1939–45 [database online]. Lehi, UT: Ancestry.com Operations, Inc., 2019.

———. 1946. Naturalisation Certificate, October 5. HO 334/229/1742 Naturalisation Certificate: "Hans Kalmus. From Czechoslovakia. Resident in Harpenden, Hertfordshire. Wife's name Anna [Kalmus]. Children: George Ernest Kalmus, Peter Ignaz Paul Kalmus." Home Office Reference: K 15381. Certificate BZ1742 issued 5 October 1946. The National Archives, Kew, Richmond, United Kingdom.

Honour, George Butler. 1944. "Operation Neptune—Force 'S' Report Enclosure 2 to Narrative, Report on Operation 'Gambit,'" June 9. DEFE 2/419 Report by Naval Commander, Force "S." The National Archives, Kew, Richmond, United Kingdom.

———. 1987. Interview by Conrad Wood. Imperial War Museum, London. Catalogue number 9709. https://www.iwm.org.uk/collections/item/object/80009495.

Hornberger, Wilhelm. 1950. "Chapter IV-L Decompression Sickness." In *German Aviation Medicine World War II*, Vol. 1. Washington, DC: The Surgeon General, US Air Force.

Horton, Max. 1942. Letter to A. V. Alexander, First Lord of the Admiralty, with included Memorandum, September 18. ADM 1/25845 Training of Crews to Man X-Craft and Chariots (2-Man Human Torpedo). The National Archives, Kew, Richmond, United Kingdom.

Howe, Quincy. 1942. "The World Today," August 20. NARA ID 115384. US National Archives and Records Administration, Washington, DC.

Hudspeth, Ken. 1944. Letter to Captain (S) Fifth Submarine Flotilla HMS "Dolphin," January 22. ADM 179/323 Reconnaissance Report on Operation "Postage Able," from Naval Commander Force J. HMS VECTIS, Cowes. The National Archives, Kew, Richmond, United Kingdom.

Huelke, Donald F., and Guy S. Nusholtz. 1966. "Cervical Spine Biomechanics: A Review of the Literature." *Journal of Orthopaedic Research* 4(2): 232–45.

Hughes, Alexander. 1946. "History of COPP 1942–1945," February 16. DEFE 2/1116 Combined Operations Pilotage Parties: History and Reports. The National Archives, Kew, Richmond, United Kingdom.

Hughes, E. R. 1939. "Post-Mortem Examination on Ernest Mitchell," September 8. ADM 116/4429 HM Submarine *Thetis*: Post-Mortem Examinations, Submarine Escape Problems and Apparatus, Nasmith Committee and Sub-committee [Committee on Safety in Submarines (Dunbar-Nasmith): Physiological Subcommittee]. The National Archives, Kew, Richmond, United Kingdom.

Hughes-Hallett, J. 1942. "Operation JUBILEE. Naval Force Commander's Narrative," August 30. RG 331, Entry 23, Box 16. US National Archives and Records Administration, Washington, DC.

Hussey, T. A. 1942. Combined Operations Headquarters Docket Demolition of Underwater Obstacles, September 23. DEFE 2/963 Clearance of Underwater Obstacles: Organisation of Landing Craft Obstruction Clearance Units. The National Archives, Kew, Richmond, United Kingdom.

———. 1942. Combined Operations Headquarters Docket Demolition of Underwater Obstacles, October 19. DEFE 2/963 Clearance of Underwater Obstacles: Organisation of Landing Craft Obstruction Clearance Units. The National Archives, Kew, Richmond, United Kingdom.

———. 1942. Combined Operations Headquarters Docket Demolition of Underwater Obstacles, October 26. DEFE 2/963 Clearance of Underwater Obstacles: Organisation of Landing Craft Obstruction Clearance Units. The National Archives, Kew, Richmond, United Kingdom.

Huxen, Keith. 2017. "Operation Husky: The Allied Invasion of Sicily." nationalww2museum .org.

Innes, Charlton Thomas. 2001. Interview by Harry Moses. Imperial War Museum, London. Catalogue number 22615. https://www.iwm.org.uk/collections/item/object /80021225.

Irwin, Hugh Michael. 1987. Interview by Peter M. Hart. Imperial War Museum. Catalogue number 9956. https://www.iwm.org.uk/collections/item/object/80009739.

James, Rear Admiral. 1944. "Remarks of Commander-in-Chief," October 21. ADM 1/15494 Foreign Countries (52): Port Mine Clearance Parties for Clearance of Underwater Obstructions in Captured Ports. The National Archives, Kew, Richmond, United Kingdom.

James, R. S. 1944. "Operation 'Neptune,'" June 7. DEFE 2/420 Report by Naval Commander, Force "S." The National Archives, Kew, Richmond, United Kingdom.

Jameson, Commander W. S. 1939. Letter to J. B. S. Haldane, August 1. HALDANE/5/6/3/1/6 1 August 1939. University College London Special Collection, College Archives, London.

Jameson, William. 2004. *Submariners VC*. Minneapolis: Periscope Publishing, Limited.

Jebb, Geraldine. 1934. Letter of Acceptance to Joan Elizabeth Jermyn, March 20. BC AR.200.80 Joan E. Jermyn student record. Royal Holloway, University of London.

Jenkins, J. 1940. Letter to Divisional Detective Inspector, Metropolitan Police Hornsey Station, June 8. KV 2/1561 Hans Kahle: "German. An officer in the German army in the First World War, Kahle went to . . ." The National Archives, Kew, Richmond, United Kingdom.

Jermyn, Joan Elizabeth. 1937. Letter to Miss Jebb, September 17. BC AR.200.80 Joan E. Jermyn student record. Royal Holloway, University of London.

———. 1941. Letter to W. Barnicot, Esq., November 7. HALDANE/3/5/1/5/4 Letter from J. B. S. Haldane's Secretary to W. Barnicot. University College London Special Collection, College Archives, London.

———. 1943. Letter to E. L. Tanner, Esq., Secretary, University College London, May 14. HALDANE/3/5/1/3/8 Letters from J. B. S. Haldane's Secretary to the Secretaries of UCL and Rothamsted Experimental Station. University College London Special Collection, College Archives, London.

———. 1944. Letter to the Manager, Harpenden Branch, Ministry of Labour & National Service, July 29. Private collection of Andrew Bamji.

Jones, E. Bell. 1942. Letter to Professor Haldane, August 2. Private collection of Andrew Bamji.

Jones, R. V. 1941. Letter to Cameron Wright, March 14. Box 19, Folder 38, Institute of Naval Medicine, Gosport, United Kingdom.

Kagan, Neil, and Stephen G. Hyslop. 2016. *The Secret History of World War II: Spies, Code Breakers, and Covert Operations*. Washington, DC: National Geographic.

Kahle, Hans. 1939. Application for Certificate of Identity, June 4. KV 2/1561 Hans Kahle: "German. An officer in the German army in the First World War, Kahle went to . . ." The National Archives, Kew, Richmond, United Kingdom.

———. 1941. Letter to J. B. S. Haldane, May 5. HALDANE/5/6/3/1/37 5 May 1941. University College London Special Collection, College Archives, London.

Kalmus, George. 2020a. Interview by Rachel Lance.

Kalmus, Hans. 1952. *Genetics*. Middlesex, UK: Penguin Books.

———. 1984. "50 Years of Exiles Working at University College London, A Public Lecture," March 21.

———. 1991. *Odyssey of a Scientist*. London: George Weidenfeld and Nicolson Ltd.

Kalmus, Miriam. 2021. Interview by Rachel Lance.

Kalmus, Peter. 2020b. Interview by Rachel Lance.

Katalinić, V., M. Milos, D. Modun, I. Musić, and M. Boban. 2004. "Antioxidant Effectivess of Selected Wines in Comparison with (+)–Catechin." *Food Chemistry* 86(4): 593–600.

Keen, George. 2006. Interview by Peter M. Hart. Imperial War Museum. Catalogue number 28887. https://www.iwm.org.uk/collections/item/object/80026738.

Kersey, John. 2012. "Members of the San Luigi Orders: Bishop Sidney E. P. Needham." The Abbey-Principality of San Luigi. https://san-luigi.org/2012/08/19/members-of-the -san-luigi-orders-bishop-sidney-e-p-needham/.

Keyfitz, Nathan. 2004. *Notes of a Wayfarer*. Victoria, BC: Canadian Population Society.

Keys, Ancel, Josef Brožek, Austin Henschel, Olaf Mickelsen, Henry Longstreet Taylor, Ernst Simonson, Angie Sturgeon Skinner, Samuel M. Wells, J. C. Drummond, Russell M. Wilder, Charles Glen King, and Robert R. Williams. 1950. *The Biology of Human Starvation*. Minneapolis: University of Minnesota Press.

Kilvert-Jones, Tim. 1999. *Omaha Beach: V Corps' Battle for the Beachhead*. Barnsley, UK: Leo Cooper.

Knowles, Sydney. 2009. *A Diver in the Dark: Experiences of a Pioneer Royal Navy Clearance Diver*. Bognor Regis, UK: Woodfield Publishing.

Konstam, Angus. 2019. *Tirpitz in Norway: X-craft Midget Submarines Raid the Fjords, Operation Source 1943*. London: Bloomsbury Publishing.

Kraut, Alan M. 2021. *Goldberger's War: The Life and Work of a Public Health Crusader*. New York: Farrar, Straus and Giroux.

Kumar, Anil, Henning Pontoppidan, Konrad J. Falke, Roger S. Wilson, and Myron B. Laver. 1973. "Pulmonary Barotrauma During Mechanical Ventilation." *Critical Care Medicine* 1 (4): 181–86.

Kvam, Wayne. 1983. "Ernest Hemingway and Hans Kahle." *Hemingway Review* 2 (2): 18–22.

Lance, Rachel M., Bruce Capehart, Omar Kadro, and Cameron R. Bass. 2015. "Human Injury Criteria for Underwater Blasts." *PloS One* 10 (11).

Lance, Rachel M., and Cameron R. Bass. 2015. "Underwater Blast Injury: A Review of Standards." *Diving Hyperb Med* 45 (3): 190–99.

Landesarchiv Berlin. 1908. *"Berlin, Deutschland; Personenstandsregister Geburtsregister; Laufendenummer: 777,"* digital image s.v. "Anna Ursula Philip," Ancestry.com.

Lanker, Albert. ca. 1944. Shown Here Are Gun Positions in Caves and Barbed Wire Entanglements Along the Top of Cliffs Northest of Dieppe, France (U.S. Air Force Number 57354AC). RG 342, Series 342-FH, NAID: 204889853. US National Archives and Records Administration, Washington, DC, USA.

———. May 1944. Element "C" A Steel Gate-Like Boat Barricade, Failed to Stop Landings (U.S. Air Force Number 57358AC). RG 342, Series 342-FH, NAID: 204889859. US National Archives and Records Administration, Washington, DC, USA.

Larsen, Leif. 1942. "Report by Quartermaster Leif Larsen, K.K., D.S.M., R.N.N.," November 28. HS 2/203 Operation Title: Abortive Chariot Attack on *Tirpitz*. The National Archives, Kew, Richmond, United Kingdom.

Leach, R. M., P. J. Rees, and P. Wilmshurst. 1998. "Hyperbaric Oxygen Therapy." *British Medical Journal* 317 (7166): 1140–43.

Leatherwood, Jeffrey M. 2012. *Nine from Aberdeen*. Newcastle upon Tyne, UK: Cambridge Scholars Publishing.

Leggatt, W. R. C. 1944. "Enclosure (No. 1) Report of Naval Officer in Charge, Sword Area," June 18. No.S42/195/1. DEFE 2/420 Report by Naval Commander, Force "S." The National Archives, Kew, Richmond, United Kingdom.

Le Poer Trench, W. 1901. "The Congress on Tuberculosis (Letter to the Editor of the *Times*)." *Times* (London), November 13, 7.

Levine, Alan J. 1992. *The Strategic Bombing of Germany, 1940–1945*. Westport, CT: Praeger.

Lewis, D., and D. M. Hunt. 1984. "Hans Grüneberg, 26 May 1907–23 October 1982." *Biographical Memoirs of Fellows of the Royal Society* 30 (November): 226–47.

Lightman, Bernard, Gordon McOuat, and Larry Stewart. 2013. *The Circulation of Knowledge Between Britain, India and China*. Leiden, Netherlands: Brill.

Lin, Mosi, Maleka T. Stewart, Sidorela Zefi, Kranthi Venkat Mateti, Alex Gauthier, Bharti Sharma, Lauren R. Martinez, Charles R. Ashby, and Lin L. Mantell. 2022. "Dual Effects of Supplement Oxygen on Pulmonary Infection, Inflammatory Lung Injury, and Neuromodulation in Aging and COVID-19." *Free Radical Biology and Medicine* 190: 247–63.

Lindblom, Ulrika, and Carl Tosterud. 2021. "Pulmonary Barotrauma with Cerebral Arterial Gas Embolism from a Depth of 0.75–1.2 Metres of Fresh Water or Less: A Case Report." *Diving and Hyperbaric Medicine* 51 (2): 224–26.

Liochev, Stefan I. 2013. "Reactive Oxygen Species and the Free Radical Theory of Aging." *Free Radical Biology and Medicine* 60: 1–4.

Liow, M. H., B. H. Ho, S. J. Kim, and K. C. Tang. 2014. "MRI Findings in Cervical Spinal Cord Type II Neurological Decompression Sickness: A Case Report." *Undersea & Hyperbaric Medicine* 41 (6): 599–604.

Little, Charles. 1944. "Operation 'Postage Able' Appendix I—Information No. 0/0302/45/15," January 15. ADM 179/323 Reconnaissance Report on Operation "Postage Able" from Naval Commander Force J. HMS Vectis, Cowes. The National Archives, Kew, Richmond, United Kingdom.

———. 1944. "Operation 'Postage Able' Ref. No. 0/0302/45/15," January 15. ADM 179/323 Reconnaissance Report on Operation "Postage Able" from Naval Commander Force J. HMS Vectis, Cowes. The National Archives, Kew, Richmond, United Kingdom.

———. 1944. "Report on Operation 'Postage Able' No. 0/0302/45/15 to Allied Naval Commander, Expeditionary Force," February 7. ADM 179/323 Reconnaissance Report on Operation "Postage Able" from Naval Commander Force J. HMS Vectis, Cowes. The National Archives, Kew, Richmond, United Kingdom.

Lloyd, Helen. 2005. "Witness to a Century: The Autobiographical Writings of Naomi Mitchison." PhD, Department of Scottish Literature, University of Glasgow, 38.

Loadsman, J. A., and T. J. McCulloch. 2017. "Wielding the Search for Suspect Data—Is the Flood of Retractions About to Become a Tsunami?" *Anaesthesia* 72 (8): 931–35.

London School of Medicine for Women. 1948. Annual Reports 1945–48. H72/SM/A/02/01/008 London School of Medicine for Women and Related Collections. London Metropolitan Archives.

London SOE HQ. 1943. Title (starting file page 78), January 2. HS 2/203 Operation Title: Abortive Chariot Attack on *Tirpitz*. The National Archives, Kew, Richmond, United Kingdom.

Love, Christine. 2020. Interview by Rachel Lance.

Lovelace, William Randolph. 1944. "High Altitude Flight Testing of Flyers Personal Equipment in Specially Engineered B-17E, ENG-49-697-1H, 3/20/44," March 20. Curtiss-Wright Corporation Records NASM.XXXX.0067, Series 2, Box 75, Folder 19, Smithsonian National Air and Space Museum Archives, Washington, DC.

Lunn, Robert Alfred George. 2004. Interview by Peter M. Hart. Imperial War Museum, London. Catalogue number 27730. https://www.iwm.org.uk/collections/item/object/80024868.

Lyne, Lionel G. 1943. C.O.P.P. Depot Progress Report Number 39 (Week Ending 31st October 1943), November 3. DEFE 2/1111 Combined Operations Pilotage Party: Progress Reports and Honours and Awards. The National Archives, Kew, Richmond, United Kingdom.

———. 1944. C.O.P.P. 9 Progress Report Number 2 (Month Ending 31st March 1944), April 11. DEFE 2/1204 Combined Operations Pilotage Parties: Progress Reports. The National Archives, Kew, Richmond, United Kingdom.

———. 1944. C.O.P.P. 9 Progress Report Number 4 (Two Months Ending 30th June 1944), July 14. DEFE 2/1204 Combined Operations Pilotage Parties: Progress Reports. The National Archives, Kew, Richmond, United Kingdom.

———. 1944. "Operation Neptune—Force 'S' Report Enclosure 2 to Narrative, Report on Operation 'Gambit' (C.O.P.P. 9 in 'X' 23) No. Z/01," June 7. DEFE 2/419 Report by Naval Commander, Force "S." The National Archives, Kew, Richmond, United Kingdom.

MacIntosh, F. C., and W. D. M. Paton. 1950. "Notes on the 'Tauchretter.'" RNP 45/171 UPS 55. ADM 298/531 Notes on the "Tauchretter" (German Submerged Escape Apparatus). The National Archives, Kew, Richmond, United Kingdom.

MacLeod, N. 1940. Letter to Juan Negrín, October 31. HALDANE/1/5/3/39 Photograph of a letter from N. MacLeod to J. Negrín. University College London Special Collection, College Archives, London.

Magnello, M. Eileen. 1999. "The Non-Correlation of Biometrics and Eugenics: Rival Forms of Laboratory Work in Karl Pearson's Career at University College London, Part I." *History of Science* 37 (1): 79–106.

———. 1999. "The Non-Correlation of Biometrics and Eugenics: Rival Forms of Laboratory Work in Karl Pearson's Career at University College London, Part II." *History of Science* 37 (1): 123–50.

Mann, C. C. 1942. "Answers to General Operational Questions," Appendix A to HF 00/934/G(Plans), September 9. RG 331, Entry 23, Box 16. US National Archives and Records Administration, Washington, DC.

———. 1942. "Dieppe Raid—1942. Naval Lessons. Lessons Learnt," September. Item C, RG 331, Entry 23, Box 16. US National Archives and Records Administration, Washington, DC.

———. 1942. "Dieppe Raid—1942. Operation Instructions," August 5. Item F, RG 331, Entry 23, Box 16. US National Archives and Records Administration, Washington, DC.

———. 1942. Lecture Notes: "The Combined Services Raid on Dieppe, 19 Aug 42," August. RG 331, Entry 23, Box 16. US National Archives and Records Administration, Washington, DC.

———. 1942. "Report on Various Equipment 'Jubilee,'" October 18. Issued with OPS 3-3-1-2 Div. RG 331, Entry 23, Box 16. US National Archives and Records Administration, Washington, DC.

Mann, Chris. 2012. *British Policy and Strategy Towards Norway, 1941–45*. London: Palgrave Macmillan.

Marie, Jennifer. 2004. "The Importance of Place: A History of Genetics in 1930s Britain." PhD, History and Philosophy of Science, University College London, 52–54.

Marler, Peter. 1985. "Hark Ye to the Birds: Autobiographical Marginalia." In *Leaders in the Study of Animal Behavior*, edited by Donald A. Dewsbury, 322. Lewisburg, PA: Bucknell University Press.

Mass Observation. 1938–45. Observations Gathered August 1940; AIR RAIDS 1938–45, January 1938–December 1945. Collection 23-5-B. Mass Observation Online.

MASTER. 1940. Mass Obs Tube Report. 23-5-B Observations Gathered August 1940; Air Raids 1938–45, January 1938–December 1945. Mass Observation Online.

Matthey, R. 1938. Letter to Wright, October 25. Box 19, Folder 45, Institute of Naval Medicine, Gosport, United Kingdom.

McCartney, Innes. 2013. *British Submarines of World War I*. London: Bloomsbury Publishing.

McCullough, David. 1972. *The Great Bridge: The Epic Story of the Building of the Brooklyn Bridge*. New York: Simon & Schuster.

McKetin, Rebecca. 2018. "Methamphetamine Psychosis: Insights from the Past." *Addiction* 113: 1522–27.

Mellanby, Kenneth. 1942. Medical Research Council Form of Application for a Research Grant: Scabies Investigation, January 13. FD 1/6673 Research on scabies: "proposed nutrition experiments, on conscientious objectors, Sheffield." The National Archives, Kew, Richmond, United Kingdom.

Merritt, Alfred G. December 1943. Land mines, booby traps and parts. NH Series 80-G-264000, Catalog # 80-G-264411. Naval History and Heritage Command, Washington, DC, USA.

MI5. 1924–53. John Burdon Sanderson Haldane: "British. A well-known scientist, he announced his membership of the CPGB in 1942, having been an undisclosed member for some years previously." KV 2/1832. The National Archives, Kew, Richmond, United Kingdom.

———. 1938. Cross-Reference P.F. 45762 Subject J. B. S. Haldane, March 13. KV 2/1832 John Burdon Sanderson Haldane: "British. A well-known scientist, he announced his membership of the CPGB in 1942, having been an undisclosed member for some years previously." The National Archives, Kew, Richmond, United Kingdom.

———. 1938. Report, February 14. KV 2/1832 John Burdon Sanderson Haldane: "British. A well-known scientist, he announced his membership of the CPGB in 1942, having been an undisclosed member for some years previously." The National Archives, Kew, Richmond, United Kingdom.

———. 1939–42. Memo. KV 4/222 Policy and Procedure for the Imposition of Home Office Warrants for the Interception of Mail and Telephone Communications in the UK 1939–1945. The National Archives, Kew, Richmond, United Kingdom.

———. 1940. Copy, Original Filed in P.F. 48682 Vol. 6 123x, November 11. KV 2/1561 Hans Kahle: "German. An officer in the German army in the First World War, Kahle went to . . ." The National Archives, Kew, Richmond, United Kingdom.

———. 1940. Cross-Reference of Form P.F. 47192 Extract from Suggested Report for Lord Swinton re.Reorganisation of Czech Hostels, July 18. KV 2/1561 Hans Kahle: "German. An officer in the German army in the First World War, Kahle went to . . ." The National Archives, Kew, Richmond, United Kingdom.

———. 1940. Cross-Reference of Form P.F. 47192 Regarding Hans Georg Kahle, April 23. KV 2/1561 Hans Kahle: "German. An officer in the German army in the First World War, Kahle went to . . ." The National Archives, Kew, Richmond, United Kingdom.

———. 1940. Letter Reference Number K. 14233/3, November 5. KV 2/1561 Hans Kahle: "German. An officer in the German army in the First World War, Kahle went to . . ." The National Archives, Kew, Richmond, United Kingdom.

———. 1940. Letter to Home Office, Paddington District Office, London, June 29. KV 2/1561 Hans Kahle: "German. An officer in the German army in the First World War, Kahle went to . . ." The National Archives, Kew, Richmond, United Kingdom.

———. 1941. Re Hans Georg Kahle of No 10 North Hill Court, North Hill, N6, June 8. KV 2/1562 Hans Kahle: "German. An officer in the German army in the First World War, Kahle went to . . ." The National Archives, Kew, Richmond, United Kingdom.

———. 1942. Letter to Deputy Assistant Coordinator, Special Branch, Reference 0F104/1/ F.2.e/WO, February 4. KV 2/1562 Hans Kahle: "German. An officer in the German army in the First World War, Kahle went to . . ." The National Archives, Kew, Richmond, United Kingdom.

———. 1942. Telephone Check Glasgow Cen. 2976.27.7.42 Conversation Between McIlhone and MacShane, July 27. KV 2/1832 John Burdon Sanderson Haldane. The National Archives, Kew, Richmond, United Kingdom.

———. 1942. Timeline of Entries, Reflecting Tel. Check. Glasgow Cen. 2976.27.7.42, July 27. KV 2/1832 John Burdon Sanderson Haldane: "British. A well-known scientist, he announced his membership of the CPGB in 1942, having been an undisclosed member for some years previously." The National Archives, Kew, Richmond, United Kingdom.

———. 1945. Telephone Transcript of Monitored Conversation, April 17. KV 2/1832 John Burdon Sanderson Haldane: "British. A well-known scientist, he announced his membership of the CPGB in 1942, having been an undisclosed member for some years previously." The National Archives, Kew, Richmond, United Kingdom.

Miller, Harry M. 1934. Diary June 27–December 7. Officers' Diaries, RG 12, M–R; Rockefeller Foundation Records. Rockefeller Archive Center, Sleepy Hollow, NY, https://dimes .rockarch.org.

———. 1936. Diary May 1–December 16. Officers' Diaries, RG 12, M–R; Rockefeller Foundation Records. Rockefeller Archive Center, Sleepy Hollow, NY, https://dimes.rockarch.org.

———. 1937. Entry January 28, 1937. Diary 1937 January 6–May 28. Officers' Diaries, RG 12, M–R; Rockefeller Foundation Records. Rockefeller Archive Center, Sleepy Hollow, NY, https://dimes.rockarch.org.

———. 1939. Diary June 10–December 16. Officers' Diaries, RG 12, M–R; Rockefeller Foundation Records. Rockefeller Archive Center, Sleepy Hollow, NY, https://dimes .rockarch.org.

———. 1942. Entry April 23, Diary 1942–1944. Officers' Diaries, RG 12, M–R; Rockefeller Foundation Records. Rockefeller Archive Center, Sleepy Hollow, NY, https://dimes .rockarch.org.

Minchin, Leslie. 1997. *My Footsteps Through the Century*. London: Pioneer Arts.

Ministry of Labour. 1946. "Disabled Persons (Employment) Act, 1944 Certificate of Registration for J. B. S. Haldane," June 3. HALDANE/3/5/1/1/20 Letters between UCL and J. B. S. Haldane's Secretary. University College London Special Collection, College Archives, London.

Mitchell, Pamela. 1993. *The Tip of the Spear*. Slovenia: Gorenjski tisk p.o.

Mitchison, N. A. 2017. "J. B. S. Haldane, As I Knew Him." *Journal of Genetics* 96 (5): 729.

Mitchison, Naomi. 1968. "Beginnings." In *Haldane and Modern Biology*, edited by Krishna Dronamraju. Baltimore: Johns Hopkins Press.

———. 1988. *As It Was: Autobiography 1897–1918; All Change Here*. Glasgow: Richard Drew Publishing.

———. 1988. *As It Was: Autobiography 1897–1918; Small Talk: Memories of an Edwardian Childhood*. Glasgow: Richard Drew Publishing.

———. 2000. *Among You Taking Notes . . . : The Wartime Diary of Naomi Mitchison 1939–1945*. London: Phone Press.

Morland, Andrew. 1942. "Report on J. M. Rendel, Esq., by Physician," April 10. ADM 1/17701 Admiralty (5) and Compensatioin (9) and Civil Power and Legal Matters (19) and Estimates . . . The National Archives, Kew, Richmond, United Kingdom.

Morris, Desmond. 2006. *Watching: Encounters with Humans and Other Animals*. London: Max Publishing.

Morris, Joan. 1944. Letter to Joan Elizabeth Jermyn from Ministry of Labour, Reference No. A.D.S.359, March 15. Private collection of Andrew Bamji.

Mountbatten, Louis. 1942. "The Dieppe Raid" (Combined Report) C.B. 04244, October. RG 331, Entry 23, Box 17. US National Archives and Records Administration, Washington, DC.

———. 1943. Letter Ref. CR 4204/43 from Chief of Combined Operations, August 2. DEFE 2/1116 Combined Operations Pilotage Parties: History and Reports. The National Archives, Kew, Richmond, United Kingdom.

MPB. 1942. Operation Title (starting file page 43), October 10. HS 2/202 Operation Title: Abortive Chariot Attack on *Tirpitz*. The National Archives, Kew, Richmond, United Kingdom.

Multiple. 1940. Letters. HALDANE/5/6/3/1/22 29 July 1940. University College London Special Collection, College Archives, London.

———. 1940. Letters Regarding James Rendel's Lungs. HALDANE/5/6/3/1/48 26 January 1942. University College London Special Collection, College Archives, London.

Nasmith Committee. 1939. "Report of Admiral Nasmith's Committee of Methods of Saving Life in Sunken Submarines," December 18. ADM 116/4460 Methods of Lifesaving in Sunken Submarines: Investigations by Professor R. R. MacIntosh and Sir Robert Davies, Reports of Admiral Dunbar-Nasmith Committee, etc. The National Archives, Kew, Richmond, United Kingdom.

National Archives Prague. 1942. Entry and Related Documents for Arnošt Kalmus born 17.06.1869. *Institutem Terezínské iniciativy* (Terezín Initiative Institute), *Židovským muzeem v Praze* (Jewish Museum in Prague), and *Památníkem Terezín* (Terezín Memorial), https://www.holocaust.cz/en/database-of-victims/victim/97069-arnost -kalmus/.

———. 1942. Entry and Related Documents for Elsa Kalmusovà (Elsa Kalmus) born 19/09/ 1882. *Institutem Terezínské iniciativy* (Terezín Initiative Institute), *Židovským muzeem v Praze* (Jewish Museum in Prague), and *Památníkem Terezín* (Terezín Memorial), https://www.holocaust.cz/databaze-obeti/obet/97071-elsa-kalmusova/.

Naturalization Records. 1942. "New York, Southern District, U.S. District Court Naturalization Records, 1824–1946," database with images, Family Search (https://www .familysearch.org/ark:/61903/1:1:7WYQ-LM3Z: 8 March 2021), Manfred Ernest Philip, 1942.

Nature. 1940. "The Severe Winter of 1939–1940." *Nature* 145 (March 9): 376–77.

Naval Sea Systems Command. 2016. *US Navy Diving Manual Revision 7, SS521-AG-PRO-010*. Washington, DC: US Government Printing Office.

Needham, Sidney E. P. 1942. Diploma from Intercollegiate University, January 30. Box 19, Folder 39, Institute of Naval Medicine, Gosport, United Kingdom.

———. 1942. Letter to Horace Wright re Degree in the Faculty of Philosophy, February 2. Box 19, Folder 45, Institute of Naval Medicine, Gosport, United Kingdom.

———. 1942. Letter to "Sir" (Horace Cameron Wright), ca. January. Box 19, Folder 45, Institute of Naval Medicine, Gosport, United Kingdom.

New York Times. ca. 1941. The London Necropolis Railway Station, privately owned station in Westminster Bridge Road, after London's biggest night raid of the war. *New York Times* Paris Bureau Collection. RG 306, Series 306-NT, NAID: 541896. US National Archives and Records Administration, Washington, DC, USA.

———. May 1945. Anti-aircraft guns in Hyde Park go into action as "enemy bombers" make a daylight raid on London, during giant air defence exercises in which over 20,000 men and 1,300 RAF planes are taking part. *New York Times* Paris Bureau Collection.

RG 306, Series 306-NT, NAID: 541897. US National Archives and Records Administration, Washington, DC, USA.

———. September 7, 1940. This picture, taken during the first mass air raid on London, September 7, 1940, describes more than words ever could, the scene in London's dock area. Tower Bridge stands out against a background of smoke and fires. *New York Times* Paris Bureau Collection. RG 306, Series 306-NT, NAID: 541917. US National Archives and Records Administration, Washington, DC, USA.

New-York Tribune. 1872. "Another Death in the Bridge Caisson." April 7, 8.

———. 1872. "Death in the Bridge Caisson." April 24, 8.

———. 1897. "In a Caisson's Depths." August 15, 40.

Nicholson, George. 1999. Interview by Bob Watkins. Imperial War Museum. Catalogue number 19592. https://www.iwm.org.uk/collections/item/object/80018929.

Nissen, Jack Maurice. 1989. Interview by Conrad Wood. Imperial War Museum. Catalogue number 10762. https://www.iwm.org.uk/collections/item/object/80010539.

Norton, Trevor. 1999. *Stars Beneath the Seas: The Extraordinary Lives of the Pioneers of Diving.* London: Century.

Nossum, Rolf. 2012. "Emigration of Mathematicians from Outside German-Speaking Academia 1933–1963, Supported by the Society for the Protection of Science and Learning." *Historia Mathematica* 39 (1): 84–104.

Nowlan, Paul. 2020. Interview by Rachel Lance.

Oakley, Kenneth George. 1997. Interview by Conrad Wood. Imperial War Museum. Catalogue number 17294. https://www.iwm.org.uk/collections/item/object/80016853.

Observer (unnamed). 1940. Air Raids (No Clue to Sender), August 15. 23-5-B Observations Gathered August 1940; Air Raids 1938–45, January 1938–December 1945. Mass Observation Online.

Observer C.F. 1940. Indirect Report Dated September 25. F25C F30C Inv. 23-5-E Tube Shelters: Observations in the London Underground, September–November 1940; Air Raids 1938–45, January 1938–December 1945. Mass Observation Online.

———. 1940. Indirect Report Dated October 22. 23-5-E Tube Shelters: Observations in the London Underground, September–November 1940; Air Raids 1938–45, January 1938–December 1945. Mass Observation Online.

Observer D.H. 1940. Stations/Tube Shelters (Holland Park Tube). 23-5-B Observations Gathered August 1940; Air Raids 1938–45, January 1938–December 1945. Mass Observation Online.

Observer H.P. 1940. Air Raid (Night of 28.8.40). 23-5-B Observations Gathered August 1940; Air Raids 1938–45, January 1938–December 1945. Mass Observation Online.

———. 1940. Air Raid (Night of 30.8.40) Belgravia. 23-5-B Observations Gathered August 1940; Air Raids 1938–45, January 1938–December 1945. Mass Observation Online.

———. 1940. Air Raid Shelterers in Underground Railways, September 23. 23-5-E Tube Shelters: Observations in the London Underground, September–November 1940; Air Raids 1938–45, January 1938–December 1945. Mass Observation Online.

———. 1940. Air Raid Warning Dated October 17, 1940. 23-5-D Observations Gathered October–November 1940; Air Raids 1938–45, January 1938–December 1945. Mass Observation Online.

Observer K.B. 1940. Air Raid Alarm—Battersea, August 15. 23-5-B Observations Gathered August 1940; Air Raids 1938–45, January 1938–December 1945. Mass Observation Online.

Observer L.E. 1940. Evacuees at Oxford, Report Dated September 19. File Report 414. Mass Observation Online.

——. 1940. Tube Shelter Observations, Dated September 25. 23-5-E Tube Shelters: Observations in the London Underground, September–November 1940; Air Raids 1938–45, January 1938–December 1945. Mass Observation Online.

Observer R. 1940. Air Raid Warning in Battersea on Thursday, August 15 (by male student, age 20). 23-5-B Observations Gathered August 1940; Air Raids 1938–45, January 1938–December 1945. Mass Observation Online.

Observer S.S. 1940a. Maida Vale Air Raid, September 6. 23-5-C Observations Gathered September 1940; Air Raids 1938–45, January 1938–December 1945. Mass Observation Online.

——. 1940b. Maida Vale Air Raid, September 9. 23-5-C Observations Gathered September 1940; Air Raids 1938–45, January 1938–December 1945. Mass Observation Online.

Observer T.H. 1940. Untitled Investigator Report, September 10. 23-5-C Observations Gathered September 1940; Air Raids 1938–45, January 1938–December 1945. Mass Observation Online.

Observer Vaugan. 1940. "Untitled Report of Tube Station Shelters, Dated September 23, 1940." 23-5-E Tube Shelters: Observations in the London Underground, September–November 1940; Air Raids 1938–45, January 1938–December 1945. Mass Observation Online.

Office of Naval Commander. 1944. "Operation Neptune—Force 'S' Orders" (Short Title ONEAST/S), May 21; ONEAST/S.7B—The Assault. DEFE 2/403 Force "S" Orders. The National Archives, Kew, Richmond, United Kingdom.

——. 1944. "Operation Neptune—Force 'S' Orders" (Short Title ONEAST/S), May 21; ONEAST/S.1—General. DEFE 2/403 Force "S" Orders. The National Archives, Kew, Richmond, United Kingdom.

——. 1944. "Operation Neptune—Force 'S' Orders" (Short Title ONEAST/S), May 21; ONEAST/S.18—Navigational Data. DEFE 2/403 Force "S" Orders. The National Archives, Kew, Richmond, United Kingdom.

——. 1944. "Operation Neptune—Force 'S' Orders" (Short Title ONEAST/S), May 21; ONEAST/S.8B—Orders for Close Support. DEFE 2/403 Force "S" Orders. The National Archives, Kew, Richmond, United Kingdom.

——. 1944. "Operation Neptune—Force 'S' Orders" (Short Title ONEAST/S), May 21; ONEAST/S.7A—The Passage. DEFE 2/403 Force "S" Orders. The National Archives, Kew, Richmond, United Kingdom.

O'Hara, Vincent P., and Enrico Cernuschi. 2015. "Frogmen Against a Fleet: The Italian Attack on Alexandria 18/19 December." *Naval War College Review* 68 (3): 119–37.

O'Keefe, David. 2020. *One Day in August: Ian Fleming, Enigma, and the Deadly Raid on Dieppe.* London: Icon Books Ltd.

Olson, Kent, and Craig Smollin. 2008. "Carbon Monoxide Poisoning (Acute)." *BMJ Clinical Evidence* 7 (2103): 1–12.

Owen, James. 2012. *Commando: Winning World War II Behind Enemy Lines.* London: Abacus.

Pals, Laurens Klass. 1980. Interview by Conrad Wood. Imperial War Museum. Catalogue number 4642. https://www.iwm.org.uk/collections/item/object/80004602.

Passenger List. 1940. "New York, New York, Passenger and Crew Lists, 1909, 1925–1957," database with images Family Search (https://familysearch.org/ark:/61903/1:1:24LS -GY9: 2 March 2021), Georg Orkin Philip, 1940; citing Immigration, New York, New York, United States, NARA Microfilm Publication T715 (Washington, DC: National Archives and Records Administration, n.d.).

——. 1941. "New York, New York Passenger and Crew Lists, 1909, 1925–1957," database with images Family Search (https://familysearch.org/ark:/61903/1:1:24LX-JHT: 2

March 2021), Dagobert Philip, 1941; citing Immigration, New York, New York, United States, NARA Microfilm Publication T715 (Washington, DC: National Archives and Records Administration, n.d.).

Patterson, T. C. 1940. Letter from the Medical Director General to the Secretary, Dunbar-Nasmith Sub-Committee, Royal Naval Medical School, June 15. HALDANE/5/6/3/1/16 19 June 1940. University College London Special Collection, College Archives, London.

Philip, Ursula. 1934. *"DiePaarungderGeschlechtschromosomen von Drosophila melanogaster."* PhD, Abteilung R. Goldschmidt, Kaiser Wilhelm-Institut für Biologie.

———. 1934. General Information/*Allgemeine Auskunft*, August 27. M.S. S.P.S.L. 203/1-5 Ursula A. Philip; Guido P. A. Pontecorvo; Ernst G. Pringsheim; Felix Rawitscher; Carl F. Robinow. Date not recorded at time of cataloguing. Weston Library, Bodleian, Oxford.

———. 1946. Letter to Ms. Ursell, Secretary, Society for the Protection of Science and Learning, December 13. Council for At-Risk Academics (CARA), MS. S.P.S.L. 203/1-5. Bodleian Library, Oxford University, Oxford.

———. 1964. "Genetics, and Eugenics in Post-War Germany." *The Eugenics Review* 56 (2): 91–92.

———, and John B. S. Haldane. 1939. "Relative Sexuality in Unicellular Algae." *Nature* 143 (February 25): 334.

Phillips, G. C. 1945. SM 04349/820 in Reply to Register No. C.W.(C)40/43 Part III Minute Sheet No. 1, June 3. ADM 358/1373 Temporary Lieutenant H. F. Cook, RNVR, Temporary Lieutenant C. E. Bonnell, RCNVR, etc. The National Archives, Kew, Richmond, United Kingdom.

Pike, W. S. 2004. "Rainfall Records of the East Anglian Blooms, 1911–88." *Weather* (Royal Meteorological Society) 59 (4): 103–6.

Pincher, Chapman. 1947. "Man to Swim Against Whale, Frog Suit Attack by Scientist." March.

Pirie, Norman Wingate. 1966. "John Burdon Sanderson Haldane." In *Biographical Memoirs of Fellows of the Royal Society*, 218–49. London: The Royal Society.

Place, Basil Charles Godfrey. 1988. Interview by Conrad Wood. Imperial War Museum. Catalogue number 10431. https://www.iwm.org.uk/collections/item/object/80010209.

Polacchini, Romolo. 1943. Message Titled "Naval Command, Palermo," May 8. ADM 223/579 Operation Principal: Chariot (Two-Man Submarine) Attacks. The National Archives, Kew, Richmond, United Kingdom.

Porteous, Patrick Antony. 1987. Interview by Conrad Wood. Imperial War Museum, London. Catalogue number 10060. https://www.iwm.org.uk/collections/item/object/80009842.

Portsmouth Bomb Map, The 2020. Portsmouth Museum and Art Gallery, Portsmouth, United Kingdom. Accessed January 2022. https://portsmouthmuseum.co.uk/collections-stories/80th-anniversary-of-the-blitz/the-portsmouth-bomb-map/.

Pratt, C. Lucan G. 1943. Letter to J. B. S. Haldane, February 22. HALDANE/1/5/3/30 Letter from C. L. G. Pratt to J. B. S. Haldane. University College London Special Collection, College Archives, London.

———. 1943. "The Organization of Naval Physiological Research and Instruction," October 14. FD 1/6069 Royal Naval Personnel Research Committee: Royal Naval Physiological Laboratory: Plans. The National Archives, Kew, Richmond, United Kingdom.

Prenatt, Jamie, and Mark Stille. 2014. *Axis Midget Submarines 1939–1945*. London: Osprey Publishing Limited.

Preston, Paul. 2007. *Spanish Civil War: Reaction, Revolution, and Revenge.* New York: W. W. Norton.

Public Record. "United States Public Records, 1970–2009," database Family Search (https://familysearch.org/ark:/61903/1:1:KP7C-H3Q: 21 November 2019), George O. Philip, 1997–1997.

Pugh, James. 2018. "The Royal Air Force, Bomber Command and the Use of Benzedrine Sulphate: An Examination of Policy and Practice During the Second World War." *Journal of Contemporary History* 53 (4): 740–61.

Pye, D. R. 1943. Letter Addressed "My Dear Haldane," August 31. FD 1/6068 Royal Naval Personnel Research Committee: Underwater Physiology. The National Archives, Kew, Richmond, United Kingdom.

———. 1946. Letter to Hans Kalmus, February 25. HALDANE/3/5/1/6/3 Letter from the UCL Secretary to H. Kalmus. University College London Special Collection, College Archives, London.

Quick, Dan. 1970. "A History of Closed Circuit Oxygen Underwater Breathing Apparatus." School of Underwater Medicine, HMAS Penguin, Balmoral, New South Wales.

Rainsford, Seymour Grome. 1940. Letter to J. B. S. Haldane, June 4. HALDANE/5/6/3/1/ 13 7 May 1940. University College London Special Collection, College Archives, London.

———. 1940. Letter to J. B. S. Haldane, July 6. HALDANE/5/6/3/1/19 6 July 1940. University College London Special Collection, College Archives, London.

Rainsford, Seymour Grome, and Robert C. Frederick. 1939. Letter to the Secretary, Admiral Dunbar-Naismith's [*sic*] Committee, July 27. HALDANE/5/6/3/1/5 27 July 1939. University College London Special Collection, College Archives, London.

Ramsay, B. H. 1944. Memo Ref. No. X/0562/21 Titled "Operations 'K.J.H.' and 'Bellpush,'" January 17. ADM 179/354 Operation "K.J.H.": Reference to Reconnaissance: Minutes of Meeting. The National Archives, Kew, Richmond, United Kingdom.

R. E., Lieutenant Colonel. May 24, 1943. "Some Notes on Underwater Obstacles," May 24, EVD/HJAF. DEFE 2/963 Clearance of Underwater Obstacles: Organisation of Landing Craft Obstruction Clearance Units. The National Archives, Kew, Richmond, United Kingdom.

Reese, H. C. 1942. "Dieppe Raid—Outline Plan," September 28. Item A. RG 331, Entry 23, Box 16. US National Archives and Records Administration, Washington, DC.

Reinold, Marcel, and John Hoberman. 2014. "The Myth of the Nazi Steroid." *The International Journal of the History of Sport* 31 (8): 871–83.

Rendel, Joan. 1939. Letter to J. B. S. Haldane, April 14. HALDANE/3/5/1/3/2 Letters between J. B. S. Haldane and Joan Rendel. University College London Special Collection, College Archives, London.

———. 1940. Letter to Naomi Mitchison, October 8. Private collection of Celia Robertson.

———. 1940. Letter to Naomi Mitchison, October 25. Private collection of Celia Robertson.

———. 1940. Letter to Naomi Mitchison, December 5. Private collection of Celia Robertson.

———. 1941. Letter to Naomi Mitchison, January 13. Private collection of Celia Robertson.

———. 1941. Letter to Naomi Mitchison, March 31. Private collection of Celia Robertson.

———. 1941. Letter to Naomi Mitchison, April 19. Private collection of Celia Robertson.

RNPRC. 1945. Minutes of the Twenty-First Meeting of the Royal Naval Personnel Research Committee, June 20. ADM 298/331 Underwater Physiology Sub-committee: 21st Meeting. The National Archives, Kew, Richmond, United Kingdom.

———. 1948. "Memorandum on Human Blast Trials Which It Is Proposed to Carry Out This Year." ADM 298/22 Human Blast Trials: Memorandum. The National Archives, Kew, Richmond, United Kingdom.

Roberts, John Hamilton. 1942. "Comd 1 Cdn Corps Report Operation JUBILEE," August 27. RG 331, Entry 23, Box 16. US National Archives and Records Administration, Washington, DC.

Robertson, Celia. 2008. *Who Was Sophie? The Lives of My Grandmother, Poet and Stranger.* London: Virago Press.

Robertson, Jane. 2020. Interview by Rachel Lance.

Rodwell & Sons. 1942. Estimate, August 8. HALDANE/6/2/4/1 Papers re Rented Homes in Harpenden. University College London Special Collection, College Archives, London.

Roe, Elsa. 2020. Interview by Rachel Lance.

Rosenhead, Jonathan. 1991. "Swords into Ploughshares: Cecil Gordon's Role in the Post-War Transition of Operational Research to Civilian Uses." *Public Administration* 69 (Winter): 481–501.

Roxby, Thomas Reginald. 1992. Interview by Conrad Wood. Imperial War Museum. Catalogue number 12722. https://www.iwm.org.uk/collections/item/object/80012454.

Royal Courts of Justice. 1939. Testimony of Professor John Burdon Sanderson Haldane, July 20. HALDANE/1/5/2/23 Minutes of Tribunal of Inquiry into the Loss of HM Submarine *Thetis*. University College London Special Collection, College Archives, London.

Royal Navy. 1940–45. Military Records for Ronald Henry Saull, born 7.Jan.1913, Official Number JX238164. Navy Command Secretariat Disclosures.

S.5. May 6, 1942. Confidential Message 2058B/6th May from S.5 to Admiralty, May 6. ADM 358/691 Lieutenant P. C. A. Brownrigg, RNVR: Report of Death Whilst Diving. The National Archives, Kew, Richmond, United Kingdom.

S.10. 1943. Message 1938A/4th January to C. in C. Mediterranean, N.C.X.F., January 4. ADM 223/579 Operation Principal: Chariot (Two-Man Submarine) Attacks. The National Archives, Kew, Richmond, United Kingdom.

———. 1943. Report Subject C.O.P.P. Reconnaissance No. 3357/05 to Commander-in-Chief Mediterranean, January 8. DEFE 2/1116 Combined Operations Pilotage Parties: History and Reports. The National Archives, Kew, Richmond, United Kingdom.

———. 1943. Report Subject C.O.P.P. Reconnaissance No. 3357/05 to Commander-in-Chief Mediterranean, July 18. DEFE 2/1116 Combined Operations Pilotage Parties: History and Reports. The National Archives, Kew, Richmond, United Kingdom.

S.10, Captain. 1943. Message 1111A/5th Jan to F.O.S. Repeated Admiralty. ADM 223/579 Operation Principal: Chariot (Two-Man Submarine) Attacks, January 6. The National Archives, Kew, Richmond, United Kingdom.

———. 1943. Message 1643A/14th January Record 2330, January 14. ADM 358/1373 Temporary Lieutenant H. F. Cook, RNVR, Temporary Lieutenant C. E. Bonnell, RCNVR, etc. The National Archives, Kew, Richmond, United Kingdom.

Sakamaki, Kazuo. 2017. *I Attacked Pearl Harbor: The True Story of America's POW #1.* Translated by Toru Matsumoto. Honolulu: Rollston Press.

Salvesen, N. A. C. 1940. Letter to J. B. S. Haldane, "Crossways," Harpenden, October 17. HALDANE/6/2/4/1 and 2 Financial Records. University College London Special Collection, College Archives, London.

———. 1941. Letter to J. B. S. Haldane, "Crossways," Harpenden, April 1. HALDANE/6/2/4/2 Financial Records. University College London Special Collection, College Archives, London.

———. 1942. "Schedule of Dilapidations Accrued During the Furnished Tenancy of Professor J. B. S. Haldane at 'Crossways' Salisbury Avenue, Harpenden," May. HALDANE/6/2/4/2 Papers re Rented Homes in Harpenden. University College London Special Collection, College Archives, London.

Sams, H. W. H. 1943. Letter P.F.45762/C.3a ADNI for Lieutenant Anderson, February 23. KV 2/1832 John Burdon Sanderson Haldane: "British. A well-known scientist, he announced his membership of the CPGB in 1942, having been an undisclosed member for some years previously." The National Archives, Kew, Richmond, United Kingdom.

Sarkar, Sahotra. 1992. "Centenary Reassessment of J. B. S. Haldane, 1892–1964." *BioScience* 42 (10): 777–85.

Scott, Dixon. 1942–1945. Houses of Parliament and "Big Ben"—London. World War II Foreign Posters. RG 44, Series 44-PF, NAID: 44266642. US National Archives and Records Administration, Washington, DC, USA.

Scott-Bowden, Logan. 1943. Letter Addressed "Dear Colonel Dove," November 29. DEFE 2/971 Combined Operations Pilotage Parties: Brief on Information Required, Establishment, Function and Use. The National Archives, Kew, Richmond, United Kingdom.

Secretary to Haldane (unnamed). 1947. Letter to HM Inspector of Taxes, October 7. HALDANE/3/5/1/1/42 Letters between J. B. S. Haldane and HM Inspector of Taxes. University College London Special Collection, College Archives, London.

Selective Service System. 1943. Registration Card for George Orkin Philip, Order Number 11394. HO 405/28087 Kalmus, H. Date of birth: 11/01/1906. The National Archives, Kew, Richmond, United Kingdom.

SHAEF. 1944. SHAEF/315GX/INT ("Report on Underwater Obstacles"), ca. May. RG 331, Entry 12, Box 13. US National Archives and Records Administration, Washington, DC.

Shelford, William O. 1942. Letter to J. B. S. Haldane from Lieutenant Commander Shelford of the Royal Navy, May 13. MS.20534 Correspondence of J. B. S. Haldane, Part 1. National Library of Scotland, Edinburgh.

———. 1944. Letter to the Chairman of the Sub-Committee on Underwater Physiology, from Shelford, Commander in the Royal Navy and Secretary of the Admiralty Diving Committee at HMS *Dolphin*, January 13. MS.20569 Miscellany of Papers, Chiefly Undated, Chiefly of J. B. S. Haldane, Including Notes, Memoranda, Reports and Mathematical Calculations Concerning Diving. National Library of Scotland, Edinburgh.

———. 1945. Letter to J. B. S. Haldane, March 4. HALDANE/5/6/3/1/65 4 March 1945. University College London Special Collection, College Archives, London.

———. 1960. *Subsunk: The Story of Submarine Escape*. Garden City, NY: Doubleday & Company, Inc.

Shepherd, Jesse Lionel. 1990. Interview by Conrad Wood. Imperial War Museum, London. Catalogue number 11626. https://www.iwm.org.uk/collections/item/object/80011373.

Sheppard, William. 1993. Interview by Peter M. Hart. Imperial War Museum, London. Catalogue number 13455. https://www.iwm.org.uk/collections/item/object/80013171.

Sherwood, James Barlow Brooks. 1987. Interview by Conrad Wood. Imperial War Museum, London. Catalogue number 9783. https://www.iwm.org.uk/collections/item/object/80009567.

Siebe Gorman & Co. Ltd. 1943. Letter to Chief of Combined Operations re Contract No. C.P.6E/28183/43/AP.2606, December 21. ADM 1/15583 Combined Operations (47): Supply of Stores to Combined Operations Pilotage Parties. The National Archives, Kew, Richmond, United Kingdom.

(signature illegible), headmistress, Benenden School. 1934. Letter of recommendation for Betty Jermyn, January 12. BC AR.200.80. Joan E. Jermyn student record. Royal Holloway, University of London.

——— for ADNI. 1941. Letter to MI5 for Captain Bennett, April 1. KV 2/1563. Hans Kahle: "German. An officer in the German army in the First World War, Kahle went to . . ." The National Archives, Kew, Richmond, United Kingdom.

——— for Director of Scientific Research. 1941. Letter to Cameron Wright, March 31. Box 19, Folder 27, Institute of Naval Medicine, Gosport, United Kingdom.

———. 1943. Letter Addressed "Dear Pye," October 16. Ref 3044/1g. FD 1/6068. Royal Naval Personnel Research Committee: Underwater Physiology. The National Archives, Kew, Richmond, United Kingdom.

——— for Chief of Staff. 1943. Underwater Obstacles, CR.12660/43, December 17. DEFE 2/963 Clearance of Underwater Obstacles: Organisation of Landing Craft Obstruction Clearance Units. The National Archives, Kew, Richmond, United Kingdom.

———. 1944. Letter to Joan E. Jermyn from the Assistant to the Academic Register at Richmond College, August 8. Private collection of Andrew Bamji.

———, Major. 1944. Letter C.M.P. to S.S. Group titled "C.O.P.P. Reconnaissance," DEFE 2/1116, January 3. Combined Operations Pilotage Parties: History and Reports. The National Archives, Kew, Richmond, United Kingdom.

———. 1945. The Work of the "P" Parties, with Letter to Sir Robert Knox, March 30. Ref H.&A.120/45. Adm 1/15494 Foreign Countries (52): Port Mine Clearance Parties for Clearance of Underwater Obstructions in Captured Ports. The National Archives, Kew, Richmond, United Kingdom.

———. 1946. Note on inside cover, June 25: "HO 405/28087 Kalmus, H. Date of birth: 11/01/ 1906." The National Archives, Kew, Richmond, United Kingdom.

———, Manager. 1946. Letter to J. B. S. Haldane from Ministry of Labour, May 17. Haldane/ 3/5/1/1/7. Letter from the Ministry of Labour to J. B. S. Haldane. University College London Special Collection, College Archives, London.

Sladen, G. M. 1942. Letter to Sir Robert Davis, Siebe Gorman, May 14. MS.20534 Correspondence of J. B. S. Haldane, Part 1. National Library of Scotland, Edinburgh.

Smith, Andrew H. 1873. *The Effects of High Atmospheric Pressure, Including the Caisson Disease.* Brooklyn, NY: Eagle Print.

Smith, John Maynard. 1993. *The Theory of Evolution.* Cambridge: Cambridge University Press.

———. 1997. Interview by Richard Dawkins. Web of Stories https://www.youtube.com /watch?v=qf1lsDsC-J0.

Smith, L. Lewis. 1941. Letter on Behalf of the Medical Director-General, Medical Department of the Navy, to the Secretary, Dunbar-Nasmith's Physiological Sub-Committee, Royal Naval Medical School, February 24. HALDANE/5/6/3/1/32 24 February 1941. University College London Special Collection, College Archives, London.

Smith, Mrs. O. 1940. Some Blitz Reactions in This District (Maida Vale). 23-5-C Observations Gathered September 1940; Air Raids 1938–45, January 1938–December 1945. Mass Observation Online.

SNL. 1942. Title (S.N.L. No. 1012) (starting file page 161), November 13. HS 2/203 Operation Title: Abortive Chariot Attack on *Tirpitz.* The National Archives, Kew, Richmond, United Kingdom.

SOE Agent 8627 in Stockholm. 1942. Title (starting file page number 83), December 4. HS 2/203 Operation Title: Abortive Chariot Attack on *Tirpitz.* The National Archives, Kew, Richmond, United Kingdom.

SOE Agent 8628. 1942. Title (starting file page 236), November 12. HS 2/202 Operation Title: Abortive Chariot Attack on *Tirpitz.* The National Archives, Kew, Richmond, United Kingdom.

SOE Agent in Trondheim. 1942. Title (starting file page 44). HS 2/203 Operation Title: Abortive Chariot Attack on *Tirpitz*. The National Archives, Kew, Richmond, United Kingdom.

———. 1943. "Translation of Report from Agent in the Field Regarding 'Title,'" January 12. HS 2/203 Operation Title: Abortive Chariot Attack on *Tirpitz*. The National Archives, Kew, Richmond, United Kingdom.

Sontag, Sherry, and Christopher Drew with Annette Lawrence Drew. 2008. *Blind Man's Bluff: The Untold Story of American Submarine Espionage*. New York: PublicAffairs.

Southworth, Helen, ed. 2012. *Leonard and Virginia Woolf, the Hogarth Press and the Networks of Modernism*. Edinburgh: Edinburgh University Press.

Spradbery, Walter E. 1942–1945. The Proud City—St. Thomas's Hospital and the Houses of Parliament. World War II Foreign Posters. RG 44, Series 44-PF, NAID: 44267211. US National Archives and Records Administration, Washington, DC, USA.

Springfield Herald (IL). 1956. "Feeling Old?" May 17, 2.

Spurway, Helen. 1938. "An Analysis of Two Free-Living Populations of *Drosophila Subobscura*." PhD, Imperial College of Science and Technology and Bedford College, University of London.

———. ca. 1940. Letter to "The Occupier." HALDANE/3/5/1/4/1 Letter from the Subdean of the UCL Faculty of Science to J. B. S. Haldane. University College London Special Collection, College Archives, London.

———. ca. 1940–44. "Personnel and Equipment." MS.20569 Miscellany of Papers, Chiefly Undated, Chiefly of J. B. S. Haldane, Including Notes, Memoranda, Reports and Mathematical Calculations Concerning Diving. National Library of Scotland, Edinburgh.

———. 1943. "Newt Larvae in Brackish Water." *Nature* 151 (January 23): 109–10.

Stanbury, R. 1943. Report on Pilotage in Operation "Avalanche," Letter No. 85/0 to the Captain(S) Eighth Submarine Flotilla, September 17. DEFE 2/1116 Combined Operations Pilotage Parties: History and Reports. The National Archives, Kew, Richmond, United Kingdom.

Stoller, Irene. 2020. Email to author, November 9.

Strand, Roald. 1942. "Title Report by Quartermaster R. Strand, R.N.N.," December 7. HS 2/203 Operation Title: Abortive Chariot Attack on *Tirpitz*. The National Archives, Kew, Richmond, United Kingdom.

Strauss, Richard H., and David E. Yount. 1977. "Decompression Sickness: Once a Little-Understood Danger, the 'Bends' Can Now Usually Be Prevented or, if the Condition Is Recognized Soon Enough, Treated Successfully." *American Scientist* 65 (5): 598–604.

Strutton, Bill, and Michael Pearson. 1958. *The Secret Invaders*. New York: British Book Centre.

Subramanian, Samanth. 2020. *A Dominant Character: The Radical Science and Restless Politics of J. B. S. Haldane*. New York: W. W. Norton & Company.

Swiderski, Richard M. 2004. *Anthrax: A History*. Jefferson, NC: McFarland & Company, Inc.

Swinfield, John. 2014. *Sea Devils: Pioneer Submariners*. Stroud, UK: History Press.

Talbot, Arthur George. 1944. "Operation Neptune—Force 'S' Report," July 22. The Narrative No. 2222/85. DEFE 2/419 Report by Naval Commander, Force "S." The National Archives, Kew, Richmond, United Kingdom.

Tanner, E. L. 1939. Letter to J. B. S. Haldane, September 7. HALDANE/3/3/4/138 Letter from the UCL Secretary to J. B. S. Haldane. University College London Special Collection, College Archives, London.

Thornberry, Cyril Joseph. 1999. Interview by Conrad Wood. Imperial War Museum. Catalogue number 19668. https://www.iwm.org.uk/collections/item/object/80018483.

Times (London). 1927. "An Anaesthetist's Experiment." September 15, 7.

———. 1932. "Charges Against Doctor." May 28, 7.

———. 1938. "Humphry Davy's Discoveries." December 30, 5.

———. 1939. "Breaking the Nazi Yoke of Tyranny." September 4, 10.

———. 1939. "Britain's Fight to Save the World." September 4, 3.

———. 1939. "Doctor's Death an Accident." January 4, 7.

———. 1939. "Last Hours in the *Thetis*: Professor Haldane's Tests." July 21, 8.

———. 1939. "Rationing in December." November 2, 8.

———. 1939. "We Stand With Britain." September 4, 9.

———. 1939. "Why Britain Goes to War." September 4, 9.

———. 1939. "Women Volunteers: Houses Wanted as Nurseries." September 4, 11.

———. 1940a. "Births." July 22, 1, People.

———. 1940b. "Births: Rendel." July 22, 1.

———. 1940. "89 Raiders Down During the Week-End." August 26, 4, News.

———. 1940. "Hitler's Proclamation to His Troops." May 11, 5.

———. 1940. "Little Damage in London: Efficiency of A.A. Defences." August 24, 4, News.

———. 1940. "London Air Raid Casualties." August 24, 2, News.

———. 1940. "More and Better Shelters: A Million Bunks on Order, Free Earplugs." September 24, 4, News.

———. 1940. "Subscriptions for War Bonds." June 28, 10, Business News,.

———. 1940. "'That Village' Raided Again: A Queen Anne House Defies a Bomb." September 4, 2, News.

———. 1945. "Personal Tribute Major F. Lennox-Boyd" (obituary). July 23, 7, People.

———. 1956. "Lecturer Stamped on Dog's Tail." August 24, 7, News.

———. 1964. "Professor J. B. S. Haldane" (obituary). December 2, 13.

———. 2006. "Richard Kendall" (obituary). February 9, 67, People.

Tredoux, Gavan. 2018. *Comrade Haldane Is Too Busy to Go on Holiday: The Genius Who Spied for Stalin*. New York: Encounter Books.

Trenowden, Ian. 1997. *Stealthily by Night: The COPPists Clandestine Beach Reonnaissance and Operations in World War II*. Manchester, UK: Crecy Publishing.

Turnbull, Maureen. 1973. "Attitude of Government and Administration Towards the 'Hunger Marches' of the 1920s and 1930s." *Journal of Social Policy* 2 (2): 131–42.

UCL Lab. MS.20548 "Diving Notes of J. B. S. Haldane and Helen Spurway" (2). National Library of Scotland, Edinburgh.

———. ca. 1940–44. MS.20569 Miscellany of Papers, Chiefly Undated, Chiefly of J. B. S. Haldane, Including Notes, Memoranda, Reports and Mathematical Calculations Concerning Diving. National Library of Scotland, Edinburgh.

———. 1942. MS.20566 "Diving Notes of J. B. S. Haldane and Helen Spurway," Part 1, May–June 1942. National Library of Scotland, Edinburgh,.

———. 1942–43. MS.20567 "Diving Notes of J. B. S. Haldane and Helen Spurway," Part 2, July 1942–May 1943. National Library of Scotland, Edinburgh.

———. 1942–44. MS.20572 Notebook, September 1942–June 1944, Containing Records of Experiments on Respiratory Problems in Diving Carried Out by J. B. S. Haldane and Helen Spurway, Recorded by Both Alternately. National Library of Scotland, Edinburgh.

———. 1943–44. MS.20568 "Diving Notes of J. B. S. Haldane and Helen Spurway," Part 3, June 1943–December 1944. National Library of Scotland, Edinburgh.

UK Immigration. March 17, 1939. Registration Card for Kalmus, Hanus [*sic*]. HO 405/28087 Kalmus, H. Date of birth: 11/01/1906. The National Archives, Kew, Richmond, United Kingdom.

UK Meteorological Office. "Historic Station Data, Oxford Station." Accessed January 8, 2021. https://www.metoffice.gov.uk/pub/data/weather/uk/climate/stationdata /oxforddata.txt.

UKNA. 1940–44. "Dr. Rendell's Claim for Compensation: Injury to Lungs While Engaged on Experiments Connected with Dunbar-Nasmith's Sub-Committee on Saving Life from Submerged Submarines." ADM 1/17701 Admiralty (5) and Compensation (9) and Civil Power and Legal Matters (19) and Estimates and Finance (69). The National Archives, Kew, Richmond, United Kingdom.

———. 1943. "Awards to 5 Officers and Men of RN and SBS for First Experimental Beach Reconnaissance by X-craft (Operation KJH)," December 31. ADM 1/29477. The National Archives, Kew, Richmond, United Kingdom.

———. 1943. "Investigations by Professor R. R. MacIntosh and Sir Robert Davies, Reports of Admiral Dunbar-Nasmith Committee, etc.," August 28. ADM 116/4460 Methods of Lifesaving in Sunken Submarines. The National Archives, Kew, Richmond, United Kingdom.

———. 1943. Royal Naval Personnel Research Committee: Minutes of Meetings and Results of Experiments, etc., November 28. ADM 218/172. The National Archives, Kew, Richmond, United Kingdom.

Ünlü, Ayşe Ezgi, Brinda Prasad, Kishan Anavekar, Paul Bubenheim, and Andreas Liese. 2017. "Investigation of a Green Process for the Polymerization of Catechin." *Preparative Biochemistry & Biotechnology* 47 (9): 918–24.

Unknown Photographer. 1888. 4. Photocopy of photograph, original negative in the possession of John R. Morison, Peterborough, New Hampshire. Photographer unknown, 14 September 1888. INSIDE CAISSON PIER IV, 60 FEET BELOW WATER LINE, 25 POUNDS AIR PRESSURE—Sioux City Bridge, Spanning the Missouri River, Sioux City, Woodbury County, IA. Survey HAER IA-96, Control Number ia0541. Library of Congress, Washington, DC, USA.

———. ca. 1940. WWII; England; "West End London Air Raid Shelter". Franklin D. Roosevelt Library Public Domain Photographs, 1882–1962, Collection FDR-PHOCO. US National Archives and Records Administration, Washington, DC, USA.

UPS. 1943. RNP 43/4 UPS Report 1, January 9; Minutes of the First Meeting; UPS Reports; Sub-Committee on Underwater Physiology; Royal Naval Personnel Research Committee of the Medical Research Council. FD 1/7026 Underwater Physiology Subcommittee; Minutes. The National Archives, Kew, Richmond, United Kingdom.

US Air Force. 1945. A North American B-25 of the 886th Operation Company, Chemical Warfare Service (attached to the 42nd Bomb Group) lays a smoke screen during tests at Palawan, Philippine Islands, in preparation for the invasion of Tarakan. 30 April 1945. (U.S. Air Force Number 76732AC). RG 342, Series 342-FH, NAID: 204950052. US National Archives and Records Administration, Washington, DC, USA.

———. June 1944. Photograph of Men and Assault Vehicles Storming the Beaches of Normandy. Photographs of Activities, Facilities and Personnel, ca. 1940–ca. 1983, RG 342, Series 342-FH, NAID: 12003984. US National Archives and Records Administration, Washington, DC, USA.

US Army. August 1944. Clearing Mines in France. RG 111, Series 111-SCA-1300, NAID: 176888168. US National Archives and Records Administration, Washington, DC, USA.

US Coast Guard. June 1944. Photograph of Stretchers Covering the Docks of a Coast Guard LCT (Landing Craft Tank). Photographs of Activities, Facilities and Personnel 1939–1967, File Unit Europe-Normandy Invasions, RG 26, Series 26-G, NAID:

205578594. US National Archives and Records Administration, Washington, DC, USA.

———. June 1944. Standing Room Only. Photographs of Activities, Facilities and Personnel 1939–1967, File Unit Europe-Normandy Invasions, RG 26, Series 26-G, NAID: 205578819. US National Archives and Records Administration, Washington, DC, USA.

US Navy. ca. 1943–1945. L51-03.07.01 Underwater Demolition Team (UDT) Two. L51 Underwater Activities, Catalog # L51-03.07.01. Naval History and Heritage Command, Washington, DC, USA.

———. December 1941. 80-G-32680 Japanese "Type A" midget submarine HA-19. Japanese Midget Submarines Used in the Attack on Pearl Harbor, Catalog # 80-G-32680. Naval History and Heritage Command, Washington, DC, USA.

Vogt, Annette B. 1999. "Ursula Philip," 31 December. The Shalvi/Hyman Encyclopedia of Jewish Women. Accessed 2020. https://jwa.org/encyclopedia/article/philip-ursula.

———. 2008. *Wissenschaftlerinnen in Kaiser-Wilhelm-Instituten, A–Z*. Berlin: Archiv zur Geschichte der Max-Planck-Gesellschaft.

Wakely, Cecil. 1977. Letter to the Royal College of Surgeons of England, January 21. Box 19, Folder 24, Institute of Naval Medicine, Gosport, United Kingdom.

Waldron, T. J., and James Gleeson. 1955. *The Frogmen: The Story of the Wartime Underwater Operators*. London: Pan Books Ltd.

Walker, Arthur Harold Colyear. 1998. Interview by Conrad Wood. Imperial War Museum, London. Catalogue number 17977. https://www.iwm.org.uk/collections/item/object/80017313.

Wallace, Bruce. 1998. *The Environment: As I See It, Science Is Not Enough*. Omaha: Elkhorn Press.

Wallich, Walter. 1979. Interview by Lyn E. Smith. Imperial War Museum, London. Catalogue number 4431. https://www.iwm.org.uk/collections/item/object/80004391.

War Office. 1944. "Amphibious Tanks." H 35178. Imperial War Museum, London.

Warren, C. E. T., and James Benson. 1953. *Above Us the Waves: The Story of Midget Submarines and Human Torpedoes*. London: George G. Harrap & Co. Ltd.

Webb, Gordon Geoffrey Henry. 1989. Interview by Conrad Wood. Imperial War Museum, London. Catalogue number 10694. https://www.iwm.org.uk/collections/item/object/80010472.

Weiner, H. 1939. Letter to J. B. S. Haldane, September 19. HALDANE/3/5/1/2/11 Letters between the Central Office for Refugees and J. B. S. Haldane. University College London Special Collection, College Archives, London.

Welch, Lilian Emmeline. 2000. Interview by Conrad Wood. Imperial War Museum, London. Catalogue number 20611. https://www.iwm.org.uk/collections/item/object/80019428.

Wheeler, Charles. 1994. *D-Day: Turning the Tide*. BBC Worldwide.

White, F. 1943. Ref. No. W/11 Subject Italian 2-Man Torpedo Boat, November 4. ADM 1/14789 Reports on Italian Miniature Torpedoes and Two-Man Motorboats. The National Archives, Kew, Richmond, United Kingdom.

White, Michael. 2015. *Australian Submarines: A History*, Vol. 2. St. Kilda West, Victoria: Australian Teachers of Media.

Wild, Hans. 1941. Haldane at Siebe Gormans, Original ID TimeLife_image_116004374. LIFE Photo Collection. LIFE Photo Collection, New York City, NY, USA, online via Google Arts & Culture.

Williams, David John. Imperial War Museum, London. Catalogue number 11291. https://www.iwm.org.uk/collections/item/object/80011047.

———. 1943. C.O.P.P. Progress Report No. 12 (Week Ending 25th April 1943), April 27. DEFE 2/1111 Combined Operations Pilotage Party: Progress Reports and Honours and Awards. The National Archives, Kew, Richmond, United Kingdom.

———. 1943. C.O.P.P. Progress Report No. 29 (Week Ending 21st August 1943), August 25. DEFE 2/1111 Combined Operations Pilotage Party: Progress Reports and Honours and Awards. The National Archives, Kew, Richmond, United Kingdom.

———. 1943. C.O.P.P. Progress Report No. 30 (Week Ending 29th August 1943), August 29. DEFE 2/1111 Combined Operations Pilotage Party: Progress Reports and Honours and Awards. The National Archives, Kew, Richmond, United Kingdom.

———. 1943. C.O.P.P. Progress Report No. 32 (Week Ending 11th September 1943), September 22. DEFE 2/1111 Combined Operations Pilotage Party: Progress Reports and Honours and Awards. The National Archives, Kew, Richmond, United Kingdom.

———. 1943. C.O.P.P.'s for "OVERLORD" No. 33/3, August 28. DEFE 2/1101 Combined Operations Pilotage Parties: Composition and Movement of Teams, Complement, Requirements for Beach Reconnaissance, etc. The National Archives, Kew, Richmond, United Kingdom.

———. 1943. Letter No. Z/448 Subject: First C.O.P.P. Operations from Malta, April 29. DEFE 2/1116 Combined Operations Pilotage Parties: History and Reports. The National Archives, Kew, Richmond, United Kingdom.

———. 1943. "War Diary: Outline History of the Combined Operations Beach and Pilotage Reconnaissance Party Known as 'C.O. Pilotage Party' or 'C.O.P.P.,'" December 15. DEFE 2/1116 Combined Operations Pilotage Parties: History and Reports. The National Archives, Kew, Richmond, United Kingdom.

———. 1944. Enclosure No. 1 to O.C., C.O.P.P. 1's No. Z/76, January 2. ADM 179/354 Operation "K.J.H.": Reference to Reconnaissance: Minutes of Meeting. The National Archives, Kew, Richmond, United Kingdom.

———. 1944. Enclosure No. 2 to O.C., C.O.P.P. 1's No. Z/76, January 2. ADM 179/354 Operation "K.J.H.": Reference to Reconnaissance: Minutes of Meeting. The National Archives, Kew, Richmond, United Kingdom.

———. 1944. Enclosure No. 4 to O.C., C.O.P.P. 1's No. Z/76, January 2. ADM 179/354 Operation "K.J.H.": Reference to Reconnaissance: Minutes of Meeting. The National Archives, Kew, Richmond, United Kingdom.

———. 1944. "Reconnaissance Report Operation 'Postage Able' 17–21st Jan 1944, Part II: Detailed Report, Swimmers Narrative," January. ADM 179/323 Reconnaissance Report on Operation "Postage Able" from Naval Commander Force J. HMS Vectis, Cowes. The National Archives, Kew, Richmond, United Kingdom.

———. 1990. Interview by Conrad Wood.

———. 1943. C.O.P.P. Depot Training, Book Section I, May 27. DEFE 2/748 Combined Operations Pilotage Parties Depot Training Book. The National Archives, Kew, Richmond, United Kingdom.

Wilson, John Skinner. 1942. Operation "Title" (starting file page 112), October 1. HS 2/202 Operation Title: Abortive Chariot Attack on *Tirpitz*. The National Archives, Kew, Richmond, United Kingdom.

———. 1942. Title (SN/2048), November 16. HS 2/203 Operation Title: Abortive Chariot Attack on *Tirpitz*. The National Archives, Kew, Richmond, United Kingdom.

———. 1943. Telegram M.F.V. "Arthur" (JSW/429) to Lieutenant Colonel B. Øen, February 21. HS 2/203 Operation Title: Abortive Chariot Attack on *Tirpitz*. The National Archives, Kew, Richmond, United Kingdom.

Wood, A. B. 1944. Testimonial, June 17. Box 19, Folder 14, Institute of Naval Medicine, Gosport, United Kingdom.

Woolf, Virginia. 1975. Letter to Hugh Walpole, dated April 12, 1931. In *The Letters of Virginia Woolf*, Vol. 4, edited by Nigel Nicolson and Joanne Trautmann, 59. New York: Harcourt Brace Jovanovich.

Wright, Catherine. 2020. Interview by Rachel Lance.

Wright, Horace Cameron. 1943. "Underwater Blast Trials—H.M.S. *Reclaim*," September 7. Your Ref. 929/49/P.6/1. ADM 218/172 Royal Naval Personnel Research Committee: Minutes of Meetings and Results of Experiments, etc. The National Archives, Kew, Richmond, United Kingdom.

———. 1947. "Investigations by RN Physiological Laboratory on Underwater Blast Problems." ADM 213/1113 Investigations by RN Physiological Laboratory on Underwater Blast Problems. The National Archives, Kew, Richmond, United Kingdom.

———. 1947. "Subjective Effects of Distant Underwater Explosions" RNPL 3/47. ADM 213/1110 Subjective Effects of Distant Underwater Explosions. The National Archives, Kew, Richmond, United Kingdom.

———. 1949. "Underwater Blast Trials—Series 'A,'" October 18 RNP 49/563. ADM 218/172 Royal Naval Personnel Research Committee: Minutes of Meetings and Results of Experiments, etc. The National Archives, Kew, Richmond, United Kingdom.

———. ca. 1950. "The Injurious Effects of Underwater Explosions." Box 19, Folder 31, Institute of Naval Medicine, Gosport, United Kingdom.

———. 1951. "Review of Underwater Blast Research by the Royal Naval Physiological Laboratory." Box 19, Folder 31, Institute of Naval Medicine, Gosport, United Kingdom.

———. 1957. "Review of Underwater Blast Research in the Royal Naval Physiological Laboratory." Box 19, Folder 31, Institute of Naval Medicine, Gosport, United Kingdom.

———. ca. 1972. "Concerning Dr. H. C. Wright OBE, PhD, FICS." Box 19, Folder 24, Institute of Naval Medicine, Gosport, United Kingdom.

———, W. M. Davidson, and H. G. Silvester. 1950. "Effects of Underwater Explosions on Shallow Water Divers," RNP No 50/639. ADM 298/102 Effects of Underwater Explosions on Shallow Water Divers. The National Archives, Kew, Richmond, United Kingdom.

Yarbrough, O. D. 1943. Letter to J. B. S. Haldane from Experimental Diving Unit, Navy Yard, Washington, DC, October 6. MS.20534 Correspondence of J. B. S. Haldane, Part 1. National Library of Scotland, Edinburgh.

Yoganandan, Narayan, Frank A. Pintar, Anthony Sances Jr., John Reinartz, and Sanford J. Larson. 1991. "Strength and Kinematic Response of Dynamic Cervical Spine Injuries." *Spine* 16 (10): S511–17.

Zamet, John. 2007. "German and Austrian Refugee Dentists: The Response of the British Authorities 1933–1945." PhD, Oxford Brookes University.

Zetterling, Niklas, and Michael Tamelander. 2009. *Tirpitz: The Life and Death of Germany's Last Super Battleship*. Newbury, UK: Casemate Publishers.

Zuehlke, Mark. 2012. *Tragedy at Dieppe: Operation Jubilee, August 19, 1942*. Vancouver, BC: Douglas & McIntyre.

Credits

147   Destroyed Necropolis station (*New York Times*, Necropolis Photo, ca. 1941)

160   Diver on Chariot (Reproduced with permission, © IWM, Catalog #A 22119)

168   Lung map (Reproduced with permission from Wright, Davidson, and Silvester, Effects of Underwater Explosions, 1950, page 28) (Note: This image was taken from a test performed in 1949. Wright 1950 100 ft water experiments.)

188   Brothers Martin Case (right) and Ralph Case (left) with friend Pat Morris (center) in uniform (courtesy Christine Love)

199   COPP swimsuit (Reproduced from Combined Operations Headquarters, Bulletin T/18, June 1945, page 1)

216   *HA-19* on the beach (US Navy, *HA-19* Beached, December 1941)

217   Americans hauling in *HA-19* (F. A. Daubin, COMSUBRON 4—Japanese Midget Submarine, No 19—Navigational Charts. RG 38, Series P 17, UD-WW 46, NAID: 133891335, December 30, 1941. US National Archives and Records Administration, Washington, DC, page 34)

218   Smart riding X-craft (Reproduced with permission, © IWM, Catalog #A 22899)

219   Robinson inside X-craft (Reproduced with permission, © IWM, Catalog #A 26933)

229   Trainee underwater (Reproduced with permission, © IWM, Catalog #A 13873)

250   LCOCU at Appledore, 1945 (Reproduced with permission, © IWM, Catalog #A 28997)

251   US Navy frogman working on hedgehog (US Navy, UDT2, ca. 1943–1945)

252   Element C (Lanker, Element C, May 1944)

281   Bomber laying smoke screen (US Air Force, North American B-25, 1945)

290   Number 4 approaches Sword (Reproduced with permission, © IWM, Catalog #BU 1181)

292   LCOCU clearing Gold (Reproduced with permission, © IWM, Catalog #A 23993)

294   Marauder over Sword (NARA, ID #12003988)

296   LCOCU relaxing on Gold (Reproduced with permission, © IWM, Catalog #A 23994)

297   Explosives types (Alfred G. Merritt, Training Explosives, December 1943)

297   Minesweeping activities on the beach (US Army, Clearing Mines, August 1944)

298   Overview of landings (US Air Force, Clearing Mines, June 1944)

298   LCM on way to beaches (US Coast Guard, Standing on LCT, June 1944)

298   LCM carrying troops on stretchers (US Coast Guard, Stretchers on LCT, June 1944)

311   Joan Elizabeth Jermyn Bamji and son, Andrew (courtesy Andrew Bamji)

Index

Note: Italicized page numbers indicate material in tables or illustrations.

About the Author

Rachel Lance is a biomedical engineer and blast-injury specialist who works as a scientific researcher on military diving projects at Duke University. Before returning to graduate school to earn her PhD, Dr. Lance spent several years as an engineer for the United States Navy, working to build specialized underwater equipment for use by navy divers, SEALs, and Marine Force Recon personnel. A native of suburban Detroit, Dr. Lance lives with her husband in Durham, North Carolina. Her first book, *In the Waves*, was published by Dutton.